GEOTECHNICAL ENGINEERING FOR THE PRESERVATION OF MONUMENTS AND HISTORIC SITES III

INVITED PAPERS

This book contains the invited lectures presented at the 3rd International Symposium on Geotechnical Engineering for the Preservation of Monuments and Historic Sites (IS NAPOLI 2022, Naples, Italy, 22-24 June 2022). It collects the opening address, the third Kerisel Lecture, four keynote lectures and eleven panel lectures, and provides a broad impression of the current state of knowledge and the techniques used worldwide for the preservation of built heritage. When confronted with structures relevant to local and global history, there is only one way to select the best possible conservation solution: the multidisciplinary approach. Therefore, the invited speakers have been selected with different pertinent skills, to represent this complexity from the points of view of geotechnical engineers, structural engineers, architects and conservation experts.

The book will be useful to researchers, practitioners, administrations and all those working or interested in the preservation of built heritage.

PROCEEDINGS OF THE THIRD INTERNATIONAL ISSMGE TC301 SYMPOSIUM, NAPOLI, ITALY, 22–24 JUNE 2022

Geotechnical Engineering for the Preservation of Monuments and Historic Sites III

Invited Papers

Edited by

Renato Lancellotta
Department of Structural, Geotechnical and Building Engineering
Politecnico di Torino
Turin, Italy

Carlo Viggiani, Alessandro Flora, Filomena de Silva & Lucia Mele
Department of Civil, Architectural and Environmental Engineering
University of Naples Federico II
Naples, Italy

CRC Press is an imprint of the
Taylor & Francis Group, an **informa** business
A BALKEMA BOOK

CRC Press/Balkema is an imprint of the Taylor & Francis Group, an informa business

Typeset in Times New Roman by MPS Limited, Chennai, India

Library of Congress Cataloging-in-Publication Data

A catalog record has been requested for this book

First published 2022

Published by: CRC Press/Balkema
Schipholweg 107C, 2316 XC Leiden, The Netherlands
e-mail: enquiries@taylorandfrancis.com
www.routledge.com – www.taylorandfrancis.com

ISBN: 978-1-032-35998-4 (Hbk)
ISBN: 978-1-032-35999-1 (Pbk)
ISBN: 978-1-003-32975-6 (ebk)
DOI: 10.1201/9781003329756

Geotechnical Engineering for the Preservation of Monuments and Historic Sites III – Lancellotta, Viggiani, Flora, de Silva & Mele (Eds)
© 2022 Copyright the Editor(s), ISBN 978-1-032-35998-4

Table of contents

Geotechnical Engineering for the Preservation of Monuments and Historic Sites III – Lancellotta, Viggiani, Flora, de Silva & Mele (Eds)

Preface

The International Symposium on the Preservation of Monuments and Historic Sites of 22–24 June 2022 in Napoli (IS NAPOLI 2022) is the third of a series, the first one having been held in 1996 and the second in 2013.

Drawing from the Preface of the latter, we underline that 'TC301 is intended to provide a forum for interchanges of ideas and discussion, to collect case histories and to promote and diffuse the culture of conservation within the geotechnical community. More specifically, it focuses on geotechnical factors affecting historic sites, monuments, cities. TC301 searches for design criteria and construction methods of our ancestors and reports on specific techniques adopted to preserve ancient sites and constructions'.

TC301 is currently supported by AGI, the Italian member society of ISSMGE, that also organized this Symposium in cooperation with the University of Napoli Federico II, and is therefore acknowledged. Special thanks are extended to the AGI Secretary, Mrs. Susanna Antonielli, for her continuous and patient assistance. We would also like to express our appreciation to the Sponsors that helped us in making this conference sustainable.

But the true success of a Symposium is determined by the quality of the scientific contributions. After a process of peer reviewing, 80 papers have been accepted to be presented to the Symposium, written by more than 250 authors coming from all over the world. In addition, an opening address, the third Kerisel Lecture, four keynote lectures and eleven panel lectures will be presented at the Conference. This printed volume collects the seventeen invited papers. All the papers – including the invited ones - are freely available in open access on the publisher website. The authors are therefore deeply acknowledged for having shared their experience, ideas, proposals, to try and set forth a reference picture of the current state of practice in the engineering approach to heritage preservation. We are confident that the discussion at the Symposium will be fruitful and stimulating and will contribute to the advancement of the discipline.

The Nobel prize poetess Wyslava Syimborska, looking at the Milkmaid, the famous painting by Vermeer, wrote: 'So long as that woman from the Rijksmuseum, in painted quiet and concentration, keeps pouring milk day after day from the pitcher to the bowl, the World hasn't earned the world's end'. In a period of serious difficulties for people everywhere in the world, due to the Covid pandemic and to the fires of war, we believe that it is very important that there are people who do care about our past and its preservation.

Renato Lancellotta
Carlo Viggiani
Alessandro Flora
Filomena de Silva
Lucia Mele

Geotechnical Engineering for the Preservation of Monuments and Historic Sites III – Lancellotta, Viggiani, Flora, de Silva & Mele (Eds)
 ISBN 978-1-032-35998-4

Symposium Organizers

AGI Associazione Geotecnica Italiana

Under the auspices of ISSMGE TC301 "Historic Sites"

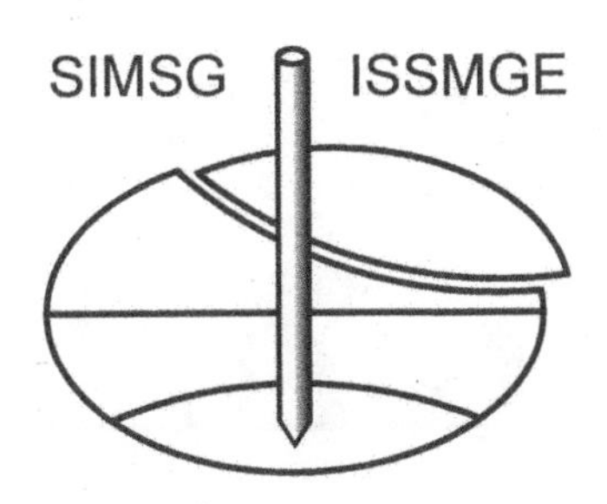

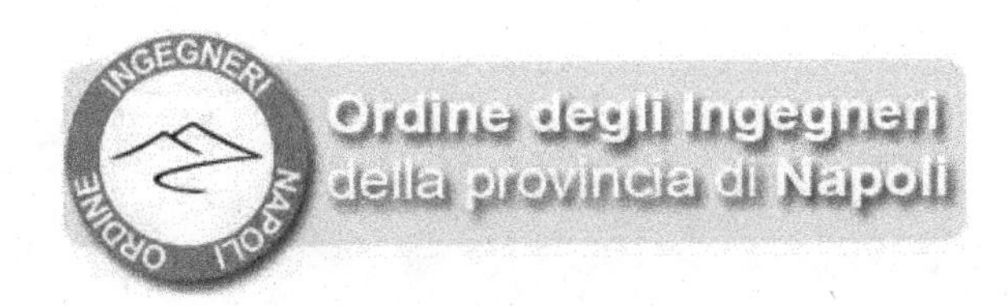

Ordine degli Architetti
della provincia di Napoli

Geotechnical Engineering for the Preservation of Monuments and Historic Sites III – Lancellotta, Viggiani, Flora, de Silva & Mele (Eds)
© 2022 Copyright the Editor(s), ISBN 978-1-032-35998-4

Committees

ORGANIZING COMMITTEE

Stefania Lirer – Chair (Guglielmo Marconi University)
Susanna Antonielli (Associazione Geotecnica Italiana, AGI)
Stefano Aversa (University of Napoli Parthenope)
Riccardo Berardi (University of Genova)
Emilio Bilotta (University of Napoli Federico II)
Luca de Sanctis (University of Napoli Parthenope)
Filomena de Silva (University of Napoli Federico II)
Guido Gottardi (University of Bologna)
Carlo Lai (University of Pavia)
Giuseppe Lanzo (Sapienza University of Roma)
Claudia Madiai (University of Firenze)
Lucia Mele (University of Napoli Federico II)
Sebastiano Rampello (Sapienza University of Roma)
Gianpiero Russo (University of Napoli Federico II)
Francesco Silvestri (University of Napoli Federico II)
Paolo Simonini (University of Padova)
Daniele Spizzichino (Institute for Environmental Protection and Research, ISPRA)
Claudio Soccodato (Associazione Geotecnica Italiana, AGI)
Fausto Somma (University of Napoli Federico II)

SCIENTIFIC COMMITTEE

Renato Lancellotta – Chair (Politecnico di Torino, Italy)
Alessandro Flora (University of Napoli Federico II, Italy)
Jitesh T. Chavda (National Institute of Technology Surat, India)
Carlo Viggiani (University of Napoli Federico II, Italy)
Giovanni Calabresi (Sapienza University of Roma, Italy)
John Burland (Imperial College London, UK)
Efraìn Ovando Shelley (Universidad Nacional Autonoma de Mexico, Mexico)
Jamie Standing (Imperial College London, UK)
Ivo Herle (Technische Universität Dresden, Germany)
Kari Avellan (University of Oulu, Finland)
Merita Guri (POLIS University Albania, Albania)
V. Ulitsky (Saint Petersburg State Transport University, Russia)
Lysandros Pantelidis (Cyprus University of Technology, Cyprus)
Panicos Papadopoulos (Frederick University, Cyprus)
Michael Bardanis (Neapolis University Pafos, Cyprus)
Christos Tsatsanifos (International Society for Soil Mechanics and Geotechnical Engineering, UK)
Rui Tomásio (JETsj Geotecnia, Lisbon, Portugal)
Guilherme Pisco (Tetraplano, Portugal)
Antonio Jaramillo (Universidad de Sevilla, Spain)
Pilar Rodríguez Monteverde (Universidad Politécnica de Madrid, Spain)
Jean Launay (Comité français de mécanique des sols et de géotechnique, CFMS, France)
Jean-David Vernhes (UniLaSalle, France)
Guido Gottardi (University of Bologna, Italy)
Stefano Aversa (University of Napoli Parthenope, Italy)

Geotechnical Engineering for the Preservation of Monuments and Historic Sites III – Lancellotta, Viggiani, Flora, de Silva & Mele (Eds)
 ISBN 978-1-032-35998-4

Sponsors

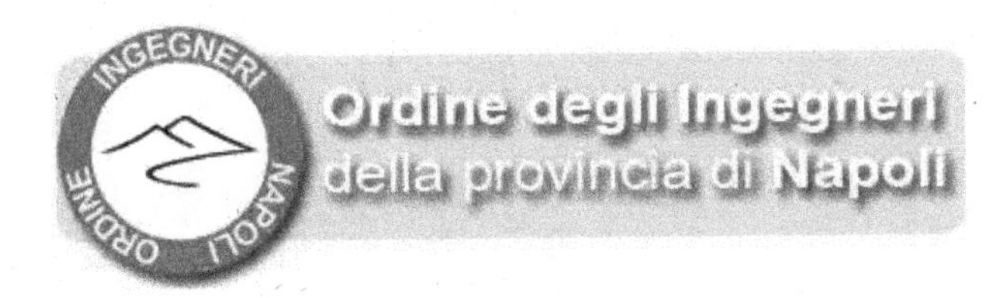

since 1990
HARPACEAS
More than BIM

CATA
COM
BEDI
NA
POLI

Opening Address

Geotechnical Engineering for the Preservation of Monuments and Historic Sites III – Lancellotta, Viggiani, Flora, de Silva & Mele (Eds)
© 2022 Copyright the Author(s), ISBN 978-1-032-35998-4

Welcome address

C. Viggiani
Emeritus Professor, University of Napoli Federico II

Dear friends, welcome to Napoli. This is the third International Symposium on Geotechnical Engineering for the Preservation of Monuments and Historic Sites organized by TC301 of ISSMGE; Napoli is an ideal location since it is itself an outstanding cultural heritage. It has a long history: it has been founded almost 3 millennia ago. Someone says that it is the only oriental city without a European district; it is indeed the only large European city whose historical center is still inhabited by people and not only by banks, offices, shopping malls. It is rich of thousands of cultural assets ranging from remains of many centuries ago to contemporary art works; many of them have geotechnical interest and/or are affected by geotechnical risks; many of them are very fascinating and will contribute making enjoyable your stay in Napoli.

A review of the history of Napoli through three millennia, as seen through the lenses of geotechnical engineering, has been presented by Aversa et al. (2013). Evangelista and Viggiani (2013) reviewed the geotechnical risks affecting the city. Borrowing from them, a bird eye view is here reported.

Greeks founded Parthenopes (also named Palaepolis, the old city) in IX century b.C. on the Pizzofalcone hill, just behind this building where we are and in front of the small Megaride island (Figure 1), where some centuries later the Castel dell'Ovo (the Castle of the Egg) was built (Figure 2). After the victory against Etruscan in 474 b.C. the city was extended in the plain area East of Palaepolis, founding Neapolis, the new city.

Neapolis was organized following a regular pattern of mutually orthogonal streets (Figure 3): three long main streets called *decumans*, oriented West to East, and many orthogonal *cardines*. The scheme is known as *Ippodameus*, from the name of the Greek architect Ippodamo of Mileto, and is found in many Greek cities of that period. It largely survives in the modern city (Figures 4, 5).

Figure 5 shows one of the old cardines, now via S. Gregorio Armeno, famous for its workshops of figures (*Pastori, i.e. shepherds)* for the Nativity scenes (*presepio*). The street is very crowded with tourists, but also Neapolitans, particularly around Christmas.

The pottery or wooden shepherds of XVIII and XIX century are authentic works of art: some instances are reported in Figures 6 and 7.

But what about geotechnical engineering? Actually, some of the most famous historical nativity scenes (presepi) are exhibited in the museum that now occupies the old S. Martino monastery. The monastery had been founded by the Angevins in 1325, at the top of a steep hill. The tavola Strozzi (Figure 8), the oldest depiction of the medieval city, shows the monastery in a dominant position with a huge retaining wall already clearly distinguishable below.

DOI 10.1201/9781003329756-1

Figure 1. The site of the Hellenic settlement of Palaepolis (the old city) (IX – V century b.C.).

Figure 2. Castel dell'Ovo on the ancient Megaride island, in a painting of A. Pitloo, 1830.

Figure 3. The network of cardines and decumans of Neapolis, superimposed to the present city.

Figure 4. Spaccanapoli (Split Naples): one of the old decumans surviving in the modern city.

Figure 5. (left) Via S. Gregorio Armeno; (right) workshops of nativity figures.

Figure 6. (left). The "Presepe Cuciniello", S. Martino Museum, Napoli; (right) The Crib.

Figure 7. Some figures of the Presepe Cuciniello. Old mandolon player (G. De Luca, XVIII Century); Angel of the Announcement (G.B. Polidoro, XVIII Century); Black woman giving a baby the breast (XVIII Century); Rustic carrying a barrel (F. Celano, 1729 – 1814).

Figure 8. The Tavola Strozzi, view of the city of Napoli from the sea. Oil on wood, 1470.

Figure 9. Bird's eye view of the S. Martino hill today.

The vineyard contains more than 7 Km of retaining walls with a height over 3 m. The system of retaining walls shows evident signs of degradation because of lack of any maintenance and uncontrolled runoff of water; the densely inhabited underlying built environment is exposed to a high landslide risk. Geotechnical engineering has thus a central role in preserving the underlying city, the S. Martino monastery and the Presepi!

Within the urban perimeter there are so many cultural assets of geotechnical interest, that we can only list some of them; for instance, underground chambers excavated in different periods for different purposes (Figures 10, 11, 12, 13) and still being excavated (Figures 14, 15, 16)

Figure 10. Subterranean tumb in ellenistic style: tomba C, ipogei dei Cristallini.

Figure 11. The S. Gennaro catacombs.

Figure 12. A Roman street beneath S. Lorenzo Maggiore.

Figure 13. The Fontanelle cemetery.

Figure 14. Excavation for a Metro station in Town Hall Square.

Figure 15. The old Roman harbour in Town Hall Square.

If we go slightly out of the urban perimeter, another world of treasures may be found. For instance, the area of the Phlegrean Fields, east of Napoli, is unique for the beauty of the landscape keeping the traces of its volcanic origin (Figures 17, 18), for the abundance of Roman remains (Figures 19, 20), for legendary events (the entrance to the underworld through the Averno Lake, Figure 21) and mythical figures like the hero Aeneas, founder of Rome, or the Cuman Sybil. The Phlegrean Fields host also the three fantastic tunnels by Lucius Cocceius Aucto (Table 1, Figures 21, 22).

If we move a little further Northwards, we find a small church in a small village, the Benedictine Basilica of S. Angelo in Formis (Figure 23), that was founded in the XI century over the ruin of a Roman temple dating back to the V century b.C. It contains an outstanding cycle of frescoes with stories of Old and New Testament (Figure 24). After the destruction of Montecassino Abbey during the World War II, the S. Angelo frescos are probably the most important document of the medieval painting in Southern Italy.

Starting in 1969 some fissures appeared in the central nave and gradually opened and extended to other parts of the Basilica. Following repeated alarms on the safety of the Basilica, geological survey of the area, subsoil investigations and geodetic monitoring of several points both inside the Basilica and outside have been carried out till present.

The subsoil of the church, as resulted from the site investigations, is schematically shown in Figure 25; it includes three horizons. The upper one is composed by made ground, for a thickness ranging from a few decimeters to some meters. Below the made ground, a layer of fractured rock (dolomite, dolomitic limestone, cemented calcareous debris) is found, with a thickness variable between 15 and 30 m. Finally, the base formation of sandstones and variegated clay shales is found-

The Basilica is located across a stratigraphic discontinuity, with the apses and the backward part of the naves founded on rock, and the front on the debris cover or even on the made ground.

It is known that some repair works have been carried out in XVIII century; in 1930 an earthquake produced the collapse of the roof, reconstructed soon after. Starting in 1969 some fissures appeared in the central nave and gradually opened and extended to other parts of the Basilica (Figure 26).

Cammarota *et al.* (2013) published the data available on settlement of the Basilica in 1980 and 2013; prof. G. Russo kindly made available further measurements in 2021.

Figure 16. Some of the stations of the new Napoli Underground lines. Someone says it is the finest underground system of the world.

Figure 17. The Phlegrean Fields.

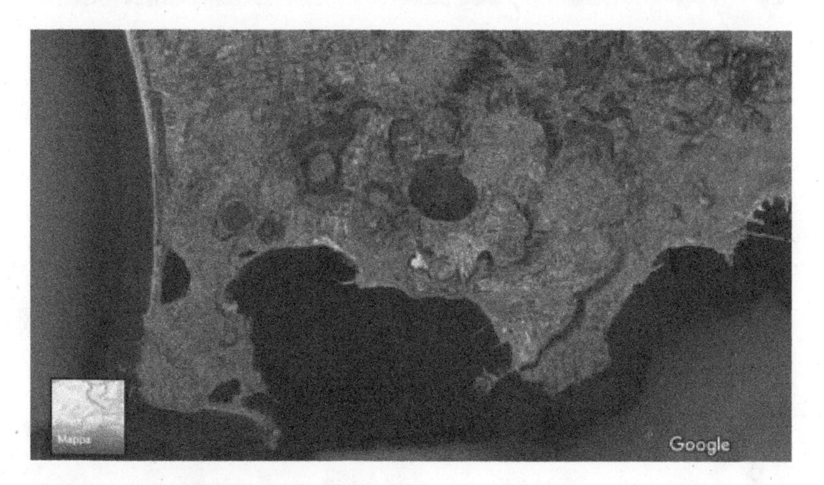

Figure 18. The volcanic origin of the Phlegrean Fields.

Figure 19. Left: the "Piscina mirabilis, an underground water reservoir for the Roman imperial fleet. Above. The remains of the Portus Julius, harbour of the imperial fleet, submerged by the sea because of a 16 m volcanic subsidence.

Figure 20. The Flavian amphitheater in Pozzuoli.

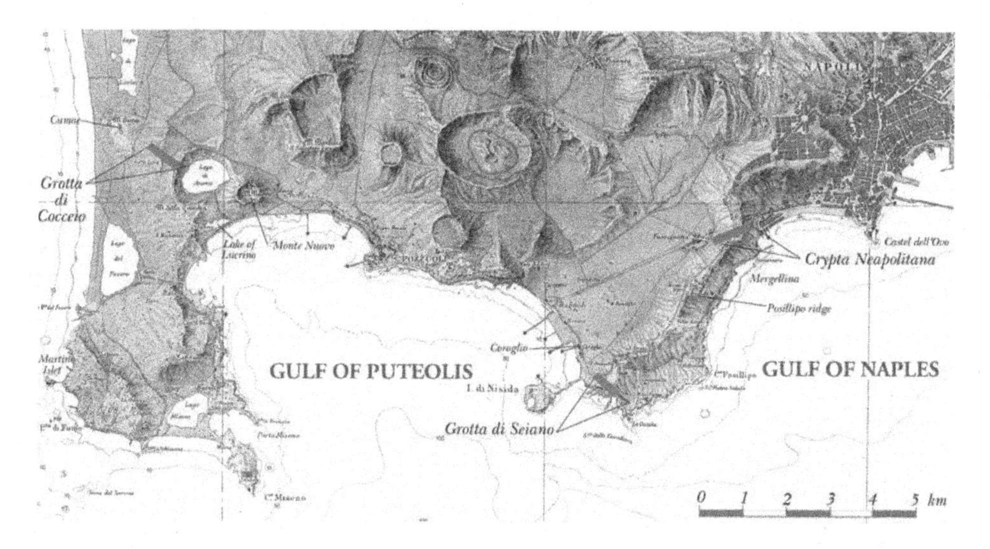

Figure 21. Map of the Phlegrean Fields showing the location of the Roman tunnels.

Table 1. Characteristics of the Roman Tunnels in the Phlegrean Fields.

Name	Length (m)	Width (m)	Height (m)	Notes
Crypta Neapolitana	711	4.5	4.6 ÷ 5.2	2 inclined ventilation shafts
Seianus Grotto	780	4.0 ÷ 6.5	5.0 ÷ 8.0	3 lateral ventilation tunnels
Cocceius Grotto	970	4.5	4.5 ÷ 8.0	5 inclined or vertical vent. shafts

Figure 22. The western intake of the Crypta Neapolitana in a painting by L. Ducros, 1793.

Figure 23. The Benedictine Basilica of S. Angelo in Formis.

Figure 24. The interior of the Basilica.

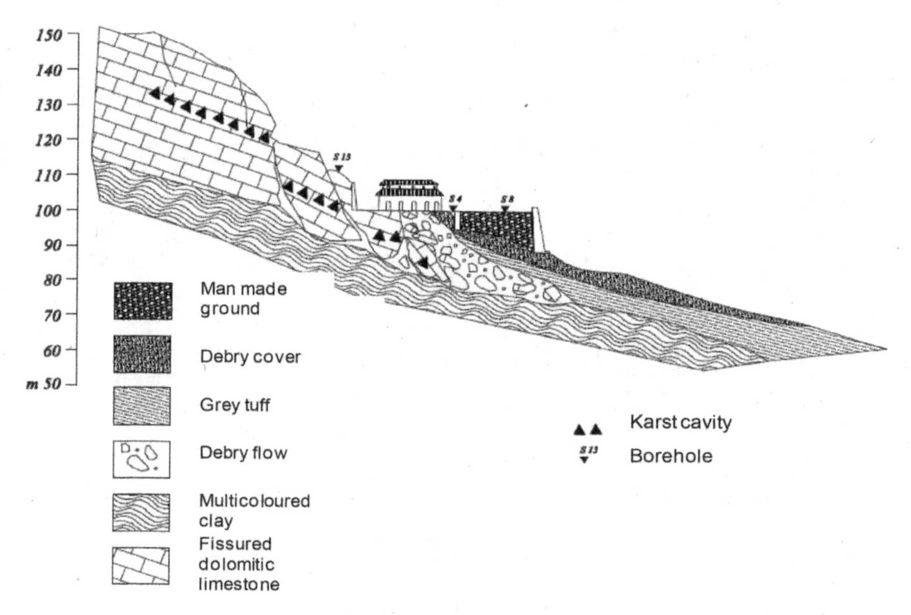

Figure 25. Schematic geological section of the area around the Basilica.

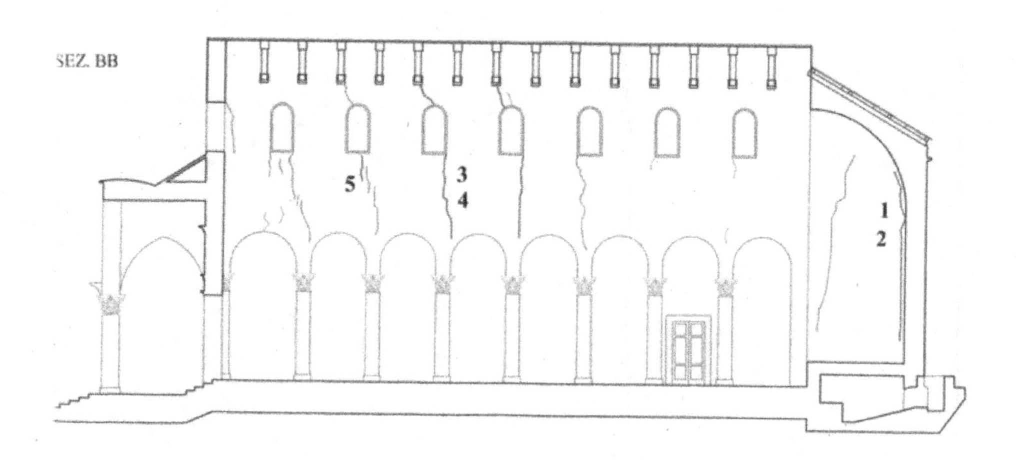

Figure 26. Main fissures of the superstructure.

The trend of differential settlement of the basilica is reported in Figures 27 and 28. The reason of the movements has not yet been understood. At present, the church is monitored and only some structural repairs have been carried out.

In a note published on a magazine some years ago (Montanari 2018) an art historian refers to the church as "…a Benedictine Basilica of clamorous beauty …, one of the most important and fascinating Italian monuments". He regrets "…the unbelievable inability of our generation to repair the structural damages threatening it" and claims that we will succeed in saving S. Angelo in Formis only if we will go deep into ourselves to really understand the problem.

This is a challenge for geotechnical engineers!

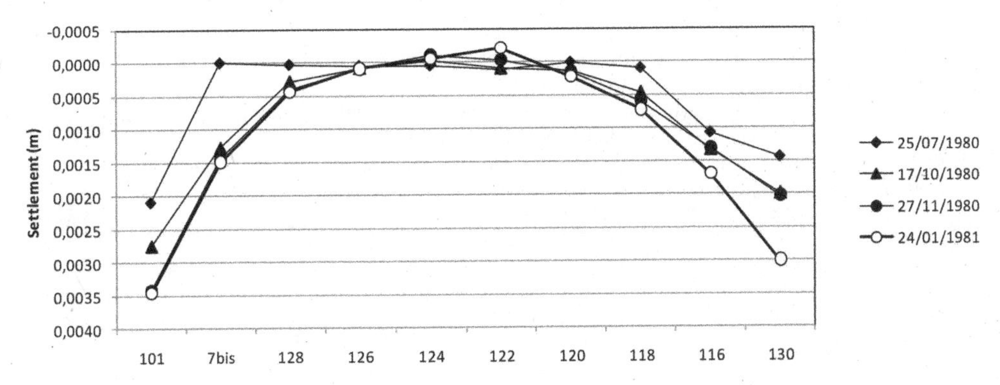

Figure 27. Survey on the internal measuring points: period February 1980-March 1981.

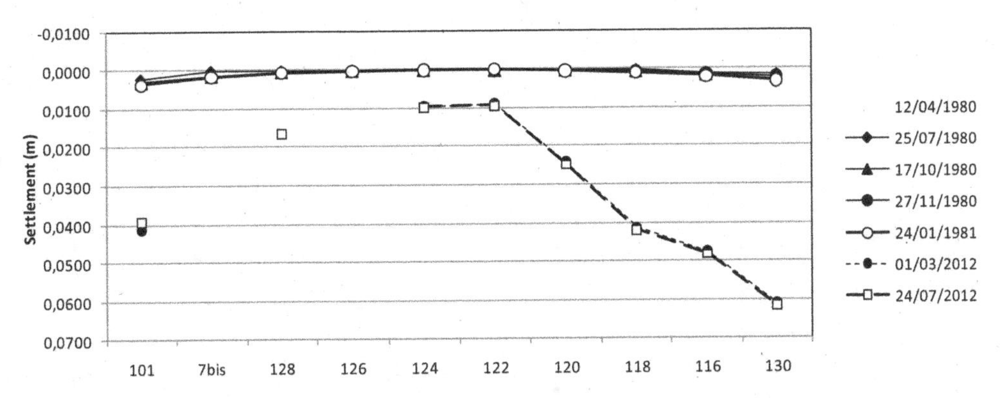

Figure 28. Survey on the internal measuring points: period February 1980-November 2012.

Finally, let me remind another aspect of Napoli: its open-minded character, its multicultural aptitude. A famous Neapolitan song says that Napoli has thousand colors; we have seen these colors in the Presepe Cuciniello. This will find, I hope, correspondence in the discussions of our Symposium, with contributions of different disciplines as structural engineering, architecture and restauration.

REFERENCES

Amato L., Evangelista A., Nicotera M.V., Viggiani C. 2000. The Crypta Neapolitana; a Roman tunnel of the early imperial age. *More than two thousand years in the history of Architecture. UNESCO-ICOMOS Internationl Congress,* Bethlehem.

Amato L., Evangelista A., Nicotera M.V., Viggiani C (2001) The tunnels of Cocceius in Napoli: an example of roman engineering of the early imperial age. *AITES/ITA World Tunnel Congress.*

Aversa S., Evangelista A., Scotto di Santolo A. (2013) Influence of the Subsoil on the urban development of Napoli. *Proc. Intern. Symp. on Geotechnical Engineering for the Preservation of Monuments and Historic Sites, 15 – 44. Bilotta et al. eds.* CRC Press/Balkema.

Cammarota A., Russo G., Viggiani C., Candela M. (2013) The Benedictine Basilica of S. Angelo in Formis (Southern Italy): a therapy without diagnosis? *Proc. Intern. Symp. on Geotechnical Engineering for the Preservation of Monuments and Historic Sites, 225 – 232. Bilotta et al. eds.* CRC Press/Balkema.

Evangelista A., Viggiani C. (2013) A paradise inhabited by devils? The geotechnical risks in the city of Napoli and their mitigation. *Geotechnics and Heritage, 75 - 96 Bilotta et al. eds.* Taylor & Francis London.

Russo G., Viggiani G.M.B., Viggiani C. (2012) Geotechnical design and construction issues for Lines 1 and 6 of Naples underground. *Geomechanics and Tunnelling,* vol. 5, n.3, 300–311.

Montanari T. (2018). Salviamo la Basilica; salveremo noi stessi. *Il Venerdì di Repubblica*, February 22, 2018.
Viggiani C. (2006) Un ingegnere romano di epoca tardo repubblicana: Lucio Cocceio Aucto. *Atti del I Convegno Nazionale di Storia dell'Ingegneria,* Napoli, vol. 2, 785–796.
Viggiani C. (2013) Portus Julius: a complex of Roman infrastructures of the late Republican age. In: E. Bilotta et al. ed., *Geotechnics and Heritage,* CRC Press/Balkema, 243–260.
Viggiani C. Le gallerie romane dei Campi Flegrei. SIG: *"1 anno dal WTC 2019: via per montes excisa, le opere in sotterraneo incontrano architettura, archeologia e arte.* Napoli.

Kerisel Lecture

Geotechnical Engineering for the Preservation of Monuments and Historic Sites III – Lancellotta, Viggiani, Flora, de Silva & Mele (Eds)

Taking care of heritage, a challenge for geotechnical engineers

Alessandro Flora
University of Napoli Federico II, Italy

ABSTRACT: When dealing with monuments or historic sites, engineers may find themselves out of the comfort zone bounded by balance and congruence, being necessary to have an approach guided not only by technical convenience and cost effectiveness, but above all from the need to preserve at the best whatever is the heritage carried by the specific structure under analysis. Such a lack of comfort has to be the guiding light in sharing the solution with experts from other fields, as clearly suggested by Article 2 of the Venice Charter. This paper reports on three case histories taken from the author's personal experience, related to the heritage of completely different cultural environments (respectively Maya, Greek-Roman and Byzantine-Ottoman), in which these constraints had to be faced from the point of view of a geotechnical engineer. The role played by geotechnical engineering differs from case to case, but the examples presented herein demonstrate that, far from being sufficient, our discipline is most times necessary. It is argued that, even though the best technical solution is always the least invasive one, geotechnical engineers should not be scared *a priori* by the possibility of interacting with historic structures, as long as their intervention is informed, necessary, respectful and above all aimed to contribute in preserving the true essence of heritage, which is not the structure in itself but the role it has in its physical and social environment.

1 THE LEGACY

What is the best possible engineering approach in the protection of monuments? This theme has been debated for a long time, and has seen a significant transformation over time. The current ruling paradigm is the result of the substantial change in the cultural approach introduced in Europe between the end of the nineteenth century and the beginning of the twentieth. Until then, in fact, the predominant tendency was to approach failure or damage of historic structures by reconstruction – total or partial – to (presumedly) make them appear as they once were. Eugène Viollet-le-Duc (1814–1879) is typically taken as an example of this old approach, summarised by the statement: '*To restore a building is not to repair or rebuild it but to re-establish it in a state of entirety which might never have existed at any given moment*'.

The new way to look at built heritage that saw the light at the beginning of the twentieth century stems from the original work of many intellectuals, like for instance John Ruskin (1819-1900) and Georg Gottfried Julius Dehio (1850–1932), whose motto '*preserve, and do not restore*' is still often quoted. A decisive contribution to the development of a new culture of preservation in Europe was given by Italian scholars (Brandi, 1963), because of the unique environment in which they were raised and educated. The peculiar and pervasive presence of archaeological sites, historic cities and villages on the Italian territory – whose long-lasting, continuous presence was guaranteed by preservation rules introduced centuries ago – has made local sensibility to the theme always extremely high (D'Agostino, 2022).

Therefore, in the twentieth century in Europe the idea that the legacy carried by a monument did not depend only on its presumed original appearance became largely dominant. Appearance did not necessarily have to be re-proposed, the wounds of time being part of the monument life and contributing to its intangible value. This cultural context explains the highly restrictive position of the Athens Charter (1933) first and of the Venice Charter (1964) subsequently.

DOI 10.1201/9781003329756-2

The Nara Document (1994), the Krakow Charter (2000) and more recent documents have added complementary information and principles to these original references, recognising that the concept of preservation and even the definition of authenticity and heritage must be referred to considering the different cultural contexts existing around the world.

Is all of this relevant for geotechnical engineering? We have learned that it is. In promoting the Technical Committee on Historic Sites of ISSMGE in 1981, Jean Kerisel and Arrigo Croce made an effort to bring to the attention of the geotechnical community the need to face all the existing cultural constraints, when dealing with cultural heritage. Since then, it has been clearly stated that the principles contained in the fundamental reference documents previously mentioned apply not only to the visible part of the structure but to the whole Ground-Monument System (Jappelli 1991). The relevance of geotechnical engineering in preservation has been highlighted many times (e.g. Jappelli & Marconi, 1997) and nicely summarised in the two previous Kerisel lectures through exemplary case histories (Calabresi, 2013, Viggiani, 2017). Clear definitions of material, iconic and historical integrity have been given (e.g. Viggiani, 2013). Historical integrity, in particular, has been dealt with in terms of authenticity from different points of view (e.g. Iwaski et al. 2013), confirming the variety of approaches related to local cultural environments, consistently with the indications of the Nara document.

Figure 1 schematically summarizes the author's personal view of what we could call the conventional relationship between the different kinds of integrity and some possible engineering actions to be carried out for preservation goals. An insight into the meaning of each action, from conservation to reconstruction, can be found in Petzet (2004). This scheme is a conceptual framework posing constraints that cannot be overlooked, even when dealing with the least visible, underground part of built heritage, and with the subsoil directly interacting with it.

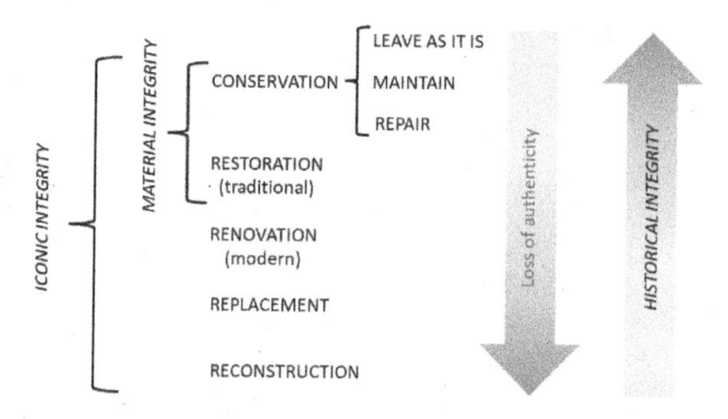

Figure 1. Conventional relationship between iconic, material and historical integrity as a function of different kinds of engineering actions.

Indeed, the scheme reported in Figure 1 may not be fully satisfactory, because historical integrity and authenticity are somehow elusive concepts, as already mentioned. An enlightening example of such elusiveness can be taken with reference to the conservation of some shrines in Japan (Flora, 2013): up to the mid of the 19th century, several wooden Shinto shrines periodically underwent complete reconstruction ever since the inception of this custom in the 7th century. Such a practice had the character of an important religious ritual, but was probably set forth to answer to the need of substituting spoiled or damaged parts. Later on, in the 19th century, all the Shinto shrines but one (Ise shrine, Figure 2) stopped the periodic reconstruction because of political changes and economic crisis. Nowadays, while the Ise shrine still keeps its ritual reconstruction every 20 years, all the other shrines are protected by law as architectural heritage, assuming as an indicator of their relevance and integrity the material value, in accordance with Figure 1.

Actually, the interruption of the periodic rebuilding process was an accident and not the norm, the Iso shrine being the only one to follow its originally conceived life cycle. So, the question is: what is authentic in this case? The frozen material situation of the 19th century or the immaterial heritage preserved by the ritual reconstruction of the Ise shrine? The answer is not easy,

Figure 2. Ise shrine (Japan), which still undergoes ritual reconstruction every 20 years ever since the 7th century.

because such problems often face the lack of unicity of the solution (Viggiani, 2017), which is certainly an uncomfortable situation for engineers.

An attempt to answer may be done complementing the list of material and immaterial values of built heritage, or better overcoming the simple distinction between tangible and intangible values, referring to the role the structure has in its social and physical environment. So, the preservation of the role – intending with it the coherence with the original scope of the construction (considered through what we may call its functional integrity, Figure 3) along with the importance it has in the perception of the local historic, physical and spatial environment – leads to the need of looking at preservation in a broader sense. Probably, the supreme value to be preserved is the message coming from the past, and therefore *continuity* may be even more important than *authenticity* (Petzet, 2004).

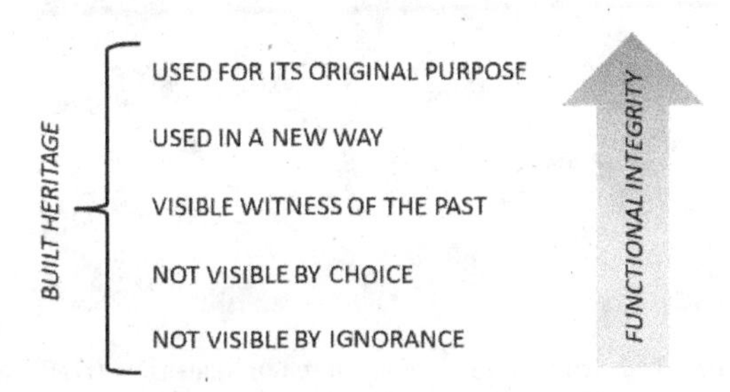

Figure 3. Schematic representation of functional integrity of the built heritage, as one of the elements defining the role it has in its environment.

Far from encouraging arbitrary reconstruction, the concept of functional integrity may certainly be a way to check if the monument or site of interest is in a good shape, alive with wounds, or just dead. In a way, these possible states have to do with the nuanced difference some scholars make between archaeology and architecture, which may seem a semantic dispute but underlays relevant differences in the preservation approach, that could be accepted to be more invasive in the latter case.

A paradigmatic example of built heritage not visible by ignorance is a part of Ercolano, a Greek-Roman town buried by the eruption of Mount Vesuvius in AD 79. Clearly, part of the town (how large? How relevant?) is still buried under modern buildings (Figure 4), built with no respect – ignorance, in its literal meaning – of the underground heritage constraints.

Figures 5 to 8 complement the information summarized in Figure 3 in terms of built heritage functional integrity.

Figure 4. The Greek-Roman town of Ercolano (Italy), still partly buried under low quality modern buildings.

Figure 5. The Samnitic-Roman town of Pompei (Italy), unique witnesses of the past.

Figure 6. Bookstore inside a 13th century Dominican church in Maastricht (Holland). Built heritage used in a new way, with respect of its original conception.

Figure 7. The Roman port of Ventotene (one of the Pontine islands, Italy), still in use (thus at the highest possible level of functional integrity), was entirely excavated in the rock bank, removing some 60.000 m^3 of material, to support emperor Augustus' (63 BC – AD 14) summer residence, as there was no natural harbour on the island.

Figure 8. The Basilica of Saint Peter in Vatican (Rome, Italy), still playing its role of church and centre of the Catholic world (thus being at the highest possible level of functional integrity).

In the framework so far depicted, the case of Ise shrine can be seen as fully respectful of its role and functional value, because able to carry the immaterial, social heritage of the ritual reconstruction.

The consideration of the social role played by the built cultural heritage – well beyond its physical features –indicate that a merely binding preservation culture, as emerging from the strongly conservative constraints posed by the different charts of the twentieth century, may be inappropriate to preserve heritage in its broader sense. Quoting Settis (2018): '*Cities are not museums: they're meant to be lived in, and that's the reason why conservation supervision should not be perceived as a way of leaving everything in a state of hibernation. I don't wish heritage protection to mean hibernation*'.

This is a crucial issue when looking at preservation from the geotechnical engineering point of view. As a matter of fact, our discipline is most often involved in preservation actions when critical mechanisms may affect the Ground-Monument System, with the structure or the site often on the verge of failure. In such cases, geotechnical contribution is typically required to solve critical static problems or to tackle a high seismic risk, often without enough time to explore in depth all the range of technical alternatives. Decisions are easier to take when the geotechnical intervention can be concentrated within the subsoil, i.e. when the action has the goal to remove the cause and not to mitigate the effects on the structure. Soft solutions can overtake on invasive ones in these cases by considering the effect of boundary geotechnical conditions and eventually acting on them, as nicely described for instance by Calabresi (2011) with reference to the Roman Milvius bridge in Rome. In other cases, taking into account the effect of dynamic soil-structure interaction may prevent from useless invasive underpinning interventions (Lancellotta, 2013) or help in better understanding the structural behaviour of monuments (e.g. de Silva et al. 2018, de Silva 2020, Flora *et al.*, 2021). The well-known case of Orvieto (Italy) (Pane & Martini, 1997) is also worth mentioning as representative of the quite frequent situations of relevant structures or sites above unstable rock cliffs, where the stabilization of the cliff is an effective, fully respectful solution.

The worldwide famous leaning tower of Pisa is another paradigmatic example of good geotechnical practice, as the preservation was successfully obtained by careful under-excavation (Burland *et al.*, 2013), i.e. just carefully removing some soil in specific zones underneath the tower, without even touching it, thus keeping the solution in the uppermost part of the conventional integrity scheme of Figure 1. However, the tower of Pisa may be also seen as a misleading example, in the sense that the successful and fully respectful solution was obtained after almost one century of careful and detailed studies, investigations and monitoring, with no economic constraints, with the support of politics and public opinion, involving in the multidisciplinary study world leading experts. Such an exceptional circumstance is rarely reproducible and cannot be considered as a routine situation, even in the case of extremely valuable historic buildings or sites. In the everyday life of geotechnical engineers, therefore, we know that a compromise is often unavoidable.

If foundation reinforcement with new technologies (to be considered as *modern renovation* only after exploring the possibility of *restoration*, Figure 1) is the only feasible solution and may solve the problem, for instance, it should not be excluded *a priori*, especially if it contributes to keep the built heritage alive (i.e. with the highest functional integrity, Figure 3). In fact, often historically valuable structures that still have a good degree of functional integrity are the result of continuous transformations that have taken place in a long time span. Therefore, modifications based on sound cultural and mechanical bases (and thus to be considered necessary) should not scare geotechnical engineers, being possible to consider them as part of the lifecycle of the structure, which should not be necessarily frozen to the present, intrinsically assumed as a reference time out of a still evolving historical pattern. Of course, the first attempt should be to use technological solutions consistent with the original structure, whose characteristics should be known (Roca et al., 2019). Apart from the formal distinction between architecture and archaeology that the concept recalls, the solution should be considered case by case, obviously taking into account all possible alternatives and privileging the least invasive ones. Hard interventions in the subsoil (for instance, underpinning) should be taken possibly avoided, considering that still undiscovered heritage may exist underneath the visible structure to protect.

The lack of a general theory, and therefore of a univocal indication of the best engineering solution to preserve built heritage, imposes the need to be extremely more cautious than with new constructions, and technical convenience or cost effectiveness must not be the guiding light in this case. Engineers have to cope with values usually out of their skills, stepping outside of their comfort zone, and have to agree on the technical solutions with archaeologists, architects, art historians and officials in charge of monuments preservation. Indeed, '*a satisfactory equilibrium between safety and conservation, between engineers and restorers, may be found only in the development of a shared culture*' (Viggiani 2013). Unfortunately, this is still to come in common geotechnical preservation practice, and extremely invasive actions on foundations are often felt acceptable just because they are not visible, with no deeper insight. Then, an effort is needed to go beyond purely academic discussion, if we want to avoid being as '*that people who give good advice if they cannot set a bad example*' (De Andrè, 1967). This is, or at least should be, the role of TC301 of ISSMGE.

The considerations reported in this introductory section clearly warn on the unusually complex task of dealing with the preservation of historic sites and buildings. Before planning any intervention, therefore, this complexity should make engineers aware of the heritage value, no other way being possible to truly perceive it than taking a humble bath into the history of the specific structure of interest, quietly listening to the silent voices of those who have imagined, designed, built, used and eventually modified it during history. Only after this empowering bath (Figure 9), balance and congruence can take the lead.

The examples reported in the next sections are taken from the author's personal experience, and have been chosen because they show situations in which the lack of unicity of the solution is clear, and the considerations reported may seem questionable and therefore worth discussing. Covid pandemic has dramatically modified the possibility to carry out site investigations in the last two years. The reported case histories have suffered this limitation, and not all the planned activities were performed. Because of this, some of the results must be considered preliminary.

Figure 9. Inner side of the Tomb of the diver's lid (Greek painting, 480/70 BC, Paestum, Italy).

2 INTERFERE WITH THE PAST HELPING TO DISCOVER

2.1 *Copan*

Copan (UNESCO site since 1980) is an ancient Maya settlement located in the extreme western highlands of Honduras, close to the Copan river (Figure 10). This city-state flourished from the 5th to the 9th century AD – a time-span known as the Classic Maya period – during which it became one of the most important sites of Maya civilization, in that period spread in contemporary southern Mexico, Guatemala, Belize, Honduras and El Salvador. Copan was the intellectual centre of the Classic Maya civilization, where important advances in astronomy and mathematics were achieved (Thompson, 1958). The site was brought to the attention of Europeans in 1576, when Diego Pedro de Palacio visited it and reported on the magnificence of its architecture and sculptures.

Copan is most famous for the so-called Main Group (Figure 10), an architectural compound exquisitely decorated with stone sculptures, comprised of a massive elevated royal complex located south, known as the Acropolis, and a series of connecting plazas and smaller structures located north. One of the most famous pieces of stonework is the superb Hieroglyphic Stairway (Figure 11), the longest known Maya hieroglyphic text, and it's incredibly high relief sculpture, some of the finest ever carved in antiquity (Sharer and Traxler, 2006). Centuries of abandonment left heavy

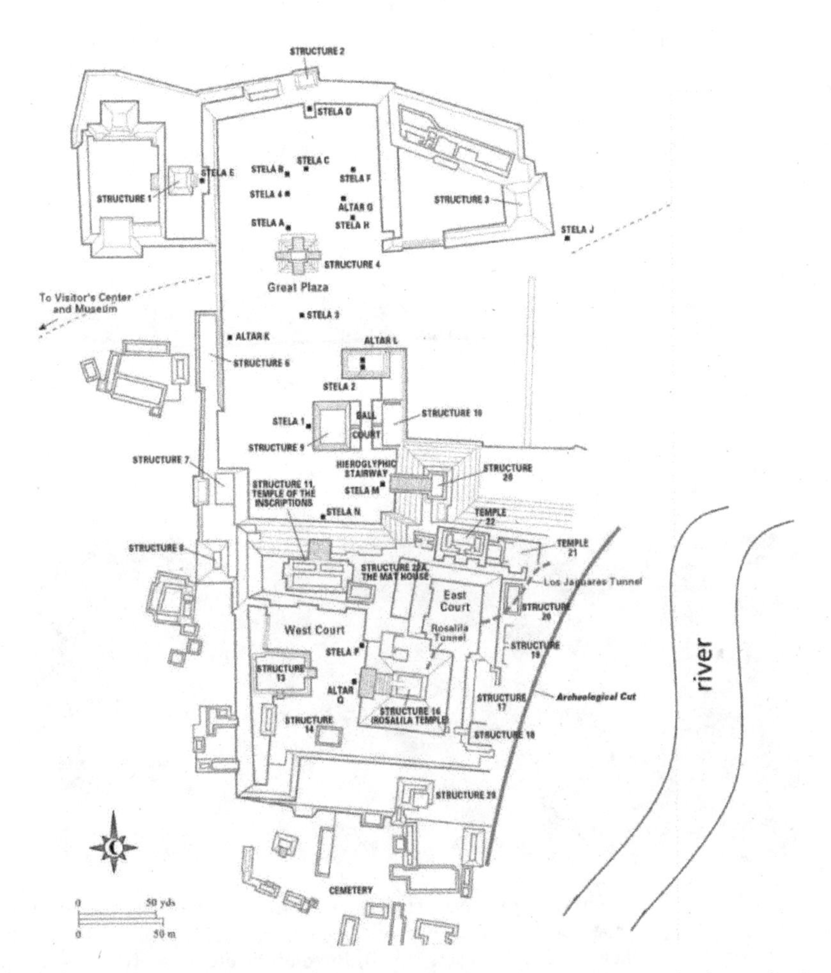

Figure 10. Plan view of the Copan archaeological area, with (sketched, non on scale) the nearby Copan river (modified after Pires, 2020).

signs on the structures, that were mostly ruined at their rediscovery in the 19th century. In 1830's, Juan Galindo was the first to systematically explore the site, bringing attention from the world to the ruins of Copan and inspiring the first official excavations to begin in the late 1880's.

When these first consistent explorations took place, the eastern side of the Acropolis had already been partially destroyed because of a large slope failure that involved part of the pyramids (Figure 12a). Such large collapse was certainly related to the action of the close by Copan river, and the cut - nowadays known as *corte* – is the result of hundreds of years of water action. In fact, at some point of the Classic Period, the Maya artificially diverted the river course. The artificial control of the river path was stopped after the city decline, letting the river meander again. Its widening floodplain eventually became responsible for the undercutting and destruction of the eastern portion of the Acropolis (Bell et al., 2004). Added to centuries of erosion, an earthquake in the 1930's threw the top of three East Court buildings into the river, partially destroying some of the structures recorded by the first archaeologists of Copan (von Schwerin, 2011).

To prevent further catastrophic collapses, in the 1930's the river was diverted again. The large slope failure partially exposed ancient layers of buildings, once hidden in the undergrounds (Figure 12b). This obviously suggested to archaeologists the existence of important buried structures. In fact, the monuments visible in the Acropolis and the enclosed upraised courtyards are just the latest components of a series of additions made by the 400-year dynasty that ruled over Copan, accumulated over the centuries as a result of built layers added by ensuing kings (Fash, 1991). As the relevance of Copan grew, so did the need for Maya rulers of higher and more impressive monuments testifying it and leaving a lasting mark of their reign. The Maya approach to this need was to use the existing pyramids and structures as a core of the new enlarged ones. Therefore, when excavated, the pyramids reveal a series of complete but smaller pyramids, often still with their original coloured stucco decoration. In some cases, individual shrines could be amalgamated into a single bigger complex over time.

Figure 11. Top: Sketch of the Main Group of Copan with its northern part marked by the low-level plazas, and its southern part constituted by the Acropolis, with emphasis in the Hieroglyphic Stairway Temple (in red) – sketch adapted from Linda Schele's drawing (Schele, 1998); Bottom: Picture of the Hieroglyphic Stairway Court, with the temple at right and part of the Ball Court at left – photo by Linda Schele (modified after Pires *et al.*, 2021).

(a) (b)

Figure 12. Left (a): looking north at the river cut (*corte*) into the 120-meter-long masonry eastern wall of the Copan Acropolis, photo by Marshall Saville 1891/92 (Peabody Museum 2004.24.66) – Right (b): a view of different entrances to tunnels in the Acropolis archaeological cut in 1989 – left (modified after Sharer *et al.*, 1992).

Once a new and bigger structure had to be built on top of an existing one, the typical erection sequence was the following: first, the ancient superstructure (i.e. the usable building) was dismantled and its construction materials kept for posterior reuse (Abrams, 1994); then, a new and usually larger superstructure and substructure (i.e. the support of the buildings) were built over and around the ancient one. In more dramatic interventions, large platforms (i.e. levelling surfaces) were built. Superstructures were generally made with three-leaf masonry walls, with external leaves composed by dressed tuff stones and an infill core with a sort of "concrete" (Loten and Pendergast, 1984). Substructures were generally composed by a mix of wet-laid earth and stone fill materials, retained peripherally by stone masonry walls.

2.2 *The archaeological tunnels*

To allow for the archaeological investigation of the buried structures under the Acropolis, the first tunnels were excavated in the 1930's. By the 1980's, a new strategy was established and more complex tunnelling started from the *corte* using the exposed layers as references (Figure 12b). In time, an incredible 4 km long and complex tunnel system developed within the monumental compound. Figure 13a reports a sketch of the tunnels network underneath the Hieroglyphic Stairway Temple, in the northern part of the Acropolis. Thanks to this underground investigation approach, extraordinary archaeological discoveries were made. A large number of decorative plasterwork reliefs provided one of the most comprehensive set of information on the origins and development of a Classic Maya complex (Lacombe et al., 2020). At the end of the 1980's, the Copan Acropolis Archaeological Project (PAAC) was established, merging past and new projects under the same direction, and allowing access to part of the tunnels by tourists.

Although most of the tunnels are apparently stable, local collapses required actions to ensure the safety conditions for researchers and visitors, and also to preserve the material cultural heritage still uncovered by the archaeologists. Moreover, in some cases the change in the environmental conditions of the buried heritage in the tunnels is also leading to the deterioration of valuable decorative materials. Since a comprehensive investigation of the conditions of the tunnels network within the site of Copan had never been carried out, a strategic plan was started by the Copan Acropolis Tunnel Conservation Program from Harvard University (Lacombe *et al.*, 2020), defining investigations, analyses and interventions to be carried out in subsequent steps. At this stage, structural and geotechnical engineers were invited to join the research group. In this framework, in fact, an important task was to obtain an insight into the stability conditions of the tunnels, starting from the available information, which in terms of the geometrical shape and position of the tunnels was well detailed. On the contrary, very little was known on the mechanical properties of the infill material. In this section, the first considerations done on the stability of the tunnels under the

Hieroglyphic Stairway Temple (Figure 13a) and on the preservation actions to be carried out are reported. Here the tunnels were excavated through Esmeralda (Figure 13b), a large embankment built around 700 AD, that would later become the supporting platform for the Hieroglyphic Stairway Temple (Sharer *et al.*, 1999).

The tunnels of Copan can be described as narrow underground passageways excavated by the archaeologists into the man-made earth fill constituting the core of the pyramids, with the goal of seeking buried structures. Because of this, they have an extremely irregular pattern (Figure 14), with sharp changes in direction and depth. When buried structures were found, tunnels usually ran aside substructures and superstructures walls (Sharer et al., 1999). From time to time, some of the tunnels have been excavated to go deeper in the embankment (and consequently to an older period of time in the development of the monumental compound). The earth fills within the pyramids have been created in a long time span, to allow subsequent enlargements of the structures. Because of this, they are not made with a homogeneous material, in terms of grain size distribution, density and even of compaction energy. However, for a certain volume related to a specific enlargement of the structure and at a certain depth, the material can be considered homogeneous, consistently with the fact that the workers were taking each time the fill material from the same pit, and were compacting it with the same tools.

Three main filling soils can be found under the Acropolis. The lowermost layers of the earth fills consist of a dark, clayey river mud (*barro*), which was mixed into a slurry to fill the earlier temples when it was time to build a new one on top (Lacombe *et al.*, 2020). The barro is a well compacted fine-grained material, and no instability problems are reported in the tunnel stretches excavated into it.

The tunnels' network was mostly excavated immediately above the barro layer, into a layer of dark reddish-brown earth (*tierra café oscuro*) mixed with construction debris, river cobbles and broken lime plaster (Lacombe *et al.*, 2020). This is a clayey silty sand with gravel. However, assuming that the larger gravel or boulder inclusions are floating in the finer matrix, the mechanical behavior of this layer corresponds to that of a sand with fines. The uppermost and outer part of the earth fills, the most recent one to be put in place, is made of a fine yellow sand (*girún*), which is the coarsest soil used for the whole earth fill. This sand was used to refine the oldest temples and to complete the latest structures such as Esmeralda. Most tunnel stretches excavated inside the *girún* layer needed to be supported at a later stage with a stone masonry lining, or suffered diffused collapses that required them to be back-filled.

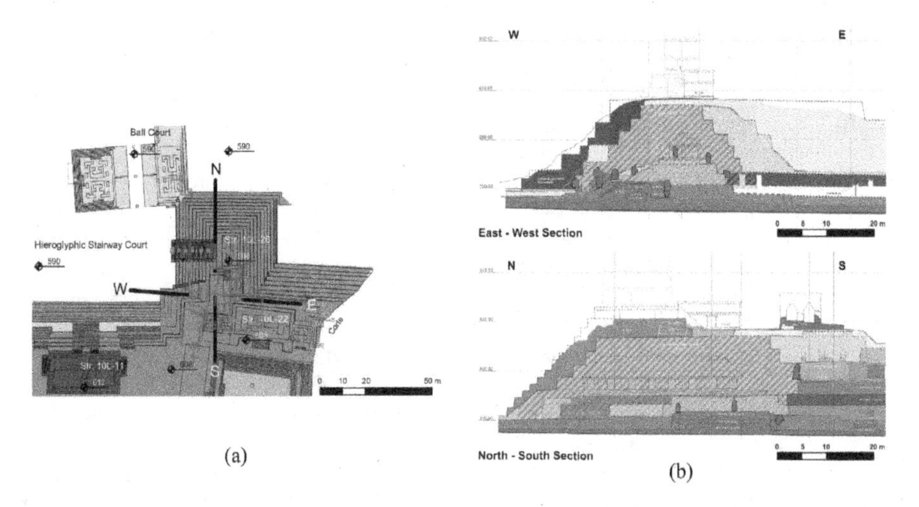

Figure 13. (a) Detail of the Acropolis plan showing the extension of currently open tunnels under the Hieroglyphic Stairway Temple (in red) – adapted from PAAC drawings; (b) Sections of the Hieroglyphic Stairway Temple showing the approximate positions of tunnels (in red) and the green hatch of the Esmeralda volume – adapted from C. Rudy Larios' PAAC drawings (modified after Pires *et al.*, 2021).

Data about tunnels position under the Hieroglyphic Stairway Temple, their internal shape, and the presence of lining, sometimes being parts of original Maya structures, were gathered by a recent 3D dimensional survey. Hence, four typical transverse sections were identified to carry out a parametric numerical analysis of their stability conditions (Figure 15) (Pires *et al.*, 2021). T1 and T2 identify the 'as excavated' tunnel sections, where no lining was applied to the walls; T3 and T4 those structurally retrofitted, that were supported by thick masonry walls as lining at a later stage. Sections T1 and T3 are fully inside the fill material, while T2 and T4 are sections of tunnel excavated adjacent to buried structures, that therefore play a role in their mechanical behavior.

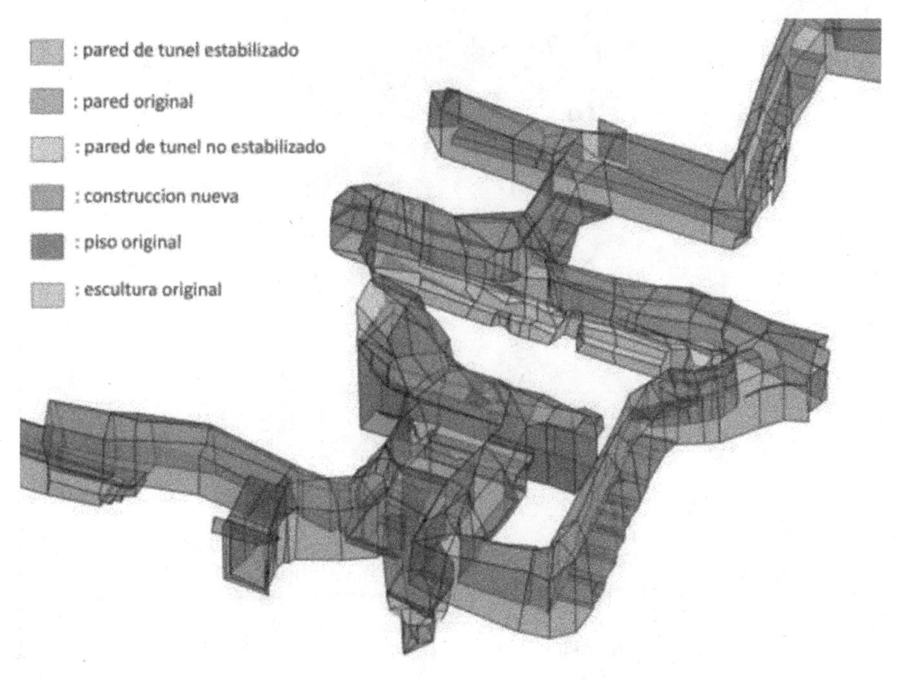

Figure 14. 3D model of part of one of the tunnels in the Acropolis (modified after Lacombe et al., 2020).

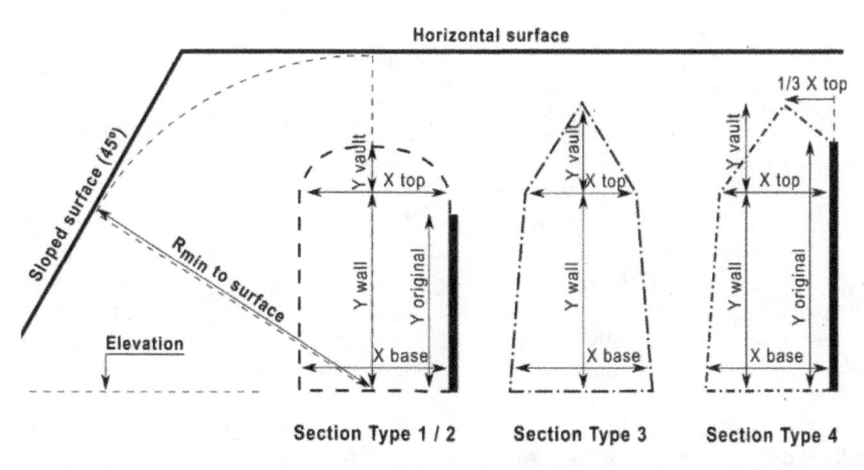

Figure 15. Shapes of typical tunnel sections: Types 1 and 2 (left), Type 3 (center) and Type 4 (right). The difference between type 1 and Type 2 is the presence of a buried Maya structure on one side for Type 2 (indicated as a thick black line) (modified after Pires et al., 2021).

The unlined tunnels in the *tierra café oscuro* layer have a typically stable arched vault (Figure 16), while the lined ones are vaulted on the top with a triangular corbelled arch. In most lined tunnels the lower walls are slanted. Section T4 is rather irregular and the final shape identified in the figure is just a rough approximation. The dimensions vary within each type but, in general, tunnels are small, serving as pathways for the archeologists, with an average width of about 1.0 m and a height ranging from 2.0 m to 3.0 m. Table 1 provides the statistical information obtained from the 3D survey. The distance between tunnels and external surfaces is represented in Figure 15 by the minimum distance R_{min}. This is measured from the axis at the base of the tunnel to the nearest ground surface.

Figure 16. Unlined stretch of a tunnel excavated in the *tierra café oscuro* (modified afterLacombe *et al.*, 2020).

Table 1. Dimensions of the four typical tunnel sections (for the definition of $Y_{original}$, Y_{wall} and Y_{vault} see Figure 15; μ = average, s = standard deviation) (Pires *et al.*, 2021).

Section type	Base width (m)		Top width (m)		$Y_{original}$ (m)		Y_{wall} (m)			Y_{vault} (m)	
	μ	s	μ	s	μ	s	median	μ	s	μ	s
T1	1.3	0.2	1.2	0.1	0.0	0.0	1.1	1.3	0.3	0.6	0.2
T2	1.2	0.3	1.2	0.2	2.0	1.1	1.3	2.0	1.1	0.4	0.3
T3	1.0	0.2	0.9	0.2	0.0	0.0	1.6	1.7	0.3	0.6	0.2
T4	1.1	0.2	0.9	0.3	2.6	1.7	1.9	2.5	1.4	0.7	0.2

The full length of the tunnels and the range of distance to the outside surface are summarized in Table 2. Here the surface is classified as sloped or horizontal. The slope of the stepped external surfaces of the pyramid-like structures is about 45°. In total, 360 m of tunnels can be found under the Hieroglyphic Stairway Temple. About 50% of these tunnels have a T3 section, while the other three types of section are equally distributed, as shown in Table 2. Most times, collapses occurred at the interface between lined and unlined tunnels or in unlined tunnels, as shown in the photos of Figure 17.

In 1942, only few years after the excavation of the first tunnels in Copan, stabilization works began by widening the tunnels (except where there were original architectural elements) and putting in place a masonry lining. The masonry walls mimic original structures from Classic Copan, and were made using mortars with Portland cement. The same cement was used also to repoint the ancient walls, making it sometimes hard to discern between additions and the original parts, even though the modern stones are usually smaller and more regularly sized. The preservation methods from the 1990's changed, keeping stabilization works but also backfilling some stretches. This time, the new masonry walls and arches did not directly touch the original walls, with the installation of plastic tarps between the new masonry and the original when necessary. With time, different local and diffused stabilization works were carried out, unfortunately without a precise strategy (Figure 18). In many sections, the stabilizing masonry structures are affected by crack patterns indicating a critical stress state.

Table 2. Total length of the four type of tunnel sections, and distance to the external surface (μ=average, s=standard deviation) (Pires *et al.*, 2021).

Section	Total Length	R_{min} to external surface (m)					
				sloped surface		horizontal surface	
Type	(m)	μ	s	min	max	min	max
T1	51.0	15.0	7.8	4.0	4.0	15.0	20.0
T2	56.0	17.0	4.4	-	-	8.5	20.0
T3	183.5	10.6	3.8	3.0	10.0	9.5	17.0
T4	66.5	7.7	4.3	2.0	11.0	9.0	12.5

Figure 17. Examples of local collapses: (left) at the interface between lined and unlined stretches; (right) in an unlined tunnel excavated in fine yellow sand (*girún*) (modified after Lacombe et al., 2020).

Figure 18. Different kinds of local stabilisation interventions (modified after Lacombe *et al.*, 2020).

2.3 *The contribution of geotechnical engineering: rational maintenance of tunnels*

Figure 19 shows in red the tunnel stretches where local collapse was observed from 2017 to 2020, and in blue the ones where the main problem was massive water intrusion during the storms of 2017. Clearly, the extension of damage is such to make the use of the tunnels network critical and dangerous. When looking jointly to the history of collapses and heavy rain or storms, a correlation is observed. Furthermore, the critical sections where collapse concentrates are in the unlined part, but close to the lined ones.

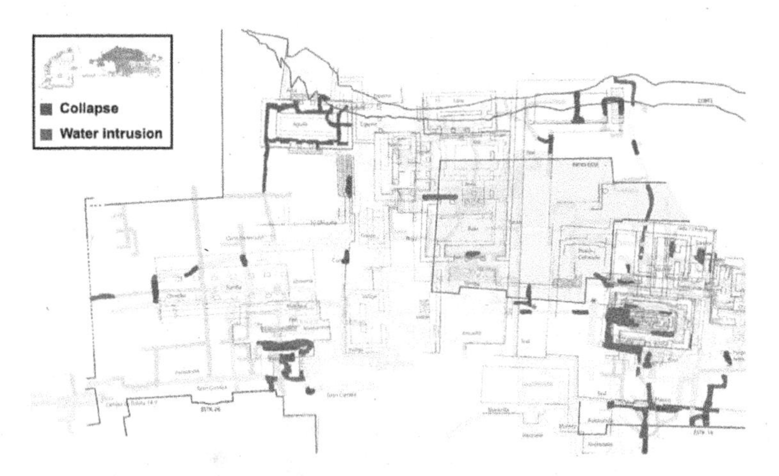

Figure 19. Map of local tunnel collapse and of water intrusion observed from 2017 to 2020 (modified after Pires, 2020).

Since the soil was originally compacted in unsaturated conditions, it has to be expected that a change in the degree of saturation has a mechanical effect, obviously more critical in the unlined sections. A main instance, happened after the installation of an impermeable geomembrane over the East Court (Patio Este) in 1998, confirms this observation: shortly after the installation (which was done without proper collection and disposal of rain water accumulated on top of the geomembrane), an hurricane (hurricane *Mitch*) struck the region, bringing intense rainfalls. Three areas along the East Court perimeter suffered tunnel collapse at the time of the hurricane, and three more the year after (Figure 20).

Figure 20. Aerial photo of the Copan Acropolis (east at top) showing approximate positions of tunnel collapses after hurricane Mitch (1998 in blue, and 1999 in yellow), with the position of the East Court membrane marked in red (modified after Lacombe et al., 2020).

In fact, the tunnel sections adjacent to the perimeter of the East Court suffered major infiltration in the earth fill from the edges of the improperly installed geomembrane because of the water build-up at ground level.

To check water effect on the mechanical behavior of tunnels, 2D FEM analyses were carried out with PLAXIS 2D on both lined and unlined sections. The earth fills in which tunnels have been excavated in the considered area is unsaturated *girún*. In this work, the shear strength of this unsaturated granular soil has been simply considered as:

$$\tau = (\sigma - u_a) \cdot tan(\varphi) + (u_a - u_w) \cdot S_r \cdot tan(\varphi) \qquad (1)$$

where $(\sigma - u_a)$ is the net normal stress, $(u_a - u_w)$ is the matric suction (s), S_r is the degree of saturation of the soil and φ is the effective shear strength angle. The second term of the second member of eq. (1) is often designated as 'apparent cohesion':

$$C_{unsat} = (u_a - u_w) \cdot S_r \cdot tan(\varphi) \qquad (2)$$

Since the apparent cohesion reduces as the degree of saturation increases, water infiltration caused by the frequent heavy rains locally reduces the available shear strength of the unsaturated soil. Therefore, the numerical analyses were carried out considering the cohesive term of the Mohr-Coulomb strength criterion as a state parameter (eq. (2)) (Figure 21), depending on the relevant soil-water characteristic curve.

The safety factor for the unlined section with an ancient wall on one side (T2) is 10% to 30% higher than for the section with no wall (T1) for low to medium degrees of saturation (Figure 22). However, approaching saturation the values converge, resulting into failure for $S_r > 80\%$.

Clearly, FS increases passing from unlined to lined sections. As long as drainage is ensured (i.e. there is no or little water pressure on the lining), the lining provides high safety margins to the tunnel, whatever the soil saturation degree, and ensures stability with high margins even in fully saturated conditions. But drainage has to be granted, which is not always the case on site. Real safety conditions can therefore be significantly lower, and have to be analyzed case by case.

The numerical results obtained (including those not reported here for the sake of brevity, see Pires et al. 2021) confirm how relevant the role of the degree of saturation in the compacted earth fill is for the stability of the unlined parts of the tunnels excavated in the Acropolis. Clearly, in the lined sections, the degree of saturation is locally higher than in the unlined ones because of the reduced exposure to draining surfaces. This explains why most collapses have taken place at the interface between the two kinds of section, where the gradient of saturation degree caused by the different boundary conditions triggers a water flow towards the closest outcome, which is the first unlined section. Eventually, this will locally result into a degree of saturation higher than in other unlined sections, further from the lined stretch, and thus in lower safety factors that may trigger the collapse of blocks.

To confirm these preliminary results, more extensive survey of the complex 3D tunnels geometrical layout is however needed, along with an adequate experimental characterization of the earth fill and of the tunnels lining masonry materials.

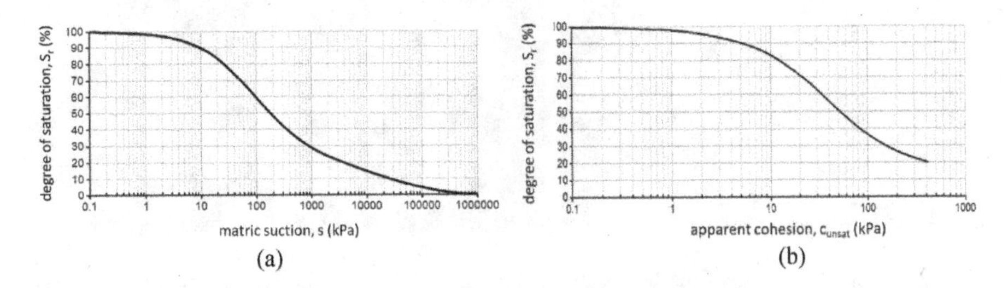

Figure 21. (a) Soil-water characteristic curve and (b) apparent cohesion versus saturation, for the silty sand infill (*'girùn'*).

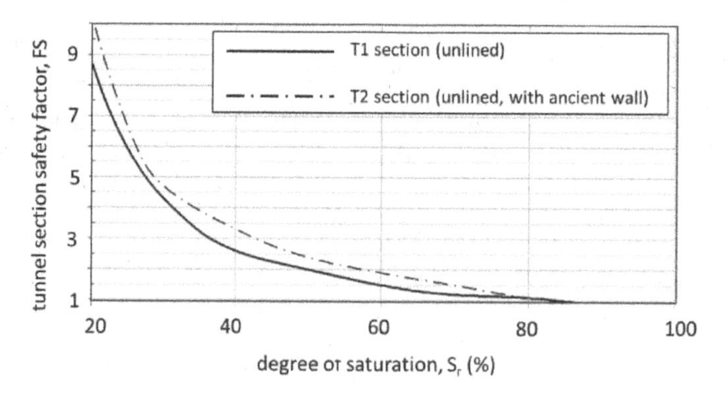

Figure 22. Safety factors for the unlined tunnel sections T1 and T2 at a depth of 20 m as a function of soil degree of saturation of soil (modified after Pires *et al.*, 2021).

These activities were planned in 2020 but never carried out because of the Covid pandemic, that dramatically reduced the possibility to operate on site and, for the author, to travel to Copan.

Based on the preliminary results herein summarized, the following recommendations have been made for the short term: correctly collect and dispose rain water at ground level, minimizing infiltration; carry out a constant monitoring of the water content in the soil around tunnels, especially at the transition between lined and unlined sections, to anticipate safety issues (instrumentation still to be put in place); ensure drainage in the lined tunnel sections to minimize water pressure on the lining; avoid the addition of new tunnels too close to the existing ones, continuously updating the 3D map of the tunnels system.

From a strategic point of view, the preservation of the invaluable heritage existing deep inside the Copan Acropolis (Figure 23) would ask for a massive application of these preservation actions.

In the medium to long term, new lining should be installed where necessary, respectful of the critical chemical conditions of the buried Maya stuccos, and possibly new tunnel stretches connecting to the outside or to other tunnels should be added as escape paths for safety reasons, to avoid extremely long blind parts. A complete lack of interaction (either mechanical or chemical) between the Maya buried structures and the new lining should be also ensured.

However, because of the length of the tunnels network and of the critical lack of funding, this whole program is unfeasible, at least in the short period. Therefore, backfilling of some parts in critical safety conditions will be required, after a more detailed analysis of the different parts of the network, sharing the decision among archaeologists, art historians and engineers (geotechnical, structural and chemical). Backfilling will in fact make not accessible sites of possible interest, and will not allow further investigations along those directions.

Figure 23. Examples of parts of the buried Maya structures discovered in Copan through archaeological tunnelling (modified after Lacombe et al., 2020).

3 INTERFERE WITH THE PAST OUT OF NECESSITY

3.1 *The Crypta Neapolitana*

This tunnel has a long-lasting story, that goes along with the development of the town of Napoli (Italy), and has been already dealt with from a geotechnical point of view in the recent past (Amato *et al.*, 2001, Viggiani, 2017). Its conception is attributed to Lucio Cocceio Aucto, a Roman architect (in a broad sense, being him a fine engineer as well) of the first century AD, highly appreciated by the emperor Octavianus Augustus and designer of the Temple of Augustus (later turned into the catholic church of San Procolo) in Pozzuoli and of the other two long Roman tunnels (Seiano Grotto and Cocceio Grotto) in the area around Naples. The Crypta Neapolitana runs close to the east-west direction, and crosses the Posillipo hill, that separates the Gulf of Pozzuoli from that of Naples (Figure 24).

According to some recent studies of the Roman tunnels in the Neapolitan area (Escalona, 2022), the hypothesis that a smaller tunnel along its longitudinal axis already existed before Cocceio's work is also considered. Well before the Romans took the area, information reported in the Odyssey of Homer (11.14) and interpreted by historians indicate that the mysterious population of Cimmerians lived in the area around the Lake Avernus (close to the later Cocceio Grotto, see Figure 26 on the left) during Ulysses peregrination on the Italian coasts. The Cimmerians were an old nomadic Indo-European people coming from the Caspian steppes, on the northern shores of the Black Sea, who probably settled in the Campania Region (as also referred by Strabo and Plinius the Old) between 1000 BC and 800 BC. During their migration, they had previously settled in Cappadocia, where they learned to excavate and live in underground spaces. Ever since they have been known as the people of darkness, living in underground houses, connected among them with tunnels. In the Phlegrean Fields, they had an oracle (a Sibyl, a sorority sister) that was venerated by the pre-Hellenic native populations. Those who lived about the oracle had an ancestral custom, that no one should see the sun, but should go outside the caverns only during the night.

Escalona (2022) claims that the existence of a small chapel within the Crypta Neapolitana may indicate a deeply seated room, originally conceived for burial (Cimmerians used to bury their Kings or leaders in the deepest heart of mountains) or religious reasons, to be reached through a small tunnel coming from the western side of the hill. Even though this is still a matter of scientific

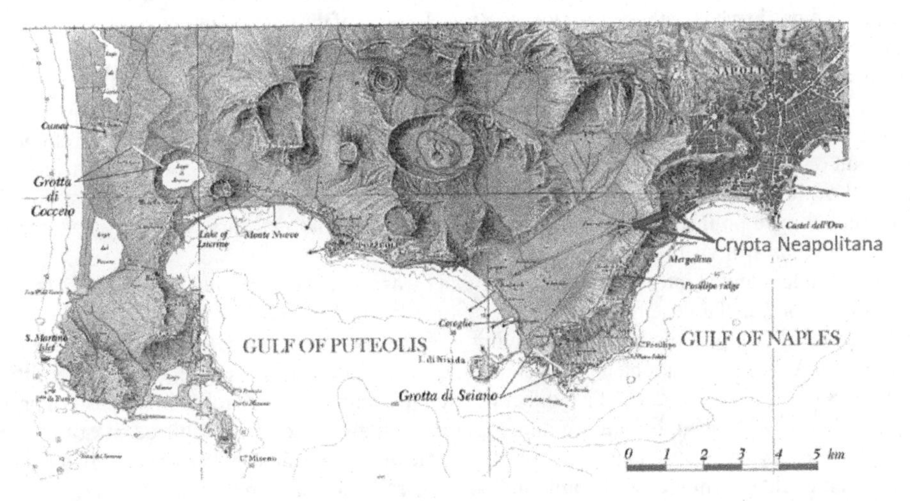

Figure 24. Map of the coastal region called *Phlegrean Fields* with in evidence the location of the Crypta Neapolitana through the Posillipo hill (The old town of Naples is on the East side). The topography refers to the second half of the XIX century, while nowadays the whole area on both sides of the tunnel is densely urbanized, the two towns of Napoli and Pozzuoli merging into a unique built environment.

discussion, it would somehow justify the mediaeval legend claiming that the tunnel was excavated by the poet Virgil in a single night: in fact, this myth could reflect a rather fast construction of the tunnel, corresponding to a possible renovation or enlargement of an originally existing one, and not to a completely new excavation. Just an hypothesis, certainly fascinating.

In any case, the Roman tunnel excavated (or enlarged) in the first century AD had the clear goal to enhance and simplify the communication between the very active town of Puteolis (nowadays Pozzuoli), where the largest commercial harbour of western Mediterranean sea was located, and the growing town of Neapolis (Napoli), and was part of a large number of public works constructed in the area under the coordination of Marcus Vipsanius Agrippa, who was a close friend and son-in-law of the emperor Augustus (Ferrari & Lamagna, 2015). The Aqua Augusta (Serino) aqueduct was built later, crossing the Posillipo hill in parallel to the Crypta Neapolitana and far only a few meters from it, letting the water flow from the far Serino springs to Puteolis. The original Roman cross section of the tunnel was between 4 and 5 m large, and likely no more than 4 m high (may be even less), with a curved vault and vertical side walls. Its shape is mostly lost, because of subsequent reshaping carried out for centuries. The tunnel is 711 m long, with two inclined ventilation shafts on the two sides of the hill. A longitudinal section with the recent shape of the tunnel is reported in Figure 25.

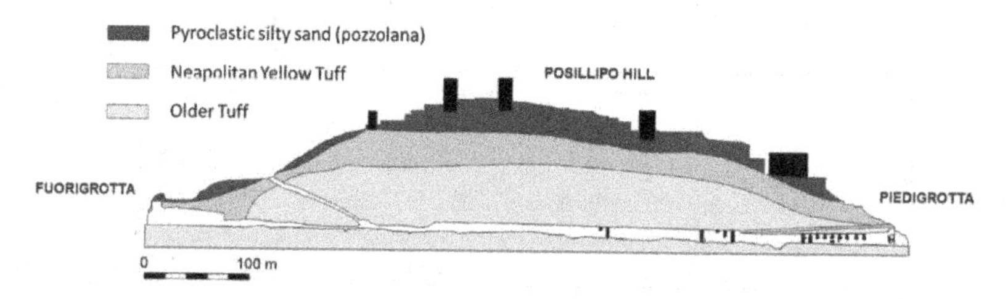

Figure 25. Cross section of the Posillipo hill along the *Crypta* longitudinal axis, with indication of the geological units.

As schematically sketched in the figure, the tunnel crosses different pyroclastic formations. The deeper one is a weakly cemented old tuff, with poor mechanical properties, that can be considered a transition material between an uncemented pyroclastic silty sand (*pozzolana*) and a tuff, with some specific sections in which cohesion is close to zero. Because of the uncomplete cementation process, this formation is not affected by the typical syngenetic cooling fractures observed elsewhere in town in well cemented tuff. Neapolitan Yellow Tuff covers the old tuff in the hill of Posillipo. As well known, it is a soft, light rock with a good degree of cementation, with a number of subvertical and sub-horizontal cooling fractures. The Crypta Neapolitana crosses the Neapolitan Yellow tuff formation only in the first 100 m on the western side, and only for few meters on the eastern one. So, most of the tunnel is excavated in the old tuff. Because of this, the eastern and western parts of the tunnel were unlined, while the central part was lined with masonry, with only few parts of *opus reticulatum* still visible.

The first documented intervention into the tunnel was carried out in 1445 by the king Alphonse of Aragòn to lower the eastern entrance. A century later, the Spanish viceroy Pedro de Toledo further lowered this part of the tunnel, paving the roadway. Other retrofitting works were done under Pedro Antonio d'Aragona (XVII century), Charles III Bourbon (1748), the Municipality of Naples (1893, with the insertion of a number of strengthening masonry arches in the eastern part of the tunnel), till 1917 when, because of ongoing local collapses and of the diffused risk of collapse of blocks, the Crypta Neapolitana was closed, its crucial connecting role between the two sides of the hill being taken from the newly built and close by rail and car tunnels.

In 1930, the eastern part of the Crypta was partly filled, and the floor raised back as much as 9 m, to accommodate the entrance to an outside green area where the supposed tombs of Virgil and of the Italian poet Leopardi are placed. Figure 26 show a reconstruction of the changes in time of floor elevation (Amato *et al.*, 2001).

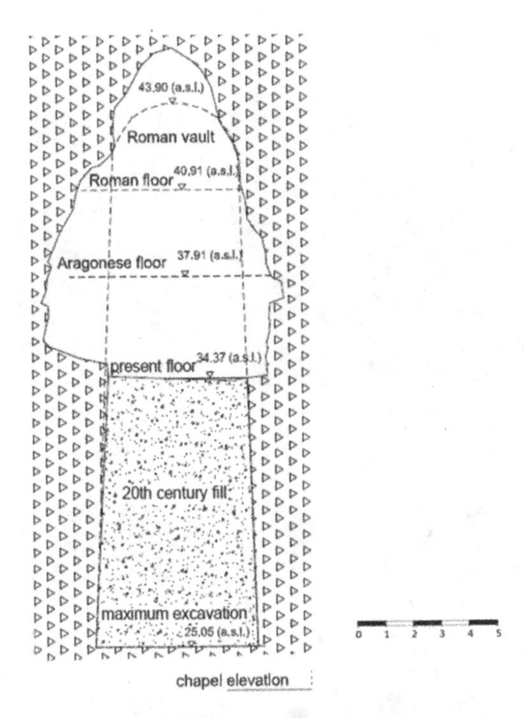

Figure 26. Cross section of the Crypta Neapolitana at 20 m from the eastern entrance in its current state, along with the reconstruction of the different elevations of the floor in time (modified after Amato *et al.*, 2001).

Figures 27 and 28 show the two entrances of the tunnel. Recently, bolted steel arches were placed to support the vault in the first 30–40 m on the western side (Figure 27).

The tunnel is actually in very bad static conditions, with diffused collapse of large blocks along its axis, especially in the central part, where site survey is now almost impossible for safety reasons. The Roman masonry lining has collapsed along the whole lined stretch, with only few parts, with no static role, still standing. A large part of the masonry supporting structures placed in the 19th century collapsed. Figures 29 and 30 show some photos of a recent survey (January 2022) from both entrances of the tunnel, giving a clear idea of the widespread state of instability.

Figure 27. Western entrance of the Crypta Neapolitana in a recent photo. On the vault, the recently placed bolted steel arches.

Figure 28. Eastern entrance of the Crypta Neapolitana in a painting (van Wittel, XVIII century, oil on copper) and in a recent photo.

Figure 29. From top left: pictures from the western entrance till ca. progressive 350 m. Collapse of blocks, tension cracks and collapse of masonry lining are clearly visible.

Figure 30. From top left: pictures from the eastern entrance till ca. progressive 150 m. Masonry arches well preserved in the first part, with complete collapse of side walls in the final part. Large subvertical rectangular blocks sometimes on the verge of instability failure.

3.2 *The contribution of geotechnical engineering: tunnels reinforcement and reuse*

To check the static conditions of the tunnel, an in situ investigation was carried out about 20 years ago. Boreholes from the ground surface and core drills from within the tunnel were executed. Samples retrieved during coring were tested to get the uniaxial compressive strength σ_c, while flat jack tests were carried out in some sections of the tunnel to quantify the in situ vertical stress (Figure 32). In the old tuff, i.e. for most of the tunnel length but the parts close to the two entrances, the compressive strength values are in the range 0,5 MPa < σ_c < 2 MPa, while in the Neapolitan Yellow Tuff σ_c is as high as 6 MPa. The comparison between σ_c and the vertical stress estimated with flat jack indicates that, for distances from the western entrance between 100 m and 200 m, the rock around the tunnel is locally on the verge of failure, consistently with the evidence of diffused collapse.

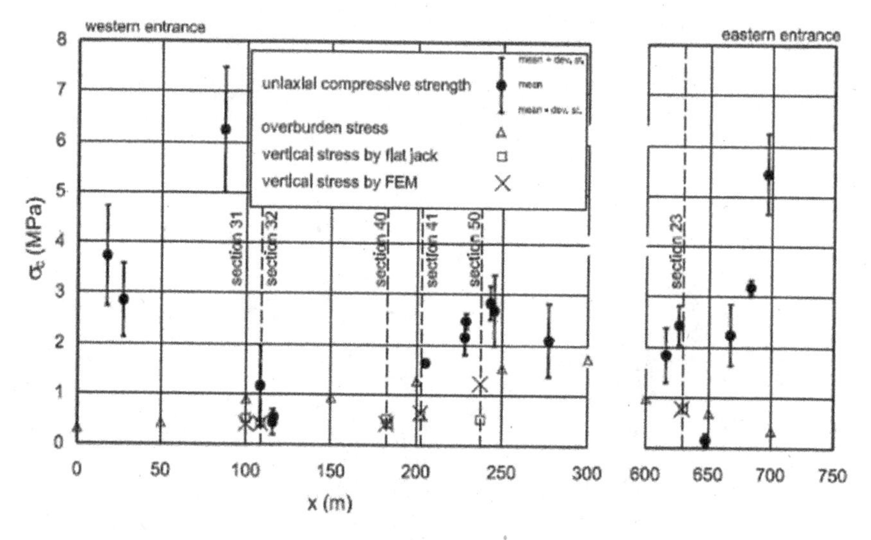

Figure 31. Values of the uniaxial compressive strength of tuff and in situ stress along the axis of the Crypta. Distances taken from the western entrance (modified after Amato *et al.*, 2001).

A detailed geometrical survey of the tunnel was also carried out with a laser scanner (Figure 32). The characterization of the rock mass was done using the Hoek & Brown failure criterion (Bilotta, 2022), then converting the parameters into the ones (c, φ) of the Mohr-Coulomb failure criterion, to be used in elastic-perfectly plastic 2D analyses carried out with Plaxis. Figure 33 briefly summarises the results in terms of plastic stress points (red and white points corresponding respectively to shear and tension failure), reporting also the values of the safety factor corresponding to a mechanism of local collapse (blocks failure). As expected, the most critical situation is in section C (see Figure 33) (FS < 1), in the part of the tunnel with the worst mechanical properties, while the other sections are globally stable, with the exception of section F (FS $\approx$ 1).

Assuming that yielded volumes progressively collapse, the numerical analyses indicate that most times the critical mechanism is the creation of tension cracks on the sides and the subsequent collapse of blocks, consistently with the experimental evidences. Since this mechanism could be perverse (i.e. local collapses, and subsequent reshaping of the sections, may trigger further collapses and reshaping, and so on), the overall indication is that the preservation of this Roman tunnel asks for a diffused reinforcement.

This brings the discussion to the constraints posed by the relevance of the historic site. The wounds of time are in this case clearly visible, and there is no reason to try and replicate the original Roman cross section, whose traces are mostly lost. The largest part of the tunnel has been naturally reshaped as a result of a stress redistribution within the rock mass or later human interventions. Then, the question arises about what should be preserved. In this case, it seems that the strongest legacy is linked to the role of the Crypta Neapolitana, which is the connection of two neighbourhoods separated by the hill, as demonstrated by their names: *Fuorigrotta* (literally, *out of the grotto*) the one on the west side, and *Piedigrotta* (literally, *at the entrance of the grotto*) the one on the east side, the grotto obviously being the Crypta Neapolitana. This path has been a living part of the city for at least two thousand years, and still keeps traces of this long lasting life (Figure 34).

Therefore, according to the author's opinion a hard interference with the iconic integrity may be justified in this case to let the Crypta Neapolitana return to its role of a connecting path. In particular, it should become a pedestrian tunnel that would correspond to the best possible preservation of its functional integrity. This is actually a matter of discussion with local authorities. Interventions should include: a new lining in the central part of the tunnel, with new materials and local openings to show details of the old masonry lining or of local connections with the close-by Aqua Augusta aqueduct; a retrofitting and underpinning of the masonry arches on the eastern side; bolting of potentially instable blocks, and bolted nets in the unlined western part of the tunnel excavated in the Neapolitan Yellow Tuff.

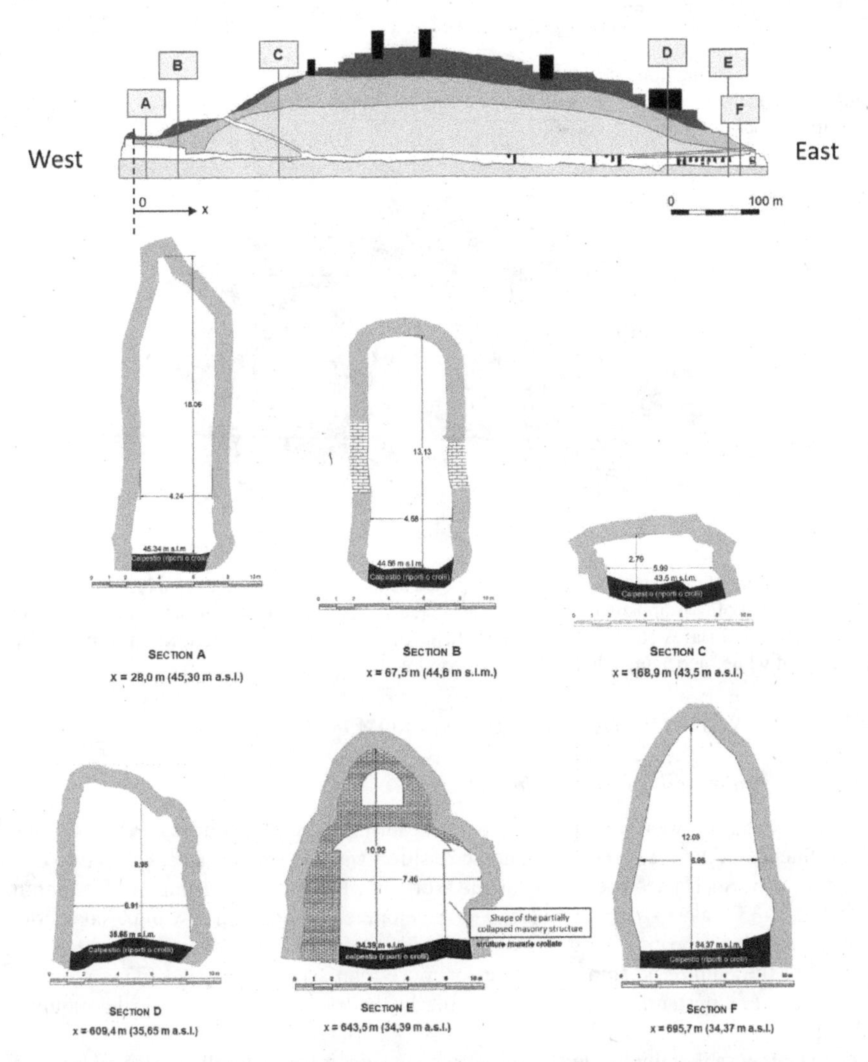

Figure 32. Cross sections (obtained by laser scanning) of the Crypta Neapolitana considered in the numerical analyses, with indication of their position (distance from western entrance) along the tunnel axis.

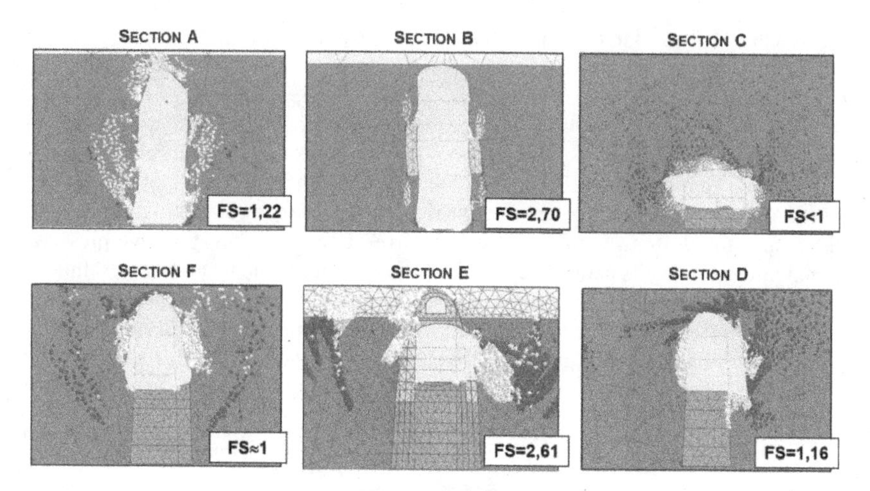

Figure 33. Results of the numerical analyses in terms of plastic stress points (red and white points corresponding respectively to shear and tension failure). FS is the safety factor corresponding to a mechanism of local collapse (blocks failure) (Bilotta, 2022).

Figure 34. Painting of the Persian God Mithra, depicted as the sun, within the Crypta Neapolitana (courtesy of L. Amato). In Napoli, Mithraism dates back to the Hellenistic period, and spread during the 5th century A.C. through prisoners and slaves coming from Cilicia. In some basements in the old town, bas-reliefs depicting the Persian God while sacrificing a bull can be still admired.

4 INTERFERE WITH THE PAST LEARNING FROM IT

4.1 *The Theodosian walls of Constantinople*

The city of Byzantium was relatively unimportant during the early Roman period, but when the Roman influence in the east grew, its strategic position was recognized, and the need to defend it became a priority. Such a need became even more urgent when the Roman emperor Constantine the Great (AD 272-337) gave a sharp impulse to its development, building a new imperial residence in the city and renaming the city Constantinople after himself. In AD 328, the city eventually became the capital of the empire, keeping this role for more than a thousand years. This move, and in general the age of Constantine, conventionally marks the transition from classical antiquity to the Middle Ages.

Since the city was initially located on the European side, it was naturally protected by the sea on the north, east and south, the major problem being to protect it from possible attacks from the inland, on the west side. Constantine arranged a wall to be built (Figure 35) from the Golden Horn to the

Sea of Marmara. As the city grew, however, this confine had to be overcome to accommodate all the people attracted from the new status of the city. An impressive double line of walls was then built on the land side during the reign of Emperor Theodosius II (408–450 AD), after whom they were named, about 2 km to the west of the old Constantinian Wall. The protective system, completed in AD 423, is the result of the skills and dedicated work of Anthemius, a Praetorian Prefect of the East, who did not see his work completed as he died in 414. In their final configuration, the Theodosian walls had a length of about 7 km, completely surrounding the city (Figure 35), thus creating a barrier not only on the west side but also along the whole coastal perimeter.

On the land side the system consisted of two closely spaced defensive lines (Figure 36): an inner wall, with a maximum height of 12 m, and an outer wall, with a lower height, each one fortified by towers placed at some tens of meters apart. The inner towers are as high as 24 m, while the outer ones are ca. 10 m high. The ground level on the terrace (*peribolos*) between the two walls is some meters (ca. 5 m) higher than that (*parateichion*) outside the outer wall, thus giving a dominant

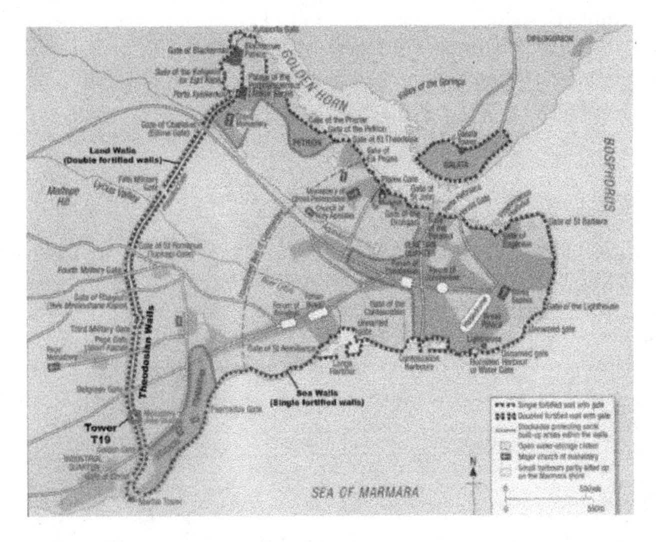

Figure 35. Map of Constantinople at the beginning of the Ottoman Period, with the location of the first defensive wall of Constantine and the location of the outer, much longer Theodosian walls (http://romeartlover.tripod.com/Murter.html). Tower T19, dealt with in this section, is indicated in the south reach of the latter wall.

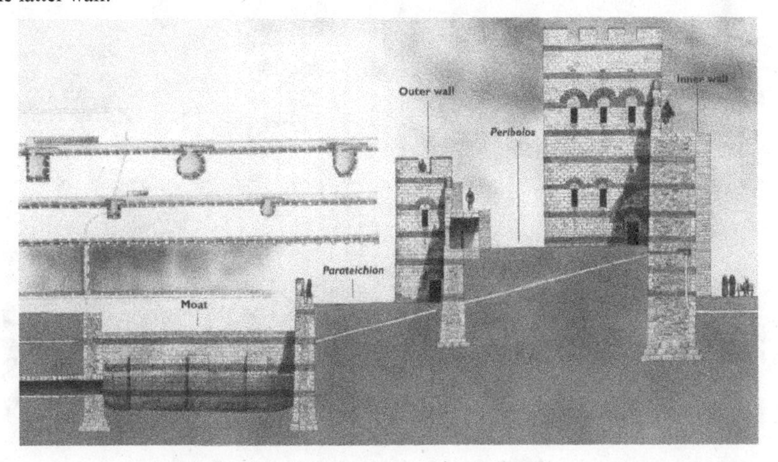

Figure 36. Typical cross section and plan of the Theodosian walls and towers. On the top left of the plan, the different shapes of the towers (modified after Turnbull, 2004).

position to the defenders in case of attack. The *parateichion* is confined on the outside by a moat, originally filled with water supplied by a sophisticated system, as a first defensive means.

Each tower had a battlemented terrace on the top. Its interior was usually divided by a floor into two chambers, not communicating with each other. The lower chamber, opened through the main wall to the city, was used for storage, while the upper one could be entered from the wall's walkway, and had windows for viewing and battling. Access to the wall was provided by large ramps along their side (Turnbull, 2004). The lower floor could also be accessed from the *peribolos* by small posterns.

This defensive system became legendary in the Middle Ages, deemed invincible and considered impregnable for any medieval besieger. The only exception was the sack of 1203-1204 during the fourth Crusade, that was a turning point in medieval history because of the decision of the Crusaders to attack the world largest Christian city. Before and after it, the Theodosian walls saved Constantinople – and the Byzantine Empire with it – during many sieges, even though with the sack of 1203 the city decadence began, and large parts of the empire were lost. The Ottomans unsuccessfully attempted to take the city in 1396 and 1422, eventually succeeding on 29 May 1453 under the guidance of sultan Mehmed the Conqueror, after a six-week siege in which a crucial role was played by the 8 m span cannons used by the besiegers. Part of the walls were largely damaged by the gunpower, and then restored by the Ottomans.

The walls were largely maintained intact during most of the long Ottoman period (1453-1922). Sections began to be dismantled only in the 19th century, as the modern city outgrew its medieval boundaries. Despite the subsequent decadence and lack of maintenance (Figure 37), many parts of the walls and towers survived and are still standing today.

Despite all the sieges they had to face throughout their history, the Theodosian walls were damaged much more by earthquakes and floods than by enemies' attacks. Earthquakes, in particular, were a major source of damage. Figure 38 reports the epicentres and years of the main historical earthquakes around the Marmara region, showing the extremely high seismicity of the area. The strong Kocaeli-Adapazari earthquake of AD 447 (only 24 years after walls completion), for instance, resulted in the partial collapse of 57 towers and large sections of the walls, and also subsequent

Figure 37. State of decadence of the Theodosian walls in the early 20th century, with clear evidence of unrepaired seismic damages to the towers (German Archaeological Institution, 1939).

major earthquakes (1509, 1719, 1754, 1766 and 1894) caused significant damages to the walls and towers (Ispir et al., 2014). Repairs were therefore undertaken on numerous occasions, as testified by the inscriptions commemorating the emperors or their servants who undertook the restoration works, and most of the surviving towers of the main wall have been rebuilt either in Byzantine or in Ottoman times.

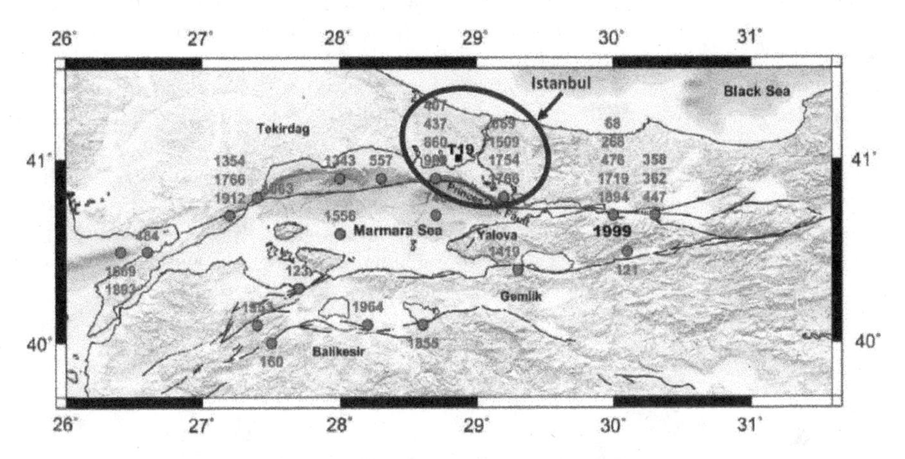

Figure 38. Historical earthquakes in the Marmara region (modified after Ambraseys, 2002), indicated with epicentral position and date. The degrees on the frame indicate North parallels (on the vertical sides) and meridians (on the horizontal sides). The area of modern Istanbul is highlighted, with the position of tower T19.

The inner wall, which is connected to the taller towers, has a thickness of ca. 5.00 m and is made of a rubble core confined between two shells built of nicely shaped squared blocks, having a thickness of 30-50 cm (Ahunbay and Ahunbay, 2000). The original, 5th century Byzantine masonry is of a fine quality, with brick bands – approximately 0.40 m high – laid at regular intervals of dressed stones. These bands run through the entire thickness of the wall, binding the structure firmly at different levels. The blocks used for the inner towers and wall in the original fifth-century construction are made of sandstone, quarried close by. The original mortar was a mixture of lime, crushed bricks and bricks powder. In subsequent restoration works, lime-based mortars were used as well.

While until the end of the 12th century the reconstructions largely replicated the original model, later modifications introduced a clear change, from both the formal and the construction point of view, and the strengthening brick bands were not used any more. Since the nearby quarries run out in the Byzantine period, after the 12th century the blocks used to restore walls and towers had different origin and had a much more irregular shape. Because of all these reasons, the resulting masonry structure was less refined and certainly less stiff.

UNESCO's designation of this defensive system as a World Heritage site in 1985 resulted into an extensive, large-scale restoration and conservation program, still under way. On the towers, the restoration works consisted in cleaning them from the vegetation and rebuilding the masonry structure, trying to replicate wherever possible the original Byzantine style. Most times, the walls and towers were over-restored and refaced rather than being repaired, possibly destroying many historical evidences.

The strong 1999 Kocaeli earthquake, whose epicentre was located about 80 km south-east of Istanbul, caused major damages to parts of the walls and to some towers. Interestingly, in some cases the damages were larger on the towers that had already undergone restoration works. As quoted by Turnbull (2004), prof. Zeynep Ahundbay (chair of Historic Preservation at Istanbul Technical University) said after the 1999 earthquake: '*The restoration campaign of the 1980s has been criticised due to its resort to the reconstruction of ruined towers and gates instead of stabilising* *and consolidating dangerous structures. The performance of the 20*th *century repairs during the recent (1999) earthquake ...omissis...constitutes a good lesson for future restoration*'.

Indeed, words to be reminded. Because of the symbolic heritage carried by this impressive, iconic defensive system, a contribution to interpret the dynamic behaviour of the walls and towers was recently attempted (Flora et al., 2021, Somma *et al.*, 2022) with the precise goal to plan better preservation actions for the future, cooling down what seems to be an endless game between men and nature. The attention was initially focused on one rectangular tower (tower T19), located on the south side of the land walls, close to the Marmara Sea and between the Belgrade and Golden Gates (Figure 35). It is one of the tallest towers connected to the inner walls, restored in the early 1990's, thus before the 1999 Kocaeli earthquake struck Istanbul. Different construction and restoration techniques are now superimposed on it (Figures 39 and 40) (Sarimese, 2018): the lower part is still made of original Byzantine masonry, while the upper part is only partially (in the centre of the west side) made of the Ottoman one. The recent restoration works completed the tower in the original Byzantine style. As a consequence, the brick bands do not completely bound the structure. The earthquake largely damaged the tower, causing the opening of cracks at the connection between the reconstructed and the Ottoman masonry (Figure 39b) on the west side, typical diagonal cracks on the windows on the south side and the collapse of the whole east side and of upper corners wedges.

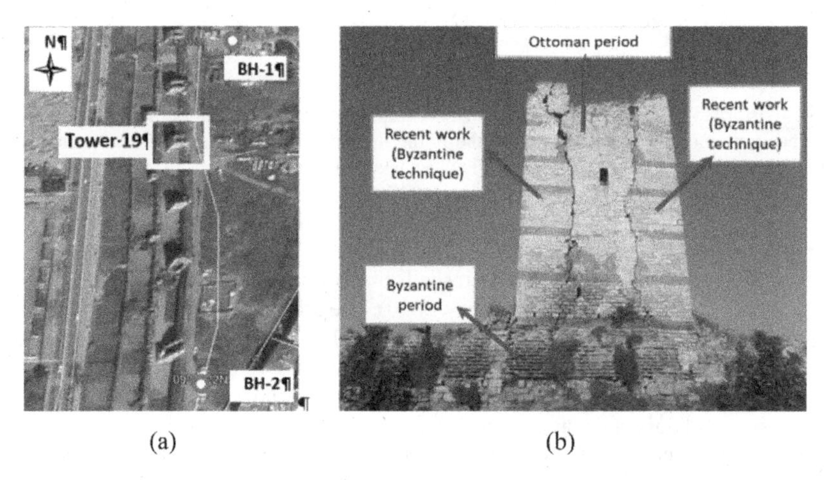

Figure 39. (a) Position of tower 19 with indication of nearby bore holes (BH-1 and BH-2) and (b) recent picture of the west side of the tower. The different, superimposed construction techniques can be clearly seen: the original Byzantine one (before 12th century) in the lower part, the later Ottoman one in the central part, without brick bands, and on the sides the restoration works carried out with the Byzantine technique, using the brick bands (Sarimese, 2018). The damages caused by 1999 Kocaeli earthquake are clearly visible.

Figure 40. Recent views of all sides of tower 19. The damages caused by 1999 Kocaeli earthquake are clearly visible.

4.2 *The contribution of geotechnical engineering: avoid preservation mistakes*

Clearly, all the restoration work was lost in what cannot be considered an unexpected natural event. To carry out a simulation of what has happened in 1999, detailed laboratory and in situ investigation was carried out (Figures 41 and 42), and the closest recorded outcrop motion (Fatih station, 4.8 km away) was deconvolved at the bedrock, using the soil profile and properties at the Fatih station site (Ince, 2008), and then considered as the seismic input at the bedrock of the tower T19 site (Flora *et al.*, 2021).

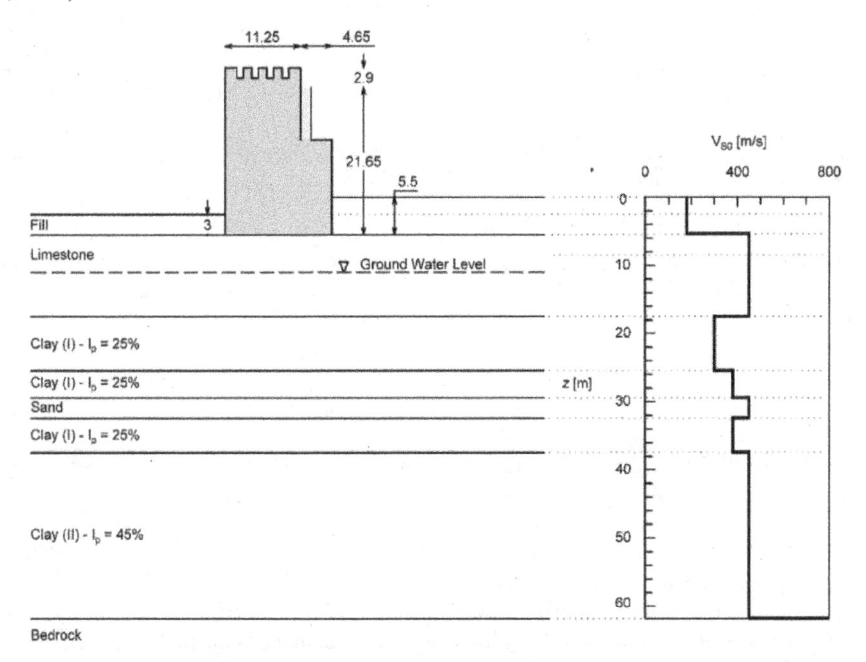

Figure 41. Soil and V_S profiles at the site of tower 19 (modified after Somma *et al.*, 2022).

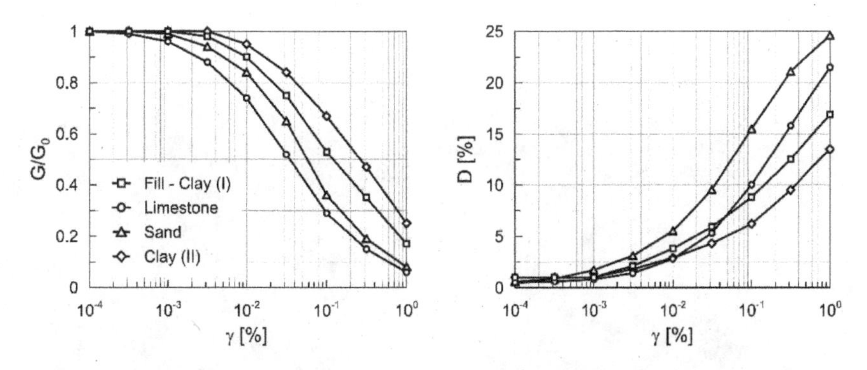

Figure 42. Shear modulus and damping ratio curves for the different soil layers (modified afterFlora *et al.*, 2021).

Seismic site response analysis shows a relevant amplification effect, with a fundamental frequency of the subsoil of 1.55 Hz (Figure 43a), close to the predominant frequency of the Kocaeli signal (1,85 Hz, Figure 43b), indicating the occurrence of resonance. Thus, the remarkable amplitude of the Fourier spectrum close to this frequency is not a startling result, with a value of the calculated PGA as high as 0.34 g.

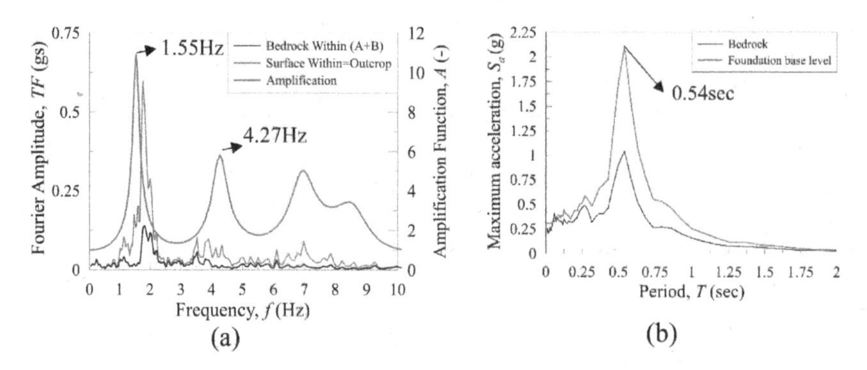

Figure 43. Fourier (a) and acceleration (b) response spectra compared with the input motion applied at the base (bedrock) of the soil profile (Somma *et al.*, 2022).

3D dynamic numerical analyses were carried out to interpret tower T19 behaviour during the 1999 earthquake (Figures 44 and 45), considering soil-structure interaction (SSI) or a fixed base, an equivalent linear elastic or a non-linear elastic-plastic behaviour for the subsoil and the tower. The tower was modelled considering two extreme scenarios, i.e. a less stiff Ottoman style masonry (M1) and a stiffer Byzantine style masonry (M2), being the real structure composed of both. The seismic input was separately applied in the x or y directions, defined respectively as the longitudinal and transversal axes of the wall (Figure 44).

The details of the analyses are reported in Somma *et al.* (2022). Figure 46 indicates that the existence of a relatively stiff layer of limestone underneath the foundation reduces the effect of dynamic soil-structure interaction; therefore, the natural period T of the tower shows relatively small increments considering the more realistic, compliant scenario (i.e. with SSI). However, being on the raising part of the response spectrum, even a small increase of T may correspond to a relevant increase of the seismic demand, as extremely evident in the x (less stiff) direction for the Ottoman (M1) masonry (top left of Figure 46), for which the seismic demand is the highest possible (peak of the spectrum). Therefore, this is a typical case in which neglecting SSI would not be conservative and would lead to an underestimate of the seismic action on the tower even if the tower is founded on a relatively stiff material.

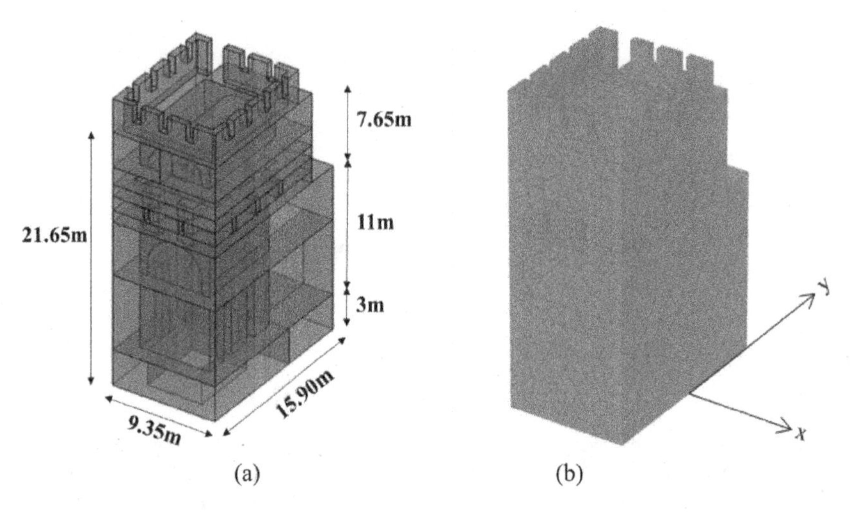

Figure 44. Tower 19: (a) geometric model and (b) Plaxis3D mesh (x=wall's longitudinal axis; y=wall's transversal axis).

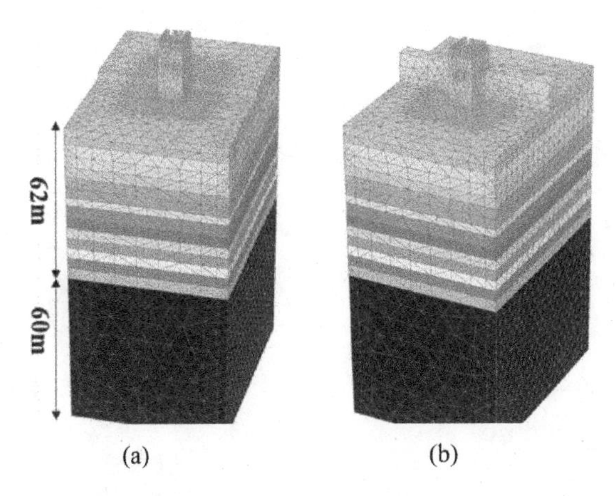

Figure 45. Plaxis3D model of the subsoil with Tower 19 considering (a) or neglecting (b) the inner wall.

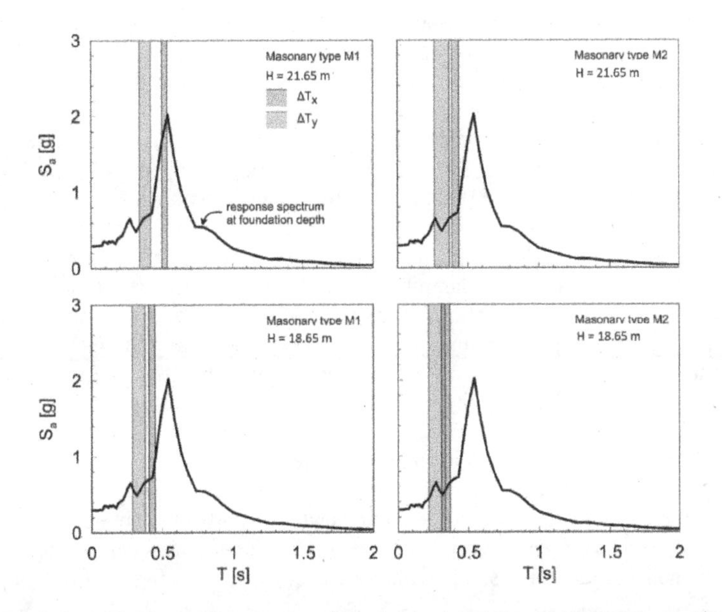

Figure 46. Equivalent elastic analyses: tower 19 seismic demand considering or neglecting dynamic soil-structure interaction in the two directions x and y, for both types of masonries (M1 and M2) and for different ways to consider the height of the mass centroid: at the foundation level (H = 21.65 m) or at ground level (H = 18,65). Period elongation indicated by the coloured bands (see legend on top left) (modified after Somma et al., 2022).

However, Figure 46 also shows that the seismic demand of tower T19 is for the Kocaeli earthquake always very high, whatever the considered scenario and earthquake direction.

The effect of seismic shaking is also shown for the non-linear analyses in terms of maximum principal tensile stress shadings in Figure 47, indicating that tensile strength (200 kPa) is reached in large areas, the red shading suggesting possible critical failure mechanisms, mostly concentrated below the first floor and at the first floor itself. These mechanisms are certainly consistent with the ones that took place in 1999 and largely damaged the freshly restored tower (Figure 40). Despite unavoidable uncertainties and simplifications (taken into account considering different scenarios), all the results demonstrate that the partial collapse of tower T19 during the Kocaeli earthquake is

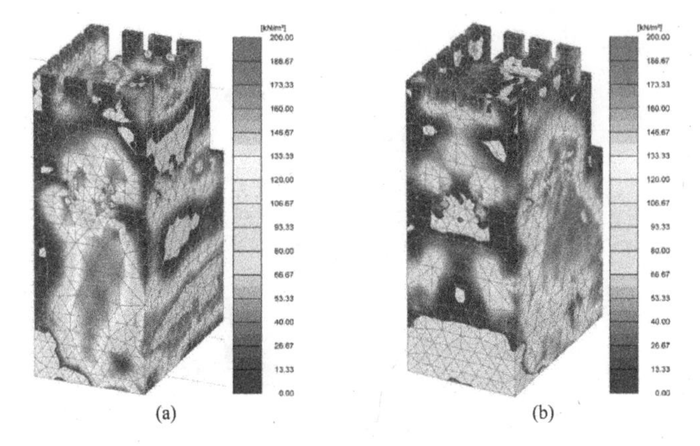

Figure 47. (a) Contours of principal maximum tensile stresses for tower 19 (compliant base, masonry M1): (a) seismic action in the x direction; (b) seismic action in the y direction.

not a surprise, being just the repetition of an event happened several times since Theodosian walls construction in the fifth century. Similar considerations could be done on the other towers damaged during the same earthquake, with differences among them linked to different site amplification effects on the seismic demand.

Then, the question to ask is what to do to step out of this reconstruction and destruction game between man and nature. From the geotechnical engineering point of view, the idea of using a Geotechnical Seismic Isolation (GSI) solution is intriguing, because it is the best way to avoid touching the structure to be protected, thus limiting to a minimum the interference with the heritage carried out by it. GSI technologies may aim to either reduce or increase soil rotational and translational stiffnesses, to increase damping or to add some soil mass to the system. All these actions modify soil-structure interaction, either reducing or increasing the natural period of the structure (the choice therefore depending on the response spectrum of interest and on the original structural period T), and can be achieved with a number of technologies (e.g. Flora et al. 2018). However, because of the high seismicity of the Istanbul area and of the poor structural qualities of tower T19 (and of the others as well), GSI by itself would not be sufficient.

This is confirmed by the results of the analyses carried out considering two simple and opposite GSI interventions: (i) lateral disconnection of the structure from the soil to the foundation base; (ii) stiffening of the soil around the tower, to be obtained with partially overlapped jet grouted columns. The previous approach is extremely simple and respectful, and has the goal to reduce the rotational stiffness of the soil-foundation system (with the trade-off of reducing also radiation damping);

Table 3. Results of the visco-elastic analyses considering the effect of 2 GSI interventions. The variation of spectral pseudo-acceleration Δa is referred to the fixed base scheme (Somma *et al.*, 2022).

		masonry M1			masonry M1		
model SSI conditions	earthquake direction	T (s)	a(T) (g)	Δa (g)	T (s)	a(T) (g)	Δa (g)
fixed base	x	0.49	1.60	–	0.39	0.69	–
compliant base		0.50	1.75	+0.15	0.41	0.72	+0.03
foundation lateral disconnection		0.56	1.86	+0.26	0.46	1.19	+0.50
lateral soil stiffening		0.40	0.70	−0.90	0.31	0.50	−0.19
fixed base	y	0.33	1.58	–	0.26	0.54	–
compliant base		0.36	1.63	+0.05	0.31	0.55	+0.01
foundation lateral disconnection		0.38	0.68	+0.10	0.33	0.56	+0.02
lateral soil stiffening		0.29	0.54	−0.04	0.24	0.50	−0.04

the latter has the opposite goal of increasing the rotational stiffness. Table 3 reports the results of the visco-elastic analyses, clearly indicating that for the Kocaely earthquake only a stiffening GSI intervention would have been of some help for the structure, because in this case (very shallow foundation, stiff layer underneath) the lateral disconnection is not sufficient to increase the period T in such a way to pass over the peak of the response spectrum, reducing the seismic action.

Similar considerations can be done looking at the results of the non-linear analyses in terms of the maximum principal tensile stress shadings on the tower, considering the two possible GSI solutions (Figures 48 and 49). Soil stiffening is beneficial, but not sufficient.

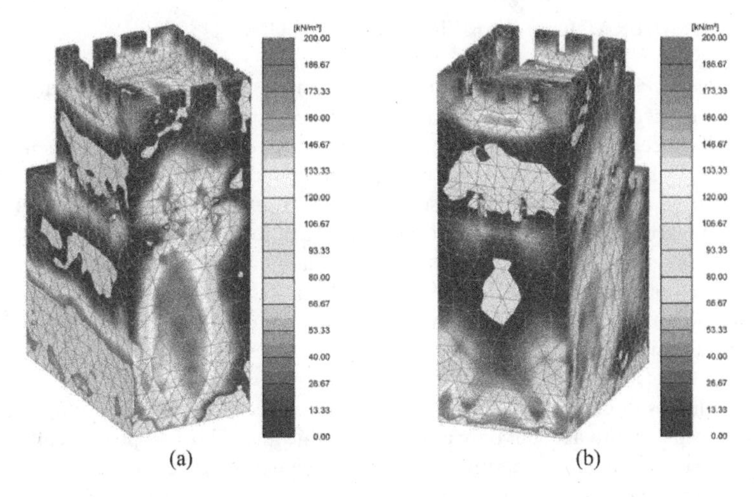

Figure 48. (a) Contours of the principal minimal tensile stresses in the laterally disconnected tower in x-direction at 5.60 s; (b) Contours of the principal minimal tensile stresses in the laterally disconnected tower in y-direction at 7.05 s.

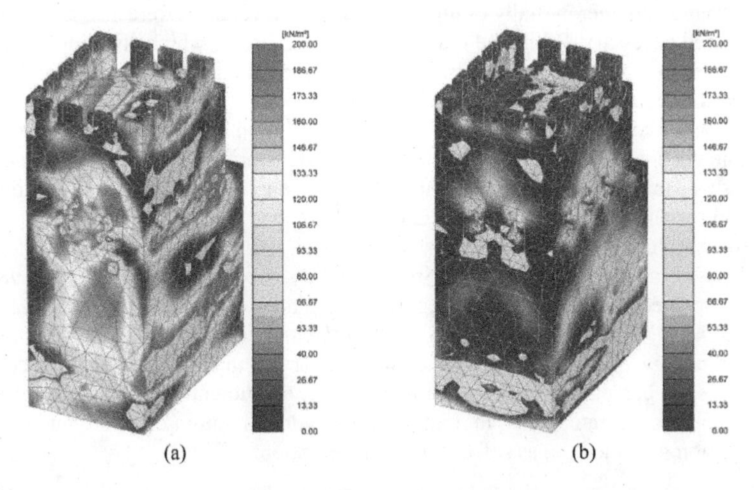

Figure 49. (a) Contours of the principal minimal tensile stresses in the laterally stiffened tower in x-direction at 7.92 s; (b) Contours of the principal minimal tensile stresses in the laterally stiffened tower in y-direction at 6.72 s.

Then, in this case there is no alternative to some strengthening of the structure. From a methodological point of view, after investigating less invasive possibilities, this is certainly consistent with preservation principles, as also explicitly suggested by Brandi himself (1963) with reference to the mitigation of seismic risk for built heritage. There are many literature indications on the best way to do so (e.g. Cosenza & Iervolino, 2007), and possible alternatives will be taken into account in the future (mitigation of the pushing effect of the vaults, improved interaction between orthogonal walls, improved connection of wall portions built in different periods, strengthening of the weaker portions), discussing about them with structural engineers but also with experts from other disciplines. Obviously, when passing from the back analysis of a single event to the design of future preservation interventions, a probabilistic approach will have to take the place of the deterministic calculations herein reported to back-analyse the effects of the 1999 Kocaeli earthquake.

5 FINAL CONSIDERATIONS (FAR FROM BEING CONCLUSIONS)

The tumultuous development in the last decades of the investigation and monitoring instruments, as well as of the theoretical and numerical tools, has certainly widened the range of options in the hands of engineers when dealing with the preservation of built heritage. When the structure shows wounds of geotechnical origin, these arrows in the quiver should inform the approach, that should concentrate as a first step on the subsoil itself, to understand the causes and possibly to restore equilibrium conditions in the soil or rock mass, lost for a number of possible changes of the boundary conditions. Only if such solutions are not available should the engineer touch the structure to protect. No action should be undertaken without thorough understanding, and the principle of necessity should always be the guiding light. This is possible only if monitoring and surveying, that should be considered as mandatory prevention measures, keep feeding information and alert against critical events.

New technologies should not be considered as enemies of cultural heritage, but their implementation should be taken into account only after a hierarchic analysis of all possible alternatives, that should start by considering the adoption of the original technological choices, consistently with the indications given in Figure 1. In any case, dogmatism should be avoided, being important to save what can be saved in the range of our possibilities, with the highest possible attention to regional traditions of preservation (Petzet, 2004).

In the examples reported in this paper, different geotechnical solutions could be found with the goal of preserving heritage. Partially or highly invasive interventions were also taken into account, when considered necessary. While the reductionistic approach in which civil engineers have been raised is a powerful tool in the solution of most practical problems, the examples show that it may not be suited to face built heritage, for which more sophisticated investigations and analyses are necessary to better catch details that may inform the solution. If it is certainly true that there is no specific geotechnical engineering of monuments and historic sites, we may conclude that there should be a specific need of consistent complexity in the approach to their preservation, for which cost effectiveness should not be the main concern.

Engineers should always face the preservation of built heritage with reference to a principle of responsibility – or rather to an imperative of responsibility, according to Hans Jonas (1984) -, which represents an updated version of Kant's categorical imperative '*act so that the consequences of your actions are compatible with the permanence of an authentic human life on earth*'. Paraphrasing this categorical imperative, we could say to the civil engineer: act in such a way that the consequences of your actions are compatible with the permanence of an authentic life of the historic structure or site on which you operate. When in doubt, we may add, do nothing; and if you have to exceed, exceed in caution, intended not as safety but as preservation.

ACKNOWLEDGEMENTS

Most of the information reported in this paper are the outcomes of team works, with the contribution of many people with different expertise and from different areas of the world, all acknowledged. During the preparation of the manuscript, ideas stemmed from fruitful discussions with (in alphabetic order) Lucio Amato, Giovanni Calabresi, Salvatore D'Agostino, Francesco Escalona, Renato

Lancellotta and Luigi Stendardo, whose contribution is highly appreciated. Emilio Bilotta and Fausto Somma are acknowledged for their valuable help in carrying out the numerical analyses. Above all, the author is indebted with his mentor and friend Carlo Viggiani, who introduced him into the magic world of ageless structures.

REFERENCES

Abrams, E. M. 1994. How the Maya built their world: energetics and ancient architecture. *Austin/University of Texas Press*.

Ahunbay, M., Ahunbay Z. 2000. Recent Work on the Land Walls of Istanbul: Tower 2 to Tower 5. *Dumbarton Oaks Papers*. 54: 227–239.

Àmato L. 2022. Personal communication.

Amato L., Evangelista A., Nicotera M.V., Viggiani C. 2001. The tunnels of Cocceius in Napoli: an example of Roman engineering of the early imperial age. *Proc. of AITES/ITA World Tunnel Congress 2001, Progress in Tunneling after 2000, Milan 10 –13 June 2001*, vol.1:.15–26, Bologna: Pàtron.

Ambraseys N.N. 2002. The seismic activity of the Marmara Sea region over the last 2000 Years. *Bulletin of the Seismological Society of America*. 92:1–18.

Bell E.E., Canuto M.A., Sharer R.J. 2004. Understanding early classic Copan. *Philadelphia: University Of Pennsylvania Museum of Archaeology And Anthropology.*

Bilotta E. 2022. Personal communication.

Brandi C. 1963. Teoria del restauro. *Edizioni di Storia e Letteratura*. Einaudi.

Burland J., Jamiolkowski M., Squeglia N., Viggiani C. 2013. The Leaning Tower of Pisa. *In Geotechnichs and Heritage, Bilotta, Flora, Lirer & Viggiani Ed.* London: Taylor & Francis Group.

Calabresi G. 2011. The soft approach to saving Monuments and Historic Sites. *Proc. Of the 15th European Conf. of ISSMGE*. Athens.

Calabresi G. 2013. 1st Kerisel Lecture: The role of Geotechnical Engineering in saving monuments and historic sites. *Proc. 18th Int. Conf. of ISSMGE*. Paris: Presse des Pontes.

Cosenza E., Iervolino I. 2007. Case study: Seismic retrofitting of a medieval bell tower with FRP. *Journal of Composites for Construction.* DOI: 10.1061/(ASCE)1090-0268(2007)11:3(319).

Escalona F. 2022. Personal communication.

de Silva F., Ceroni F., Sica S., Silvestri F. 2018. Non-linear analysis of the Carmine bell tower under seismic actions accounting for soil–foundation–structure interaction. Bulletin of Earthquake Engineering, 16(7):2775–2808. DOI: 10.1007/s10518-017-0298-0.

de Silva F. 2020. Influence of soil-structure interaction on the site-specific seismic demand to masonry towers. *Soil Dynamics and Earthquake Engineering 131*. DOI: 10.1016/j.soildyn.2019.106023

D'Agostino S. 2022. Conservation and Restoration of Built Heritage. A history of conservation culture and its more recent developments. *Built Heritage and Geotechnics Series, Lancellotta R. Ed.* London: CRC Press.

De Andrè F. 1967. Bocca di rosa. *Bluebell Records*.

Fash W.L. 1991. Scribes, Warriors and Kings: The City of Copán and the Ancient Maya. *London: Thames and Hudson*.

Ferrari G.W., Lamagna R. 2015. Crypta Neapolitana: non solo un tunnel. *Trasporti e Cultura, Rivista di architettura delle infrastrutture nel paesaggio*, 40, 88–93.

Flora A. 2013. General Report TC301: Monuments, Historic Sites and case histories. *Proc. 18th Int. Conf. of ISSMGE*. Paris: Presse des Pontes.

Flora A., Lombardi D., Nappa V., Bilotta E. 2018. Numerical Analyses of the Effectiveness of Soft Barriers into the Soil for the Mitigation of Seismic Risk. *Journal of Earthquake Engineering, 22(1)*. DOI:10.1080/13632469.2016.1217802.

Flora A., Chiaradonna A., de Sanctis L., Lignola G.P., Nappa V., Oztoprak S., Ramaglia G., Sargin S. 2021. Understanding the Damages Caused by the 1999 Kocaeli Earthquake on One of the Towers of the Theodosian Walls of Constantinople. *Int. Journal of Architectural Heritage*, Taylor & Francis. DOI: 10.1080/15583058.2020.1864512

Ince G. C., Yildirim M., Ozaydin K., Ozner P.T. 2008. Seismic microzonation of the historic peninsula of Istanbul. *Bulletin of Engineering Geology and the Environment.* 67:41–51. DOI 10.1007/s10064-007-0099-9.

Ispir M., Demir C., Alper I., Kumbasar N. 2014. An outline of the seismic damages of several monumental structures in Istanbul after historical earthquakes. *Workshop on Seismicity of Historical Structures*. Istanbul, November.

Iwasaki Y., Zhussupbekov A., Issina A. 2013. Authenticity of Foundations for Heritage Structures. *Proc. 18th Int. Conf. of ISSMGE*. Paris: Presse des Pontes.

Jappelli R. 1991. Contribution to a systematic approach. *in The Contribution of Geotechnical Engineering to the preservation of Italian historic sites*. AGI Associazione Geotecnica Italiana Ed.

Jappelli R., Marconi N. 1997. Recommendations and prejudices in the realm of foundation engineering in Italy: A historical review. *In Carlo Viggiani (ed.), Geotechnical engineering for the preservation of monuments and historical sites; Proc. Intern. Symp., Napoli, 3-4 October 1996*. Rotterdam: Balkema.

Jonas H. 1984. The Imperative of Responsibility – In Search of an Ethics for the Technological Age. *The University of Chicago Press*.

Lacombe L., Fash W.L., Fash B. 2020. Plan for the long-term Conservation of the Copan Acropolis Tunnels. *Report presented to the Instituto Hondureno de Antropologia e Historia, October*. University of Harvard.

Lancellotta R. 2013. 11th Croce Lecture: La Torre Ghirlandina: una storia di interazione struttura-terreno. *Rivista Italiana di Geotecnica, 47(2)*. Bologna: Patròn.

Loten S.H. and Pendergast D.M. 1984. A Lexicon for Maya Architecture. *Archaeology Mongraph 8*.

Pane V., Martini E. 1997. The preservation of historical towns of Umbria: the Orvieto case and its observatory. *Proc. of the 1st Int. Conf. on Geotechnical Engineering for the Preservation of Monuments and Historic Sites, Viggiani C. Ed.* Rotterdam: Balkema.

Petzet M. 2004. Principles of preservation. *ICOMOS Open Archive.*

Pires F., Bilotta E., Flora A., Lourenço P.B. 2021. Assessment of Excavated Tunnels Stability in the Maya Archeological Area of Copán, Honduras. *Int. Journal of Architectural Heritage*, Taylor & Francis. DOI: 10.1080/15583058.2021.1931730.

Pires F. 2020. Safety Assessment of the Archaeological Tunnels in Copán – Honduras. *SAHC Masters Course Thesis, Advanced Master in Structural analysis of Monuments and historical Constructions*. University of Minho.

Roca P., Lourenço P.B., Gaetani A. 2019. Historic Construction and Conservation. London: Routledge.

Sarimese F. 2018. Restoration and renovation of 18th and 19th Century Istanbul Land Walls. *Master of Science (MSc) Thesis, Turkish Research Institute of Marmara University* (in Turkish).

Schele L. 1998. Aerial View of Copan. *http://ancientamericas.org/collection/aa010001*.

Settis S. 2018. About the future: the Besieged City. *Interview by Antonio Guerriero*. Electra Vol. 2, 2018.

Sharer R. J., Miller J. C., Traxler L. P. 1992. Evolution of classic period architecture in the eastern acropolis, Copan: A progress report. *Ancient Mesoamerica*, 3(1). DOI: 10.1017/S0956536100002364.

Sharer R. J., Traxler L.P., Sedat D.W., Bell E.E., Canuto M.A., Powel C. 1999. Early Classic Architecture Beneath the Copan Acropolis. *Ancient Mesoamerica, 10(1).* DOI: 10.1017/s0956536199101056.

Sharer R. J., Traxler L.P. 2006. The Ancient Maya. *6th Ed., Stanford University Press.*

Somma F., Bilotta E., Flora A., Viggiani G. 2021. Centrifuge Modelling of Shallow Foundation Lateral Disconnection to Reduce Seismic Vulnerability. *Journal of Geotechnical and Geoenvironmental Engineering.* ASCE 148(2).

Somma F., Lignola G., Ramaglia G., de Sanctis L., Iovino M., Oztoprak S., Flora A. 2022. Earthquake damages to an historic tower: back analysis and possible mitigation measures. *Bullettin of Earthquake Engineering.* Submitted.

Thompson J. E. S. 1958. The Civilization of the Mayas. *Chicago Natural History Museum.*

Turnbull S. 2004. The walls of Constantinople AD 324-1453. *Osprey Publishing.*

Viggiani C. 2013. Cultural Heritage and Geotechnical Engineering: an introduction. *In Geotechnichs and Heritage, Bilotta, Flora, Lirer & Viggiani Ed.* London: Taylor & Francis Group.

Viggiani C. 2017. 2nd Kerisel Lecture: Geotechnics and Heritage. *Proc. Of the 19*th *Int. Conf. of ISSMGE.* Seoul.

von Schwerin, J. 2011. The sacred mountain in social context. Symbolism and history in maya architecture: Temple 22 at Copan, Honduras. *Ancient Mesoamerica, 22(2).* DOI: 10.1017/S0956536111000319.

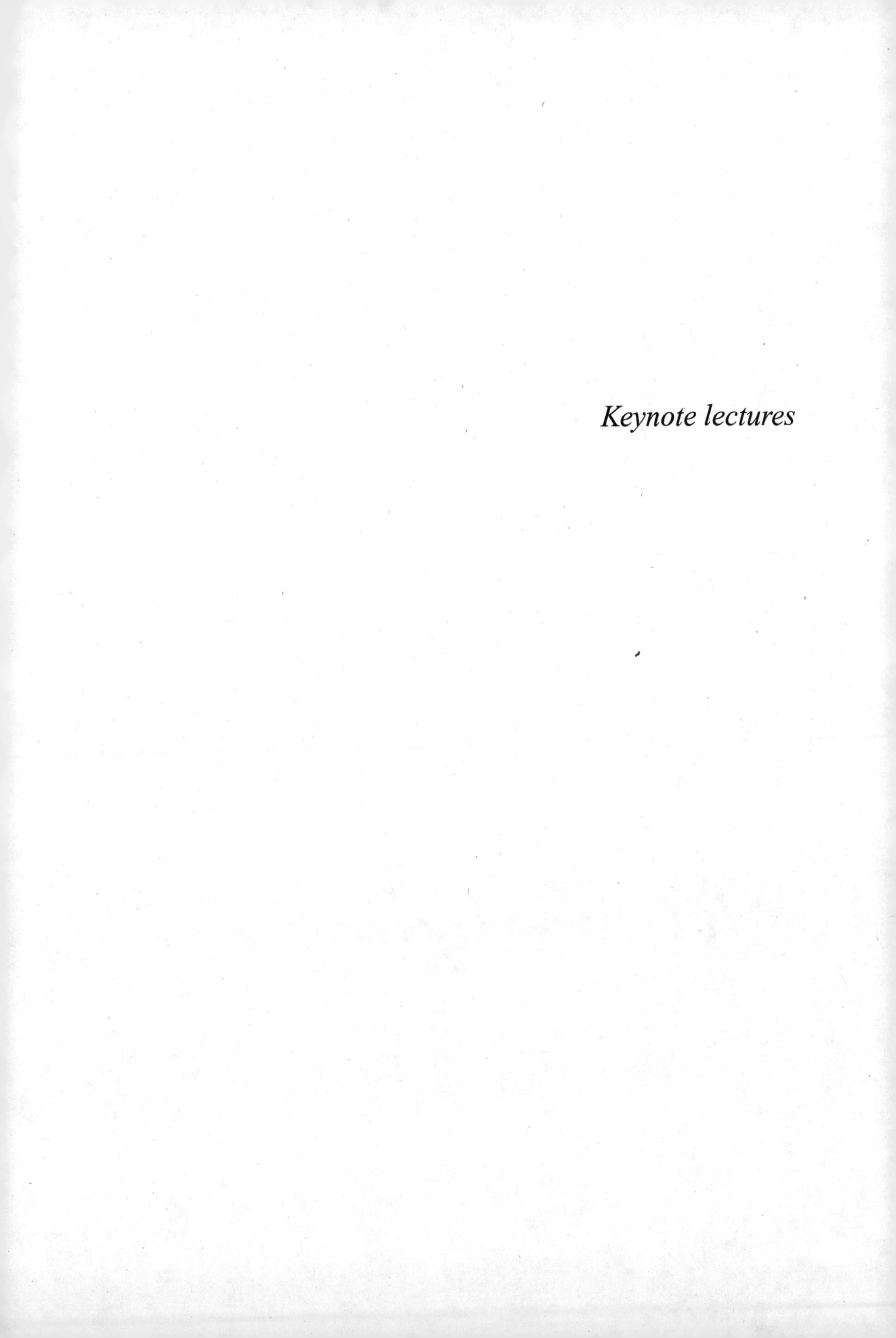

Keynote lectures

Geotechnical Engineering for the Preservation of Monuments and Historic Sites III – Lancellotta, Viggiani, Flora, de Silva & Mele (Eds)

The historical underpinning of Winchester Cathedral – Heroic or horrific?

J.B. Burland & J. Standing
Imperial College London, London, UK

J. Yu
Buro Happold, London, UK

ABSTRACT: Winchester Cathedral is not only famous for its size and magnificent Gothic architecture but also for the underpinning work that was carried out between 1905 and 1912. This work ran into a number of serious problems due to a high-water table and poor ground conditions. The former meant that the underpinning work had to be carried out by a diver, William Walker, who has become a legend for his heroic work. The need for the underpinning work has seldom been questioned and the purpose of this paper is to examine the evidence on which this key decision was taken. The lessons learned are important for civil engineers and architects called in to advise on the need for the stabilisation of historic buildings and monuments.

1 INTRODUCTION

Winchester Cathedral is one of the largest Gothic cathedrals in Northern Europe and the longest in overall length. The present building was begun in 1079 and was completed in 1532. It has a cruciform plan, with a long nave, transepts, central crossing tower, choir, presbytery, and lady chapel. The Cathedral is 170 m long, and the vaulting has a height of 24 metres. The central tower is 46 m high. Figure 1 is an aerial view of the Cathedral and Figure 2 is an isometric drawing from the same viewpoint. The various locations within the cathedral are labelled for ease of reference. Figure 3 is a plan view of the present Cathedral. As well as being famous for the magnificence of its architecture, the Cathedral is perhaps equally well known for the heroic efforts of a diver, William Walker, in underpinning virtually the whole Cathedral in the early 20th Century. The need for this underpinning appears to be seldom if ever questioned, yet it was a major undertaking fraught with problems and dangers. The purpose of this paper is to revisit the way in which the decision to underpin was made and the information on which this decision was based.

2 A BRIEF HISTORY OF CONSTRUCTION

2.1 *The Saxon Minster*

Winchester is built on an old Roman settlement with evidence of roads and buildings beneath the present city. The original Saxon church was located just north of the present Cathedral and had its origins in the 7th Century when England's pagan monarchy first became Christian. The church (later known as the Old Minster) was constructed around 648AD. Within a quarter of a century, it had achieved cathedral status. In the early 10th Century, the church was greatly enlarged, going through at least four stages of change and additions within that century. The outlines of the original church and its later extensions have been marked on the ground of the cathedral precincts and

DOI 10.1201/9781003329756-3

Figure 1. Aerial view of Winchester Cathedral looking south.

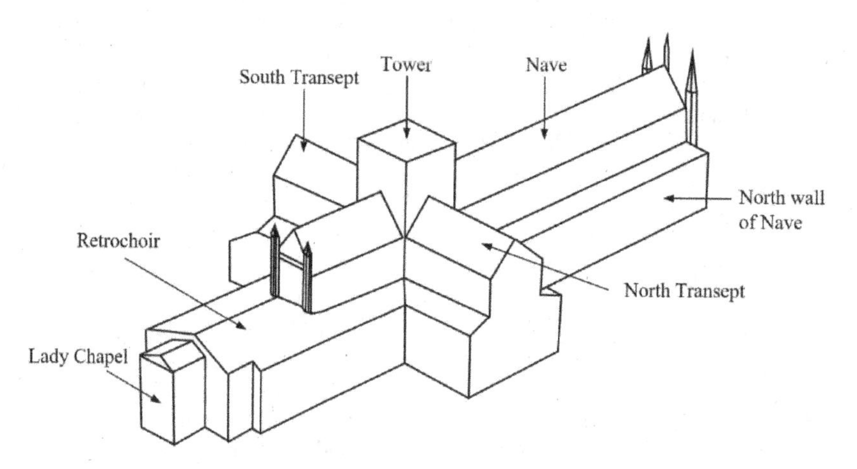

Figure 2. Isometric drawing of Winchester Cathedral looking south.

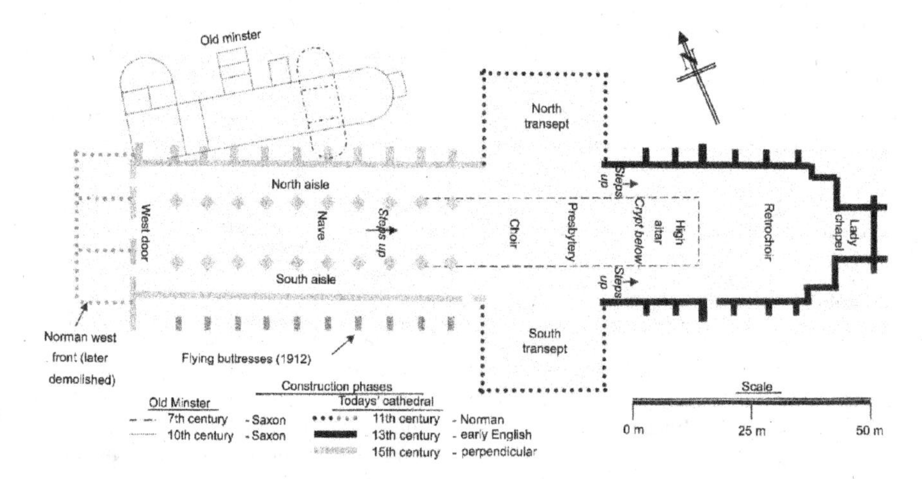

Figure 3. Plan of Winchester Cathedral (Roberts 2013).

Figure 4. Location of Old and New Minster on the north side of Winchester Cathedral.

can be seen just north of the west end of the present Cathedral as shown in Figure 4. Its location may well have been chosen because it is sited in an area where the foundation conditions are more favourable than elsewhere in the area.

2.2 *The New Norman Cathedral (1079–1120)*

Following the invasion of England by William the Conqueror in 1066, the Saxon bishop of Winchester was replaced by the Norman royal chaplain Walkelin. In 1079 he began the construction of a large new cathedral in the Norman Romanesque style, a plan of which is shown by the broken lines in Figure 5, and probably took 40 years to complete (Henderson and Crook, 1984). As mentioned above, the location of the new Cathedral was south of and immediately alongside the Saxon Minster as shown in Figure 4 and was probably chosen for several reasons. First, the site of the Old Minster was prestigious and important, being directly opposite King William's royal palace. Also, it was an important pilgrimage centre containing the shrine of St Swithun, the patron saint of Winchester Cathedral. Moreover, this arrangement meant that demolition of the Old Minster could be delayed until after the completion of the east end of new Cathedral. Therefore the site of the new cathedral was almost certainly dictated by issues other than ground conditions.

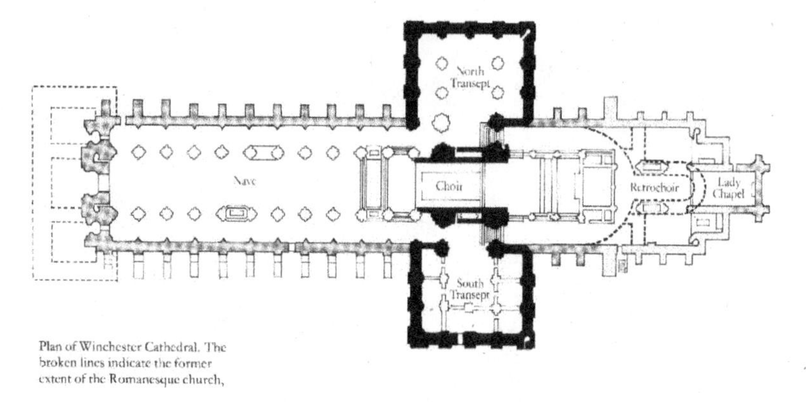

Figure 5. Plan of Winchester Cathedral with the original Norman Romanesque church shown as broken lines (Henderson & Crook 2004).

The design of the new Cathedral was based on that of contemporary Norman churches, with a cruciform shape in plan and a semi-circular aisle behind the Choir and High Altar as shown in Figure 5. Below the Choir, and accessed from the North Transept, is the Crypt which would have been constructed first. Unlike most other crypts, it is only partially below the ground – almost certainly because of the high-water table at the time of construction. Indeed, it now gets flooded most winters with up to 30 cm of ground water implying that the water table may have risen over the centuries.

The 12th Century Cathedral was completed around 1120 but not before some construction problems had been overcome. The ground profile consists of fill, overlying successively, chalky marl, a layer of peat, and dense gravel resting on chalk. The ground water level was about 2m above the top of the peat layer. In general, the bases of the Norman foundations invariably stop at the ground water level of that time, about 2.7m above the top of the gravel and are hence underlain by the peat layer. In places the foundations rest on a mat of timber logs and oak piles some of which were driven through the peat to the top of the underlying gravel layer. Figure 6 gives a typical section through the wall of the Presbytery.

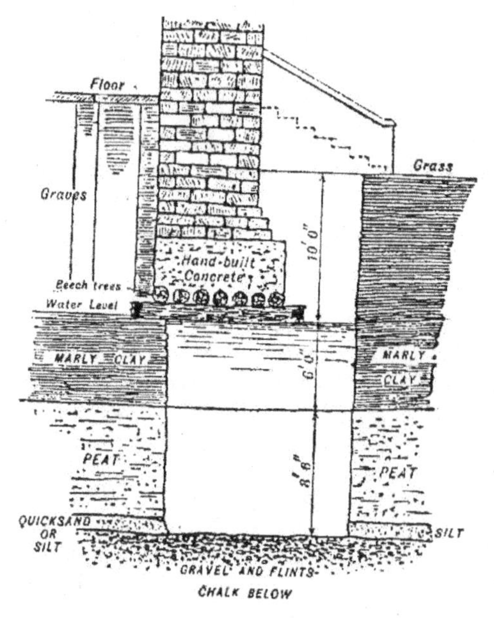

Figure 6. Section through wall of Presbytery, showing underlying soil profile (Henderson & Crook 2004).

Work quickly progressed to the transepts and central tower and was completed by about 1100. In 1107 the central Tower collapsed while work on the Nave was going on. The reason for the collapse is not known but Henderson and Crook (1984) postulate that it may have been due to unstable foundations. They note that the re-built Tower is on exaggeratedly massive piers at the four corners of the central Crossing. They also note the contrast between the exceptionally fine jointing of the replacement Tower and the thick jointing of the earlier masonry. It is therefore also possible that the failure of the original Tower may have been due to poor workmanship.

2.3 *Gothic expansions and refashioning of the Cathedral*

The first alterations to the Norman Cathedral were in 1202 when Bishop Godfrey de Luci started construction of a new Retrochoir at the east end of the Cathedral – see Figure 3. Work continued to progress westwards and ended with the creation of the new Choir and Presbytery in the first half of the 14th Century. Between 1500 and 1528, the Choir clearstory was remodelled, and the Choirisles were rebuilt. Henderson and Crook (1984) note that the architectural history of the entire eastern limb of the Cathedral is complex and remains a matter of controversy amongst architectural historians.

Also, in the 14th Century the west end of the Nave was demolished, shortening the Cathedral by 12.2m. Under Bishop Edington and later Bishop Wykeham, the Nave of the Cathedral was refashioned to the Perpendicular style. This they achieved by encasing the Norman stone in new masonry, recutting the piers with Gothic mouldings and pointed arches and reorganising the three-tier Nave into two tiers. Figure 7 is a photograph of the present refashioned Nave which is truly breath-taking.

Figure 7. The Nave of Winchester Cathedral looking East (courtesy of Michael Beckwith).

2.4 *Continued decay and the decision to underpin the Cathedral*

Henderson and Crook (1984) conclude that, from the early 16th Century to the twentieth, the story of the Cathedral is one of structural neglect and decay with occasional attempts at repair. In 1775 James Essex noted the defective vaulting of the Retrochoir aisles and attributed that to settlement during and immediately after construction. However, he doubted that subsidence was still taking place.

In 1809 a programme of inspection and repair was initiated following the appointment of the Cathedral's first architect, William Garbett. He appears to have shared James Essex's view that subsidence was complete. He did however note the inclinations of the south walls of the Retrochoir and of the South Transept. He also noted 'alarming fissures' in the South Transept walls and recommended filling them so that an eye could be kept on possible movement. At the turn of the 20th Century the structural state of the Cathedral was exceedingly worrying. There were huge cracks in the walls, some of the walls were bulging and leaning and occasional falls of stone took place from the vaults. In 1905 the Cathedral Architect, J. B. Colson, submitted a report to the Dean and Chapter describing these serious structural problems and recommending timber propping of the leaning walls. The Diocesan Architect, Thomas Jackson, then undertook a detailed inspection of the Cathedral in early March 1905. A trial excavation was made on the south side of the east end of the Cathedral and revealed dense gravel at a depth of about 7.5m underlying a thick layer of peat. Jackson in turn decided that he needed specialist advice and recommended the consulting engineer Francis Fox to the Dean and Chapter who agreed to call him in. It is important to note that much of Fox's previous experience had been concerned with bridges, tunnelling and railways. His involvement with historic buildings was minimal. He did have some very limited involvement with St Paul's Cathedral, but it seems to have been controversial.

On 27th June 1905 Fox, Jackson and Colson closely examined a second trial hole, which confirmed the ground profile revealed in the first trial hole, and they agreed unanimously that the south wall of the Retrochoir should be underpinned on to the underlying gravel. On 5th July 1905 Fox presented his recommendations to the Dean and Chapter. In his book "Sixty-three Years of Engineering", Fox (1924) summarised the recommended sequence of remedial measures for the south wall of the Retrochoir as follows.

(1) Shoring up the leaning south wall at the east-end of the Cathedral and the façade of the South Transept
(2) Centring the arched vaulting of the Retrochoir to prevent collapse.
(3) Inserting steel tie-rods between the north and south walls of the Retrochoir.
(4) Grouting with liquid cement under compressed air every portion of the relevant walls commencing at the base.
(5) Lastly, underpinning the walls down to the bed of gravel.

It is very important to note that these recommendations were made very quickly, within a few days of Fox's first visit. Moreover, nothing was included in the recommendations about the need to monitor settlement or structural movement prior to, and during, grouting and underpinning.

3 UNDERPINNING OPERATIONS

3.1 *Preliminary operations*

As described by Roberts (2013), the first operation was to timber up and strut the Retrochoir and South Transept walls followed by the centring of the roof and the installation of the tie-rods. These can still be seen in the Retrochoir tying the north and south walls together. The next stage was the cement pressure grouting, which at that time was a relatively unknown technique. A pressure grouting machine had been invented in the 1870s by James Greathead, who had specified it for use in the construction of London's Northern and Central line tube railways with which Fox was familiar. The first step of the grouting process was to force in compressed air to blow out the accumulated dust of ages. Water was then injected to wash out the cracks. The grouting was begun at the bottom and, running horizontally, filled in the interstices and cavities. This was continued progressively upwards to the top levels of the walls and tower.

3.2 *Underpinning work encounters problems*

On completion of the grouting, the underpinning operation commenced. It was proposed that the underpinning would be carried out progressively by excavating one-metre-wide trenches (known as drifts) at right angles to the wall to be underpinned. Once the base of the existing foundations was reached, the drift would be extended under the existing foundation and excavation would continue down to the gravel. A bed of concrete would then be placed on the gravel and a solid masonry wall would then be constructed up to the underside of the existing wall. Fox recognised that this would involve excavating beneath the water table, but he believed that the water level within each trench could be controlled by pumping. Work commenced in December 1905 at the southeast corner of the Retrochoir.

The underpinning work soon ran into trouble. It was discovered that the layer of peat overlying the gravel had become so compressed by the foundation pressure that it became very impermeable and difficult to excavate. As excavation of a drift progressed downwards below the water table, initially little water entered the trench. However, as the excavation progressed through the peat, the water under pressure burst through from the underlying very permeable gravel, flooding the trench with black opaque water.

A powerful pump was required to keep the trench dry enough to carry out the work. Serious doubts began to be expressed about the advisability of continuing the work with extensive pumping. These doubts were reinforced by the discovery of a thin layer of silt between the peat and the underlying gravel which could be eroded by the flow of water into the drift thereby risking damaging subsidence.

3.3 *Underpinning commences using a diver*

Fox concluded that the best way forward was to use a diver to carry out the excavation beneath the water table. In April 1906 a trial excavation was undertaken using a diver named William Walker. When the excavation was complete Fox, who was himself an accomplished diver, examined the base of the drift. He reported that the base *"proved to be a hard flinty gravel, quite excellent and, as this overlies the chalk, no better foundation could be either secured or desired"* (Fox, 1924). It was therefore decided to continue with this method of working. Little did William Walker know that he would be engaged until September 1911 in underpinning most of the Cathedral.

A detailed description of the method of working is given by Roberts (2013) and reference is made to Figure 8. Work on a drift began by labourers excavating a short 1m wide trench at right angles to the Cathedral wall and taken down to below the level of the existing footings. Excavation continued down into the peat layer leaving a thickness of peat above the gravel of about 0.6m. A drift or small tunnel up to 7m long was then excavated beneath the existing wall footing. The peat was found to be so compressed by the foundation bearing pressure that a pick was needed to excavate it. It was also necessary to remove beech logs that had been placed there when the walls had been built some seven centuries earlier – often having to resort to sawing them into short lengths. The wall footings were then shored up before the remaining peat layer was excavated, whereupon black opaque peaty groundwater poured into the excavation filling it to a depth of some 4m.

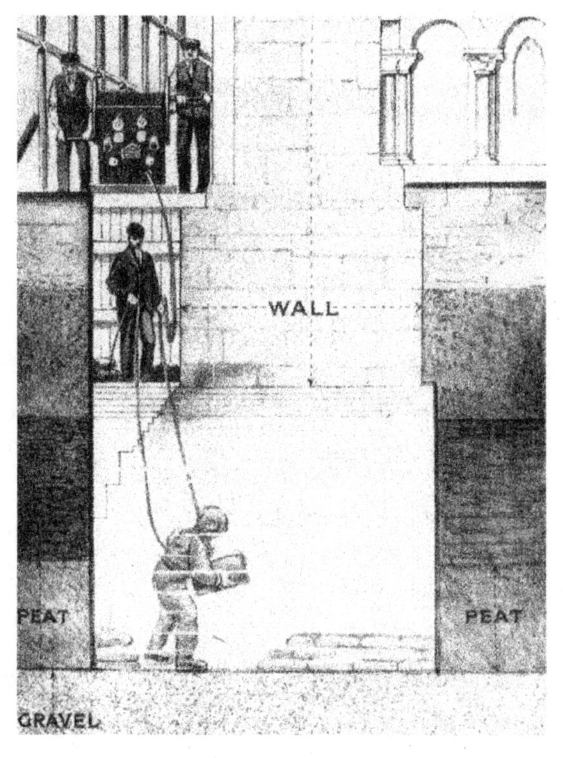

Figure 8. The diver at work (Roberts 2013).

The diver then took over and, working by touch in the opaque black water, completed the excavation down to the gravel. A layer of sacks filled with concrete was then placed on the gravel, packed tightly, trodden into place, and then slit open before being covered by the next layer. Three further layers of concrete filled bags were then similarly laid on top of the first layer. Finally, the bags were grouted to ensure they became a solid mass of concrete. All this had to be done by feel due to the black opaqueness of the water. The grouted bags of concrete made an effective seal that stemmed the flow of water from the gravel, thus enabling the trench to be pumped dry and allowing workmen to complete the underpinning of that section of the wall by means of mass concrete or engineering brick.

Originally it had been anticipated that only the Retrochoir and South Transept would need underpinning, but in July 1906 there was a small collapse of some vaulting of the Nave's south aisle. The opinion was offered that this was due to the historic subsidence of the foundations of the wall causing it to lean outwards thereby distorting the vaulting. This led to the recommendation that the wall of the Nave's south aisle would have to be underpinned. Ultimately this wall was stabilised by means of new flying buttresses and the foundations for these were constructed underwater by William Walker using the same method as was used for the underpinning work. A few months later it was reported that on the north side "every tell-tale has moved, and the whole North Transept will have to be underpinned". This underpinning work was completed in February 1910. In June 1907, Jackson surveyed the South Transept which had clearly undergone significant settlement. It was concluded that underpinning was also required, and this work was completed in April 1911. In his survey of ongoing and completed works in January 1909, Jackson identified problems with the north wall of the Nave. The buttresses which had been added during the 14th Century to support the wall had increased the danger of collapse because the buttress foundations were at a higher level than the Norman foundations. Therefore, the buttresses along the north aisle wall had to be underpinned. This work was completed in November 1910.

By the end of August 1911 virtually the whole of the Cathedral had been underpinned and a total of 235 drifts (trenches) were dug. It was estimated that William Walker had handled 25,800 bags of concrete. The original estimate of total costs was £3250 compared with the actual cost which turned out to be £113,000 which is equivalent to a sum approaching £14 million today. There can be no doubt that the work of the diver was heroic and it is universally recognised as such. But we are faced with the question "was this huge underpinning operation really necessary at that time"?

4 RECENT STUDIES

4.1 *The ground conditions beneath the Cathedral*

In 2006 and 2007 a team from Imperial College London carried out some detailed studies of the Cathedral with a view to ascertaining the following.

1. The ground conditions underlying the Cathedral precincts.
2. The total and differential settlements experienced by the Cathedral.
3. Whether subsidence was still occurring.
4. What surveys were put in place to monitor the foundation and wall movements during underpinning operations.
5. What modern technology would be used now to stabilise the foundations.

Much of the work of the Imperial College Team was carried out as MSc projects (Galdos-Ispizua (2007); Wilkinson (2006); Yu (2006)). The subsurface of the soil profile around the perimeter of the Cathedral was painstakingly pieced together using the records of the underpinning works during 1905 to 1912. These records are kept at the Winchester Cathedral Library Archives as well as the Library Archives at the Institution of Civil Engineers, London. The diver, William Walker, recorded the depths to the top of the gravel for all the drifts. Figure 9 shows the absolute levels of the top of the gravel along the north and south side of the Cathedral. On the north side the Reduced Levels (RL) of the top of the gravel varies from RL 31.4m to RL 30.0m, while the south sideshows a greater range varying from RL 31.6m to RL 29.3m. The diver's records also show that the peat thicknesses and levels vary significantly across the site. Thus, the ground conditions at Winchester Cathedral are highly variable, with appreciable changes in levels and thicknesses over short distances, which is a reflection of its fluvial and alluvial depositional environment.

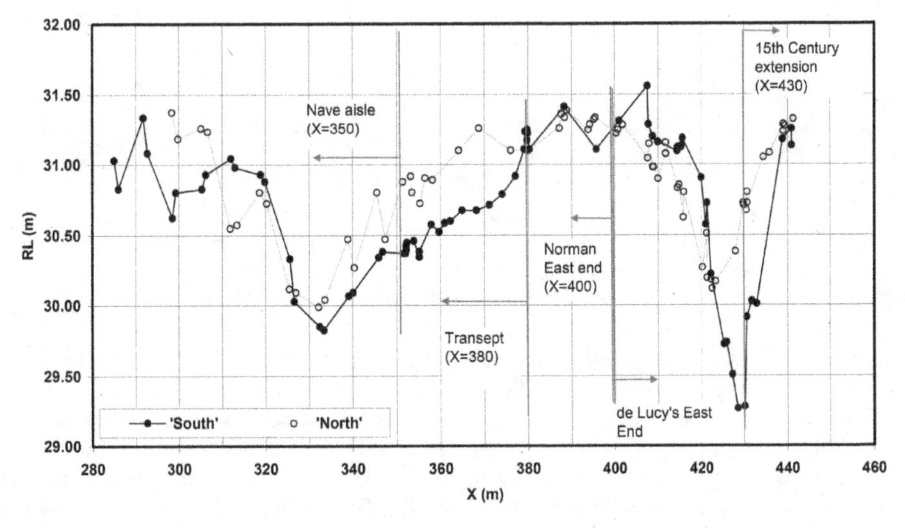

Figure 9. Levels of the top of the gravel layer along the north and south sides of Winchester Cathedral (Yu 2006).

4.2 *Settlement profiles around the Cathedral*

It is reasonable to assume that in laying down the base course or plinth of a medieval building every effort would be made to keep it horizontal. Therefore, it was considered that surveying the Norman plinth course would provide reliable information about the differential settlements of the Cathedral that had taken place during and subsequent to construction. In some locations the plinth course was not obvious and either a ledge course or string course was surveyed. Figure 10 shows

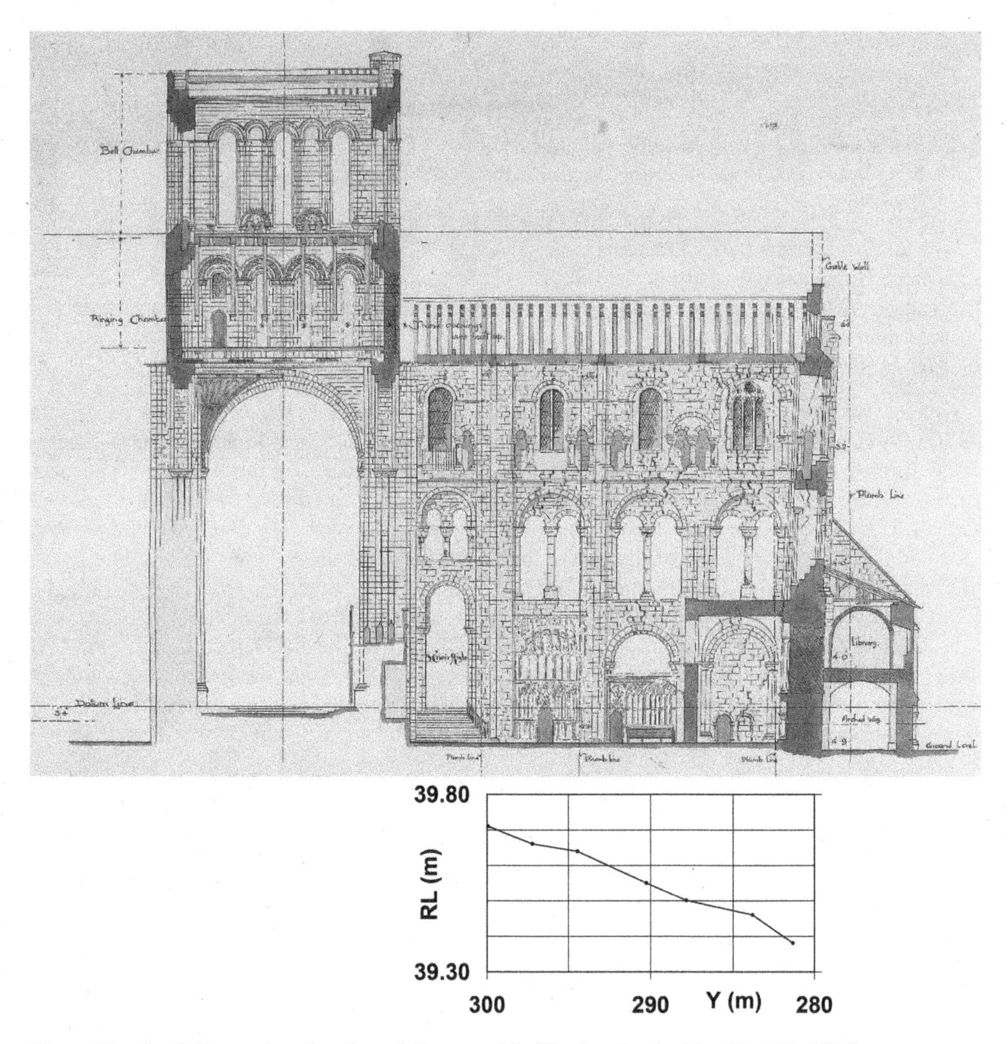

Figure 10. South Transept section through Tower and inside elevation looking East (Yu 2006).

a section through the Tower and inside elevation of the South Transept looking East and beneath it is plotted the levelling profile along the plinth. It is evident that a differential settlement of about 0.3m towards the south has taken place over a length of 18m. Figure 11 shows the elevation of the west wall of the South Transept looking East. It can be seen that the Slype at the southern end of the Transept has settled by about 0.15m and a hogging profile of settlement has taken place giving rise to significant cracking near the top of the west wall. The settlement profiles shown in Figures 10 and 11 also explain the outward tilt of the south façade wall. The value of such levelling surveys is all too apparent, and they do not require sophisticated equipment.

Henderson and Crook (1984) drew a most important conclusion from a study of the tapered masonry courses in the east facing wall of the South Transept that significant subsidence must have taken place during construction and that settlement was complete by the middle of the 14th Century.

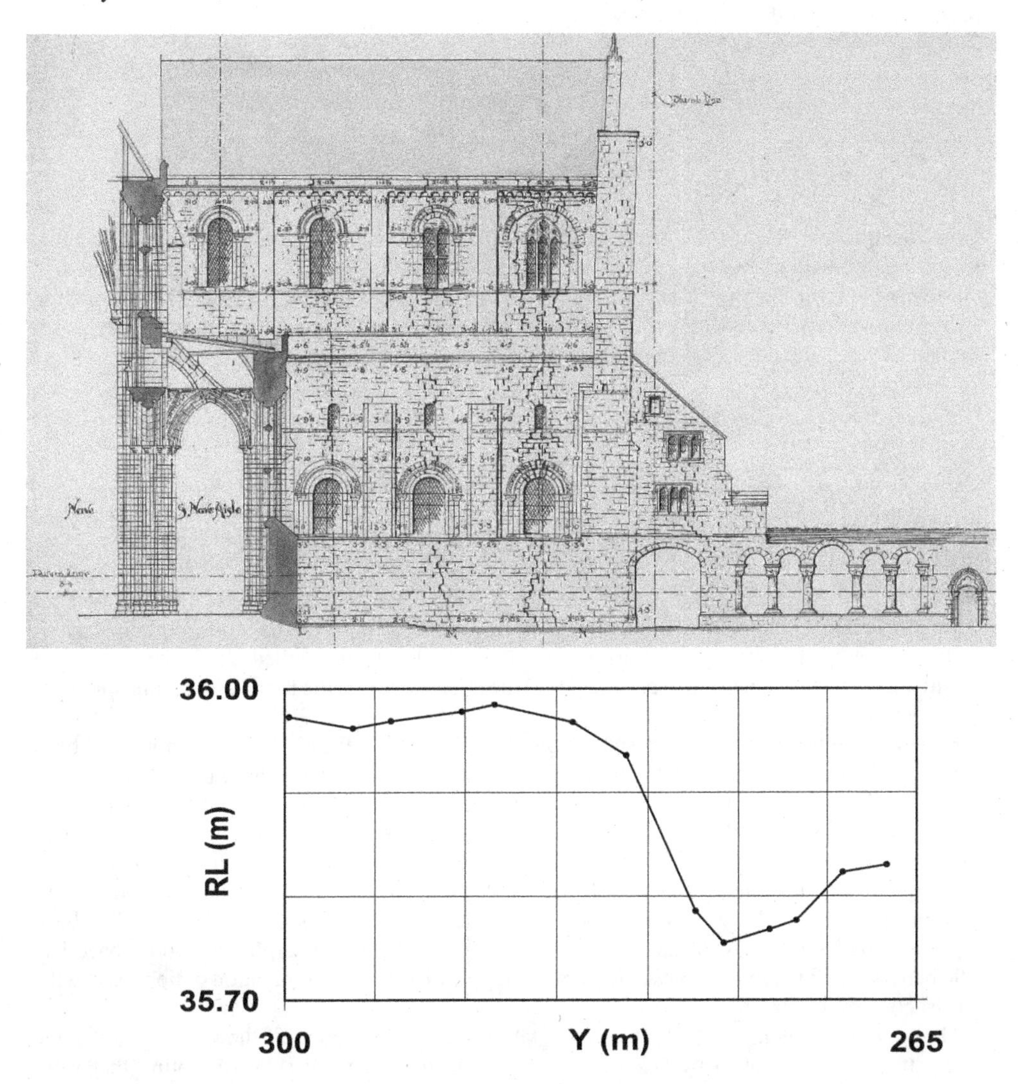

Figure 11. South Transept west elevation showing section through the south aisle of the Nave (Yu 2006).

4.3 *Settlements from precision levelling surveys*

In 1988 Ove Arup and Partners were commissioned to initiate a precision levelling survey of the whole Cathedral. A total of 42 BRE levelling points (Cheney 1973) were installed throughout the Cathedral and on 11th May 1988 initial readings were taken on most of them and referenced to a datum point installed on The Wessex Hotel some distance north of the Cathedral. The Imperial College team carried out a second survey on 24th July 2007 some 19 years after the initial survey. The results showed that over this period, the vertical movements were everywhere extremely small. The average settlement of the Tower, which was not underpinned, was less than 2mm. Elsewhere towards the western and of the Cathedral and the Transepts the settlements were generally less than about 1mm. At the eastern end small upward movements of generally less than about 1mm were recorded. For all practical purposes it can be concluded that over a period of 19 years no progressive vertical movements had taken place both for those parts of the Cathedral that had been underpinned and for those that had not.

5 DISCUSSION AND CONCLUSIONS

It was mentioned in Section 2.4 that the decision to underpin was made very quickly within a few days of Fox's first visit. It is easy to be critical and wise after the event, but it is most instructive to consider what information was needed and could have been obtained at that time, before making such a momentous, far-reaching and costly decision. It is appropriate to quote from Fox's own account of when and how the decision was taken:

> *"I was requested by Dr Furneaux, the Dean [of the Cathedral] to accompany Thomas Jackson [the consultant architect] on June 27, 1905, to Winchester. The architects had found very serious subsidence in various parts of the Cathedral, that in the Presbytery amounting to nearly 2ft 6in [0.75m]. The outer walls and buttresses had gone seriously out of the perpendicular, while the beautiful, groined arches were distorted in form, and disintegrated in character, and alarm had been caused by the fall of some stone from the roof. Thomas Jackson had sunk a trial pit some few yards distant, and had discovered a bed of peat 8ft [2.4m] below the clay and resting upon a fine solid bed of flints and gravel, into which he had bored to some depth to prove its solidity.... The problem of strengthening the foundations was, therefore, a very formidable one."*

The decision to underpin the Cathedral was evidently based purely on the following information: (1) the current visible structural distortions within the 800-year-old Cathedral; (2) that in the past considerable settlement had taken place; and (3) the knowledge that the foundations were underlain by a layer of peat.

However, the trial pits that were sunk showed that the peat beneath the foundations had been compressed into a very dense and stiff material, but no account seems to have been taken of this. Moreover, the chalk marl revealed in the trial pit overlying the peat would not have given rise to long-term progressive settlement. No attempt was made to measure what, if any, settlement was still taking place. Yet high quality surveying equipment was in wide use at that time. Indeed, it is interesting to note that precision surveying of the Leaning Tower of Pisa was commenced in 1911 (Burland et al. 2021). It seems very probable that the inclination of some of the walls had been caused by the lateral thrust of the internal arches. In which case underpinning would have had little beneficial effect. Indeed, these problems were dealt with by the installation of tie rods and the construction of new buttresses.

There can be no doubt that the structural stabilisation and grouting of the masonry walls and arches that was undertaken by Fox were essential. In this regard it is worth noting that work carried out by Professor Jacques Heyman on the stability of masonry structures has demonstrated how robust masonry arches are, even when significantly distorted. Reference can be made to a fascinating article by Heyman (2021) entitled "Why ancient cathedrals stand up".

Perhaps the most important lesson to be learned from the study described here is that before any work on the foundations of an historic building is carried out, it is essential to establish (1) what foundation distortions have actually occurred, by levelling around plinths or string lines and (2) whether on-going movements are taking place in the foundations or structure. Because of the lack of such information in the case of Winchester Cathedral it is not possible to be definitive about how necessary the underpinning works were. However there is evidence given in this paper to suggest that any progressive settlement must have been very small. There can be no doubt that the work of the diver, William Walker, was indeed heroic and his name is forever linked with Winchester Cathedral.

ACKNOWLEDGEMENTS

The authors are most grateful for the assistance received from The Revd Canon Dr Roland Riem, Vice-Dean of Winchester Cathedral. Dr John Crook, Cathedral Archaeologist, gave us freely of his intimate knowledge of Winchester Cathedral and its history. We are grateful too for the assistance given to us by Mr Carlton Bath, Cathedral Clerk of Works and the Cathedral Archivist.

REFERENCES

Burland, J.B., Jamiolkowski, M.B., Squelia, N. & Viggiani, C. 2021. *The Tower of Pisa – History, Construction and Geotechnical Stabilization.* Taylor and Francis, London, UK.

Cheney, J.E. 1973. Techniques and equipment using the surveyor's level for accurate measurements of building movement. *Proc. Symp. on Field Instrumentation*: 85-89. BGS, London, UK.

Fox, F. 1924. *Sixty-Three Years of Engineering.* John Murray, London, UK.

Galdos-Ispizua, A. 2007. *Winchester Cathedral Foundation Engineering Past and Present Approach.* MSc dissertation, Dept. of Civil and Environmental Engineering, Imperial College London, UK.

Henderson, I.T. & Crook, J. 2004. *The Winchester Diver: The Saving of a Great Cathedral. Henderson & Stirk Publishers*, Crawley, Winchester, UK.

Heyman, J. 2021. Why ancient cathedrals stand up: *Ingenia, Magazine of the Royal Academy of Engineering.* 10:19–23.

Roberts, G. 2013. How a diver saved Winchester Cathedral, UK: and today's solution? *Proc. ICE, Engineering and Heritage,* 166 (EH3): 164-176.

Wilkinson, S. 2006. *A review of the Ground Conditions below Winchester Cathedral and their Effects on the Cathedral's Structure.* MSc dissertation, Dept. of Civil and Environmental Engineering, Imperial College London, UK.

Yu, J. 2006. *Winchester Cathedral – an Engineering Look into the Past.* MSc dissertation, Dept. of Civil and Environmental Engineering, Imperial College London, UK.

Geotechnical Engineering for the Preservation of Monuments and Historic Sites III – Lancellotta, Viggiani, Flora, de Silva & Mele (Eds)
© 2022 Copyright the Author(s), ISBN 978-1-032-35998-4

Rethinking preventive conservation: Recent examples

P.B. Lourenço, A. Barontini & D.V. Oliveira
University of Minho, ISISE, Department of Civil Engineering, Guimarães, Portugal

J. Ortega
Instituto de Tecnologías Físicas y de la Información "Leonardo Torres Quevedo" (ITEFI), CSIC, Madrid, Spain

ABSTRACT: The past few decades have seen an increasing awareness of the potential socio-economical and environmental impact of investment in Cultural Heritage (CH). Preserving CH is not only an obligation to sustain and transmit it to the future generation but is also a driver of sustainable growth. Here, recently concluded projects are taken in consideration for a reflective thinking on preventive conservation, as the only viable strategy towards a sustainable and cost-effective management of CH, to face unprecedented challenges posed by increasing natural and man-made threats. Here, the main open issues for a widespread implementation of preventive conservation are identified, moreover, standardised, integrated good practices, validated over significant case studies, are presented within a multi-level replicable framework.

1 INTRODUCTION

Over the past few decades, the awareness of Cultural Heritage (CH) potential and of the benefits brought by it to society as a whole is significantly grown, leading to a strong development of national and international policies. Several examples demonstrated the significant socio-economical and environmental impact of investing in CH. CH has been recognised not only as an irreplaceable asset, but also as a driver of sustainable development and a strategic resource to promote peace, diversity, inclusiveness and participation (Jagodzińska et al. 2015). In order to preserve CH, its intrinsic fragility and the growing threats that is facing are particularly worrying and are calling for the development and enforcement of good and validated practices. To this end, preventive conservation is likely to be the most cost-effective strategy, strongly recommended by international institutions involved in preservation, as the International Council on Monuments and Sites (ICOMOS). The 2003 charter (ICOMOS 2003), for instance, while setting an ensemble of principles for conservation, recognises preventive maintenance as the best therapy for built heritage.

According to preventive conservation philosophy, damage and decay are unavoidable, however they can be tolerated as long as the affected system is fit-for-purpose, namely it meets a set of requirements related, for instance, to structural capacity, aesthetic, comfort and safety of the user, economic and market values and, in the case of historic buildings, authenticity and heritage value. The probability of failing to meet one or more of such requirements is, therefore, reduced by scheduling maintenance and interventions according to prescribed criteria based on performance and/or parameter monitoring and the consequent analysis and prognosis (CEN 2010). This approach allows timely detection of anomalies, optimized long-term allocation of the resources and prioritisation of the required measures.

A preventive conservation framework is the effective integration of condition survey and monitoring with risk assessment (Taylor 2005), where condition survey and monitoring repeat over time

 DOI 10.1201/9781003329756-4

the estimation of the building performance through qualitative and quantitative methods, whereas risk assessment forecasts the potential loss of performance due to specific hazardous events. Their integration allows the identification of the probable causes from the detected damage and the prediction of the remaining service life, based on the expected evolution of the degradation under given scenarios (Taylor 2005). This integration is boosted by the development of a reliable digital twin namely a duplicate of the asset, generated by a fusion of models and data and able to evolve replicating the physical twin evolution in time (Wagg et al. 2020). Documenting the CH asset condition and understanding its need in connection with its environment and its operation become fundamental pillars of conservation. Nonetheless, this asset investigation is complicated by several factors, as further described in the following sections. Such shortcomings should be tackled by validated, replicable and cost-effective strategies able to adapt to the specificity of each asset, without losing objectivity in the interpretation of the evidence.

Two recently concluded European projects led by the University of Minho, with an active involvement of the authors, have been an incredible opportunity to reflect further on CH preventive conservation, formulating a multi-level comprehensive methodology, built on standardised protocols and aimed at addressing the aforementioned shortcomings. These protocols were validated and tested on a large set of assets located in different context and geographical area. The main outcomes of the projects are described in the last part of this paper.

2 PREVENTIVE CONSERVATION: A MEDICAL ANALOGY

The similarities between the diagnostic process for human and building pathologies have led to a medical analogy, embraced by international recommendations and scientific literature. This analogy supports the identification of a standardised framework for conservation, drawing inspiration from a field that has theorised that prevention is better than cure, for a long time. Following this analogy, in (Della Torre 2010) and (Balen 2015), three levels of prevention were defined: (i) a primary level aimed at avoiding the causes; (ii) a second level aimed at an early detection of the symptoms; (iii) a third level aimed at preventing a further spread of the pathology and its side effects. First level encompasses mitigation strategies, ranging from simple measures (e.g. proper use of the asset and constant maintenance) to more systematic modifications of the level of hazard, exposure and/or vulnerability. Second level relies on a systematic screening. Third level consists in an urgent cure. It is clear the change of perspective: the remedial treatment (level 3) should not be the standard way, it is rather an ultimate solution when prevention (level 1 and 2) fails, namely it is a defeat of the conservation system (Balen 2015). Extending the medical analogy, preventive conservation can be, thus, seen as a process of early diagnosis and treatment, repeated across the asset lifespan, in which damage and defects are seen as symptoms of a pathology.

The diagnosis aims at defining with sufficient degree of certainty the most probable causes for these symptoms, following a differential diagnostic procedure, namely comparing multiple alternatives. These alternatives are reduced to the most likely based on the data collected by anamnesis (interview and search for medical history, presenting complaint and relevant data), examination (mainly qualitative and supported by simple tools) and testing (experimental quantitative evaluations). The causes are then tackled through a proper therapy, whose effects are object of control in time. Shifting from cure to prevention requires a change of mindset and the acquisition of a new awareness of its advantages that are often visible only in the long-term and, nowadays, are not always quantifiable yet. Moreover, such positive consequences are only achieved upon an initial investment in screening that may demand considerable economic and societal costs, especially when a proper planning and knowledge is lacking (Balen 2015). In this regard, the implementation of a preventive conservation strategy should be driven by a case-specific cost-benefit analysis, based on factors as time and resources required for a repeated assessment, costs of timely and delayed measures, significance of the asset and level of risk.

3 PREVENTIVE CONSERVATION: CONDITIONS FOR IMPLEMENTATION AND OPEN ISSUES

A wide-spread implementation of preventive conservation rests on (Balen 2015): (i) scientific knowledge; (ii) clear codes and guidelines; (iii) supportive policies; (iv) trained professionals; (v) a society aware of the importance of its heritage and the advantages of prevention. Nowadays, relevant examples for each of these factors exist in many countries, nonetheless, it is argued that there is no system with the compresence of all of them (Balen 2015).

Promoting education of society, making it aware of the benefit of regular control and maintenance and of the significance of heritage, is a paramount strategy to avoid neglection and vandalism and spread the responsibility of prevention among local communities (Balen 2015, Della Torre 2010). Increasing awareness requires consistent regulations and policies including financial supports but also the evidence of the preventive conservation benefit. This evidence consists of good examples that, in turn, requires trained professionals. The legal framework can rely on obligations and recommendations. The former are intended to enforce preventive measures, by indicating, for instance, when the assessment is mandatory or the periodicity of inspection and maintenance. The latter are indications of the steps to take and instructions, aiming at spreading good practices. In different countries, protection, conservation and actuation criteria are commonly defined at completely different scales (e.g. both national and local, only national or only local), with different levels of coordination between the involved entities. Intervention and management strategies may exist, although disperse, or lack completely (Masciotta et al. 2019). Such sparse and vague instructions induce a dangerous state of uncertainty regarding conservation policies and may result in a non-compliance.

The legal framework is also relevant in the field of testing. Lack of accreditations, regulations and guidelines jeopardise the reliability of the tests that are affected by the personal experience and judgment of the operator. Together with laws and regulations, supportive policies and financial investment through grants and incentives may play an essential role in promoting proactive conservation strategies. Indeed, most of the CH buildings and sites are financially non-self-sufficient and rely on public subsidies to invest in conservation, but recent financial crises have led to deep cuts to heritage sector funding (Marjanović 2014). Moreover, funding is mostly addressed to listed assets, neglecting a large number of historic buildings. Preventive habits may optimise resources allocation and management of the limited budget available, but many assets are already in a severe state of decay to a point where financial institutions are unwilling or unable to invest in urgent remedial measures that are preparatory to a preventive management (Marjanović 2014).

Finally, as already mentioned, the evidence of the preventive conservation benefit is essential to spread its practice. All the stages of conservation, including technical and practical activities, require expert professionals. Lack of training and education of the parties involved in conservation is likely the principal cause of inappropriate decision making. It results in interventions that do not address the causes, but just the symptoms, leading to negative consequences like a recurrence of the pathologies, a diffusion of the damage or an acceleration of the decay. Inadequate interventions, including use of incompatible materials or incompatible structural systems, is commonly driven by non-systematic documentation, limited testing, misinterpretation of the collected data and, in general, an excessive appeal to subjective judgment in the absence of conclusive evidence. Even good decision making may be hindered by the lack of skilled craftsmen to carry out the required activities (Balen 2015).

On the other hand, sometimes, the practitioners voluntarily avoid a detailed investigation claiming that it is expensive and time-consuming, reporting the dispersion of information and the inconsequential existing procedures (Gonçalves et al. 2017). This is likely due to the nature of the sought information, cumulative and dependent on the availability of time and sources. Significant data can be non-existent, unreliable or outdated. Documentation, thus, requires iterative and flexible procedures and adequate platforms to store and retrieve it. In this regard, digital technology may offer an unvaluable support to inspection and documentation. Recent advancement in software and hardware allows to collect, store, retrieve and process an unprecedented large amount of data. Potentially, advanced surveying techniques and structural health monitoring strategies are likely to reduce the time requirement to produce updated and precise information.

Nonetheless, purchasing and maintaining the required software and hardware components, including instrumentation, licences, storage platforms and processing systems, require a significant investment that should be considered in the cost-benefit analysis. More importantly, the information that is generated by such advanced tools, is indiscriminate and growing in size and complexity, is demanding in terms of data management and interrogation and requires a time-consuming processing to become significant and meaningful to the stakeholders. This not only increases the costs but also requires new expertise from fields that were not directly involved in conservation before. On the other hand, saving on the sources or on the post-processing of the data is likely leading to an insufficient level of knowledge. In both the cases, excess or lack of data, confusing and meaningless information is produced, contributing to the scepticism of the stakeholders regarding the diagnostic process.

A final issue, to be mentioned, is the multidisciplinary nature of heritage analysis and preservation that encompasses different approaches, each one with its own terms, methods and sources. A synergistic framework needs a coordination and unification that start from the terminology. Indeed, different disciplines currently use the same words with subtly different meanings or address similar concepts by means of completely different words.

4 RETHINKING PREVENTIVE CONSERVATION: HERITAGECARE METHODOLOGY

Addressing all the aforementioned open issues is not an easy task as they are strongly connected. A non-harmonised, sparse and vague legal framework, without uniform terminology and standardised methodologies for data collection and interpretation, leaves room to subjectivity, prevents interdisciplinary collaboration and hinders successful preventive conservation instances, leading, in some cases, to poor decision making that enhances decay and loss. The lack of good examples contributes to a diffuse scepticism regarding preventive strategies and building diagnosis considered inconclusive, expensive and time-consuming, therefore unworthy.

Unaware owners do not demand preventive conservation, since codes do not enforce it and policies do not provide financial support for it, moreover, they do not resort to expert professionals in case of need that, therefore, are not encouraged to invest in training and perform accurate diagnosis while assessing existing structures. Tackling such issues requires rethinking preventive conservation and the role of academia in disseminating good practices and boosting advancement in scientific knowledge. This process should lead to the development of a consistent and cost-effective preventive conservation framework, defined according to the following steps: (i) review of existing methodologies, documentation and management systems, standards and codes relevant to assessment and conservation; (ii) standardisation of terminology, protocols, recommendations and criteria, integrating the existing ones into a consistent unitary approach harmonised with the current regulations; (iii) identification of flaws or gaps in the overall process flow or in its tasks (e.g. outdated methodologies or conservation needs that are not properly addressed); (iv) development and validation of novel strategies, including testing techniques and diagnostic tools, to fill such gaps; (v) validation of the overall methodology in real scenarios.

A systematic literature review is of paramount importance. Indeed, beside policy-makers, other institutions and scholars have produced a large number of protocols, recommendations and testing strategies, often focused on specific goals within the field of documentation, inspection, diagnosis and conservation (Gonçalves et al. 2018, Kioussi et al. 2011, Pereira et al. 2021). A comparative analysis of these methods aimed at an integration and a harmonisation also with the existing codes is unavoidable. Moreover, the consistency of the framework should be improved by addressing built heritage conservation as a specific case of a wider existing building conservation discipline rather than a separated instance, allowing good practices to be generalised, irrespective of the original field of application. This beneficial integration of the methodologies should follow a holistic approach that includes in the assessment all the needs of the asset as a whole, thus related, not only to structural safety and material conservation, but also to user comfort, energy efficiency and sustainability among others. This requires a comprehensive and multidisciplinary process of harmonising the good practices towards a cost-effective management, in which any activity or intervention carried out on the building aims at fulfilling more of its needs at the same time or, at least, at minimising the negative impact in case of conflicting needs.

Upon this preliminary analysis, a set of basic requirements for a preventive conservation framework are defined as follows:

- The framework must use a clear and unified terminology. Glossaries of damages, activities, principles, concepts, assets typologies and components have been collected in national and international standards and guidelines, e.g. (EN 15898 2019, ICOMOS 2003). A standardised glossary should be built by harmonising such sources and it should be furnished with clear textual and graphical information for a univocal identification of each item.
- Informed decision-making must be supported by a set of relationship databases connecting at least: (i) symptoms and causes, defining also the most effective diagnostic tools to formulate a correct diagnosis; (ii) causes and remedial measures, based on the urgency of the intervention and the extent reached by the pathology. Such databases are built on scientific knowledge and previous experiences and should be updated upon advancement of research. Their correct use prevents the influence of subjective judgment and experience on the diagnostic process and the implementation of unnecessary or incorrect treatments. Moreover, they can help the stakeholders select the most appropriate equipment for their specific predictive conservation needs, based on ongoing and expected damage scenarios (Pereira et al. 2021).
- For each asset a database must be created to collect all the documentation produced. The database should allow a dynamic updating across the entire lifespan of the asset, adapting to the cumulative nature of information (Kioussi et al. 2011). The preventive conservation framework should aim at a comprehensive documentation of the whole investigated system, including the building envelope, the interior, the technical installations, the equipment and the integrated movable assets, as they all contribute to the value and the performance of the system.
- Documentation and data collection must be as free as possible of biases. In case of qualitative methods, especially visual inspection, subjectivity can be prevented by defining a standardised mean of recording the information, presenting clear requests through fields to be filled and unambiguous options for pre-set multiple choices, to be used according to a protocol for each method. In case of quantitative methods (e.g. on-site or lab tests) errors and uncertainties should be minimised by defining clear protocols including information as the equipment, the data storage and retrieval strategy and the tasks to be performed in preparation, during the execution and afterwards, namely, to plan the activities, apply the method and interpret its outcomes.
- Expert and trained professionals are the main actors of the diagnostic process. The protocol of each task of the framework should clearly specify the needed expertise and the accreditation when relevant. Owners and users, irrespective of their background and education, should contribute by correctly using the asset and its components and by monitoring the application of the technical recommendations. Moreover, they should participate in documentation, not only through interviews, but actively, carrying out non-expert regular inspections, aided by simple checklists or questionnaires, to report in a standardised way malfunctioning, damage and decay in the very early stage. This ensures an adequate level of maintenance and a timely identification of the anomalies, optimising the subsequent expert activity.
- To guarantee a flawless exchange of information, for each party involved (e.g. professionals, owners and managers, stakeholders, policy-makers, etc.), databases access and editing rights must be clear, defining type, amount and format of information that each category can query, produce and/or edit. This ensures the quality control and that each party interacts only with information that is meaningful for its purposes. A standard minimum quality and amount of information needed for each task of the framework should be defined, aiming at a good trade-off between costs and benefits of documentation.

- The framework must be flexible and multi-level, in order to be replicable and scalable, adapting to the expected variability and peculiarity of diverse geographical areas and target assets, with various complexity, level of performance, conservation needs, protection status, local environmental, social, economic and financial conditions. Such factors affect the extent of the information that can be collected and generated. Therefore, the granularity of information should be defined upon agreement among the parties involved, based on the pre-defined level needed for each task. A hierarchic and nested organisation of the levels, where the specific tasks of a lower level are included in the upper levels and complemented by additional activities, allows a dynamic adaptation of the service, over time, to new conditions, resources or needs. At least two levels for the assessment can be identified, in agreement with ISO 13822 standard (ISO 2010), namely a preliminary and a detailed assessment. An harmonisation of standard procedures can be attempted by integrating the preliminary inspection with the condition survey detailed in EN 16096 standard (EN 2012).
- The diagnosis should produce an indication of the recommended measures and their urgency for the asset as a whole, based on the condition grading, the risk and the recommended measures for its components (EN 2012). The criteria and the relevant features used to issue the grade should be clearly expressed as well as the aggregation formulas to estimate the overall score based on the component's values. Standardised criteria should be also defined to link condition and risk to the urgency and type of intervention needed. A colour-coded rating supports the interpretation by making reporting more user-friendly to non-technical users (Abbott et al. 2007).
- The preventive conservation framework should present a high level of digitisation. All the aforementioned relationship databases, glossary, protocols and previously generated documentation on the asset should allow online, real-time exploitation, especially to support on-site activities. This reduces the time invested by the operators in learning the methodology and performing the tasks, reduces the gap in technical knowledge between different operators and increases the accuracy of the inspection (Gonçalves et al. 2018). Moreover, a digital platform supports the definition of clear access and editing rights, automatically filtering the information and providing ad-hoc authorisations and restrictions to each category of user. The digitisation of the information is also essential to establish an effective interoperability between all the parties involved in the management and preservation of the assets. To this end, specific protocols should be defined to guarantee that the documentation is made available for other purposes, as analysing energy efficiency, managing activities within the spaces, estimating quantity take-off, allowing interactive and virtual engagement with the asset, etc.
- The databases created for each instance allow data collection and exploitation at the individual building level. However, a centralised management of the information allows the statistical analysis of an increasing group of assets, offering an invaluable tool for policy-makers to learn from experience and establishing good practices. This higher level analysis, indeed, provides a paramount insight into pathologies occurrence, reliability of the diagnostic techniques and effectiveness of remedial measures.

The recently concluded HeritageCare project (SOE1/P5/P0258) has significantly contributed to this ongoing process of rethinking preventive conservation. This multidisciplinary high-technological effort, involving eight beneficiary partners and eleven associated partners from three countries (Portugal, Spain and Southwestern France), coordinated by the University of Minho, has led to the development of a new validated methodology for heritage preventive conservation, according to the aforementioned requirements.

For a thorough description of the project and its outcomes refer to (Barontini et al. 2021; Masciotta et al. 2019, 2021). HeritageCare multi-level methodology encompasses a set of tasks organised according to a systematic workflow in three following stages, namely prior to, during and after inspection, each one with specific data categories to be collected and generated and activities to be carried out. The granularity of the documentation and information searched, stored and produced varies according to three Service Levels, SLs (Table 1).

Table 1. HeritageCare service level definition.

Service Level	Designation	Functionality
SL1	StandardCare	Provision of what is essential for the primary health and ordinary maintenance of the heritage building.
SL2	PlusCare	Provision of what is necessary for the primary health, ordinary maintenance and thorough screening of the heritage building along with its integrated and movable assets, including monitoring data to support decision making.
SL2	TotalCare	Provision of what is necessary for the primary health, ordinary maintenance, thorough screening and enhanced management of the heritage building along with its integrated and movable assets.

SL1 provides a low-cost and rapid, although complete, assessment of the historic building, harmonised with the methodology described by EN 16096 Standard (EN 2012). Prior to inspection, the off-site documentation is carried out, through historical survey and bibliographic search. The reliability of all the textual and graphical sources is assessed and all the relevant data is extrapolated and collected within the Building ID and management information, namely a series of descriptors, updated over time. These include univocal code, name, category, protection status, property, time of construction, original and current functions, localization, important historical information, architectural features, construction system, principal building materials, previous interventions, inspections, maintenance actions, test reports, number of integrated and movable assets of cultural interest with a description of their significance, age and main geometrical and material features. This documentation is furnished with bibliographic references, sketches and drawings of the main components and spaces. The inspection at SL1 is mainly qualitative and is performed by at least two experienced professionals, with complementary expertise, capable of grading the condition state of the building and its components. The main support on-site is the inspection app (Figure 1), with an e-form to be filled online, with a standardised checklist of items and sub-items to survey. During on-site activities, the surveyors have access to informative materials as the standardised glossary, the damage atlas with definitions and examples and a collection of most common damages and deterioration processes for each sub-item. Each damage affecting the sub-item is graded through a condition index and a risk index, according to a 4-point scale, from 0 to 3 (Figure 2). These indexes are then used to assess the sub-items, the items and the asset as a whole, in a bottom to top cascade. More details on the assessment criteria are provided in (Masciotta et al. 2019).

Upon completion of the inspection process the report for the asset managers and owners is automatically produced and stored on an online platform. The report encompasses an informative section on the overall condition of the building and its main components and a set of clear and schematic recommendations regarding remedial or preventive actions to undertake in the short/medium/long term based on the identified damages, their most likely causes and consequences (Figure 3). The asset managers or the owners are then invited to provide feedbacks, in order to document any following measure undertaken on the building. Building ID, management info and subsequent condition reports constitute a simple but informative attribute-based digital twin. Movable assets hosted within the building may deserve specific attention during the inspection. To this end an ad-hoc documentation protocol is defined, aimed at producing an Asset ID, namely a set of relevant data, similar to the Building ID, used to unambiguously identify any specific heritage object and allow its standardised inspection. This inspection is carried out by means of a dedicated form where damage, alteration and operational and environmental conditions that are likely to affect the asset conservation (e.g. temperature, relative humidity, illumination, etc.) are recorded.

SL2 complements and increases the level of information on the building and the integrated and movable assets, producing a virtual restitution and collecting quantitative information, through testing, monitoring and surveying techniques. Typology, location and number of tests are defined

Figure 1. HeritageCare mobile inspection app.

Class N.	Condition index
0	Good – No symptoms
1	Fair – Minor symptoms
2	Poor – Moderately strong symptoms
3	Bad – Major symptoms
NA	Not (safely) accessible

Class N.	Risk index
0	Long term No actions | preventive monitoring
1	Medium term Non-urgent actions | Monitoring
2	Short term Timely repair or additional inspection and diagnosis
3	Urgent and immediate Urgent repair or additional inspection and diagnosis
NA	Not inspected Not safely accessible or not visible

Figure 2. Condition and risk rating indexes.

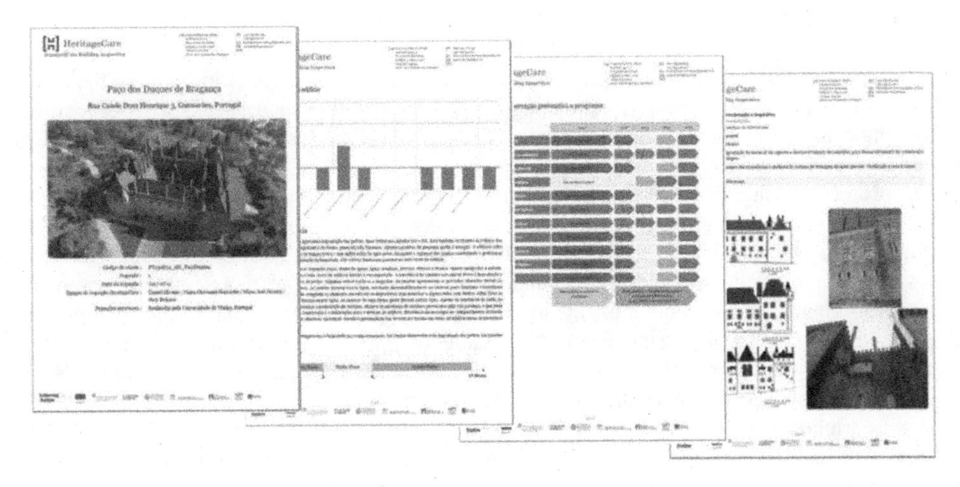

Figure 3. Excerpt from the Guimarães ducal palace inspection report.

prior to inspection, based on the condition assessment and upon agreement among the involved parties. Each test has a pre-set protocol defining the expertise requested to the operators, the equipment and the procedures for a correct execution. Testing and monitoring is intended to identify and track physical, mechanical and environmental parameters for a detailed assessment of the asset condition and the evolution of its performance. All the required techniques are summed up to build a service that is tailored to the specific asset needs and resources. This flexibility fosters an application of the methodology to any context, without requiring extra costs or specific expertise of the operators.

For instance, the methodology is independent of the surveying strategy, allowing the managers to decide whether to request a traditional metrical survey or a more advanced one, image-based (e.g. terrestrial or aerial photogrammetry) or range-based (e.g. static or dynamic laser scanning). For photogrammetry three protocols exist depending on the goal, namely reconstruction of planar objects, 360° reconstruction of movable assets or detailed reconstructions. The protocol determines the rules and the parameters of the acquisition (e.g. number of captured images, the shot overlap and path, the lens system, the focal length, the exposure triangle), based on the characteristics of the object to capture and the required level of detail. For laser scanners, protocols are more flexible. Nonetheless, essential practical rules are defined and strictly followed to optimize the outcome of the data acquisition. On-site, beside surveying and testing, the operators collect an ensemble of 360° panoramic photos of the whole buildings, recording all its main components and integrated movable assets.

The resulting digital twin integrates the alphanumerical information of SL1 with graphical information obtained by interlinking these 360° panoramic views. This simple, although clear 3D reconstruction allows a virtual tour inside and around the building and is enriched by the identification of hotspots (e.g. damage hotspot, asset hotspot, sensor hotspot, etc.), clearly recognisable through a predefined visual code. Each hotspot is a link to stored information, as SL1 condition reports, specific documents and images concerning the assets, alteration or damage detected during on-site inspections, real-time reading from the monitoring systems installed. When a point cloud is generated, this is also navigable for the stakeholders on the platform, contributing to the information content of the digital twin. Moreover, the platform allows important operations directly on the point cloud, such as slicing to produce 2D drawings (plans, sections and elevations) and segmentation to identify functional parts of the building (e.g. roof, façades and rooms).

Finally, SL3 produces a highly informative 3D model in a BIM environment. A protocol with a standardized workflow to develop and update the BIM model is further discussed in (Barontini et al. 2021). The protocol is based on a clear separation of roles and expertise, a standardisation of the documentation process and the interoperability with the e-form and other professional software. This ensures the exchange of information and its use for other preservation purposes, as for instance organising and performing an inspection, analysing the structural safety or the energy efficiency, designing interventions, managing activities within the spaces, etc., without requiring modelling expertise and holding software licence to any of the parties involved, except for the BIM modeller.

The protocol defines the extent and granularity of the information for each element of the BIM model according to the purpose of the model, to be defined in agreement by all the parties, and in compliance with EN 17412-1 standard (BS EN 2020), as a combination of geometrical data, alphanumerical data and documentation. Indeed, the model consists in an assembly of parametric objects representing the real components of the building with an acceptable level of geometrical detail. These objects are placed in their correct location in the three-dimensional space, as resulting from the existing documentation and the surveys. Accurate measurements and point clouds produced in service level 2 are a fundamental source for the model. The information related to each object is enriched by means of non-graphical attributes and linked documents as, for instance, the outcomes of the historical survey, the bibliographic research and the condition survey. Localised damage can be easily represented by patch-type objects. Similarly, in case of monitoring, the exact location of the sensors can be shown within the 3D model, enabling the real-time reading of their records.

The main purpose of the BIM model is to support on-site activities. Operators can navigate and interrogate it online by using mobile devices or even with the aid of augmented reality technology through mixed reality smart glasses. The availability on-site of an informative model that collects all the previous documentation on the assets is paramount, providing a continuous and timely interaction between virtual objects and physical counterparts. This approach allows a fast comparison of actual damages or alteration phenomena with a previously recorded condition, for a fast decision making regarding urgent measures or an optimisation of the inspection process towards the causes of the phenomena. Augmented reality is a further improvement, permitting a visualisation of the information collected in the BIM model directly on top of the real inspected objects. Smart devices allow also an efficient and rapid data collection by taking pictures or measuring distances. Beside purchasing the smart devices, the surveyors do not need to invest in licences, since navigation and query of the model can be done on free of licence model viewer software or directly on the online platform. No expertise in BIM modelling is demanded to the surveyors, since the manipulation of the model is carried out as back-office activity by an expert BIM modeller upon receiving the inspection forms and reports.

Increasing service level, from SL1 to SL3, implies an increment of the quantity and complexity of the information, requiring more advanced surveying and testing techniques, thus more sophisticated equipment, costs, time and expertise of the operators. It also demands advanced strategy to store, manage and visualise the produced documentation. The flexibility of a multi-level approach allows to provide a service tailored to the specific requirements and financial availability of the owners and the conservation needs and complexity of the specific asset. The adaptability and effectiveness of the methodology were tested over several typologies of heritage buildings (e.g. churches, chapels, palaces, castles, etc.) equally distributed over the three countries. In particular, sixty case-study buildings (twenty per country) were inspected and assessed according to the first service level. Fifteen out of these sixty (five per country) were selected for the implementation of the second service level. Finally, one case study per country was included in the third service level. Two selected case studies, namely São Torcato church and the Guimarães ducal palace, are discussed hereafter to show a complete SL2 and SL3 application, respectively.

4.1 *São Torcato church*

Located in the homonymous village, close to Guimarães, in the north of Portugal, São Torcato church's construction started in 1871, based on the original conception proposed in 1825, featuring a Neo-Manuelino, revival style, and continued in phases for more than 130 years (Ramos et al. 2013). Photos taken during the construction allow documenting the evolution of the work (Figure 4). In its actual configuration, the church has a Latin-cross plan, with a gallery entrance, a single nave (57.5 m long, 17.5 m wide and 26.5 m high) with side chapels and an apse at the north end. The transept is 37 m long and 11.5 m wide. Nave and transept are covered with barrel vaults and a dome with octagonal tambour stands at their crossing. Two towers are placed on the sides of the façade featuring a rectangular plan (7.5×6.5 m^2) and a height of 50 m. Wall thickness varies in the façade, from 2.5 m to 1.7 m. The thickness is 1.3 m in the lateral walls and 1.45 m in the towers (Ramos et al. 2013).

The succession of building phases determined the use of different materials, in particular three-leaf granite masonry for towers, nave and transept, and reinforced concrete for the dome and apse. The gabled roof is supported by timber trusses. Since 1970s the church has been subjected to inspections and regular control, due to a severe cracking in the front area of the church, especially the façade (Figure 5a), likely associated to their differential settlement and tilting, due to the poor mechanical characteristics of the soil. Between 2014 and 2015, the church underwent a structural intervention aimed at eliminating the differential settlement and restoring material continuity, by means of micro-piles, post-stressed tie rods and crack injection (Masciotta et al. 2017). To assess the impact of the construction activities on the church and validate the intervention, a monitoring system was installed and was active before, during and after the works (Masciotta et al. 2017).

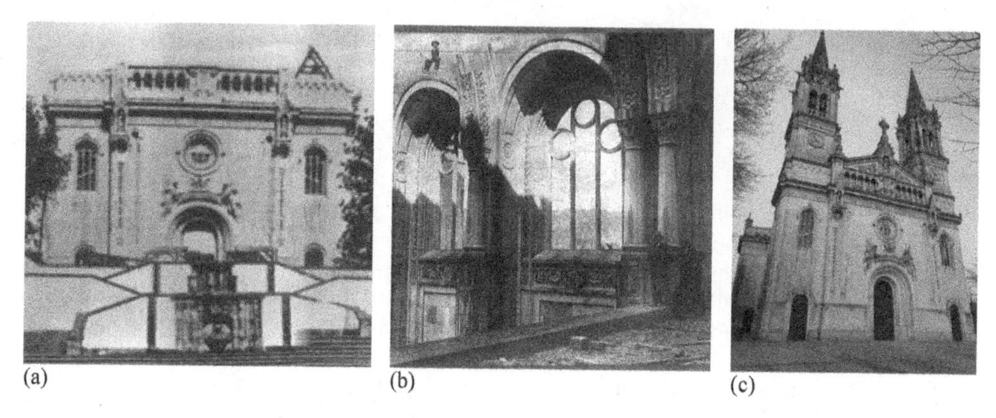

Figure 4. São Torcato church, building phases: (a) exterior view; (b) interior view; (c) actual aspect.

After few years, HeritageCare SL2 protocol was applied to the church. The large number of existing documents was collected and used to constitute the Building ID and management information. In this case, dealing with a quite recent asset that underwent several investigations and a significant intervention, the material available encompasses previous surveys, photos taken during the construction and interventions and test reports. The inspection carried out by the HeritageCare team identified several forms of degradation and damage, such as discolouration, efflorescence, biological growth and bird infestation (Masciotta et al. 2021). Most of the problems were related to water infiltration through roof, walls and openings. After the intervention, a permanent deformation was still evident in the choir, whereas new cracks appeared on the triumphal arch and along the lateral walls of the nave.

3D documentation consisted in a laser scanner survey, by means of a Leica ScanStation P20 (Figure 5). A 3D point cloud with about 3 billion of points, then reduced by processing and filtering to 17 million, was generated over 174 scan stations. The generation of the model required 3 weeks of work to two technicians. Contextually, a photographic survey was carried out in 110 minutes by means of a 360° camera Ricoh theta V at 42 locations, capturing inside and outside. The enriched virtual tour model was generated through the proprietary software Pano2VR®and ad-hoc developed plugins. A more detailed report on this case study can be found in (Masciotta et al. 2021).

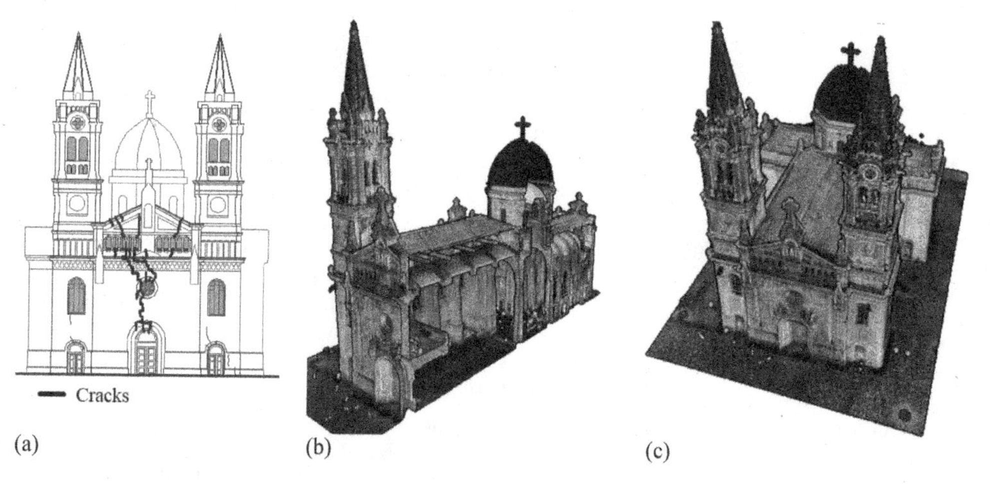

Figure 5. São Torcato church: (a) crack survey before intervention; (b) point cloud of the interior; (c) point cloud of the exterior.

4.2 *Guimãraes ducal palace*

The construction of the ducal palace of Bragança, located in Guimarães, Portugal, was more articulated than the previous case study and was affected by several vicissitudes. The construction, begun in 1420, under the first Duke of Bragança, suffered a first stop after his death. In 1478, the construction continued under the third Duke. In this period, the actual organisation of the spaces was defined. However, the palace remained incomplete, abandoned and subjected to dismantling and reuse of the materials, since the beginning of the 16th century, once the court was moved to another town. The alteration of the building continued during the 19th century, when it was turned into a military barrack. Finally, after the acknowledgment of its importance and the inclusion in the national monument list in 1910, the palace underwent a series of strong interventions, with demolition of the changes occurred in the previous century and addition and reconstruction of several of its parts, aiming at the supposed original aspect of the building. During the repair, an extensive use of reinforced concrete beams was made in the floors and roof. The actual building features a rectangular plan around a central courtyard, surrounded by a colonnade at the two lower storeys. Thirty-nine brick chimneys, among which only four are original, constitute a landmark of the city, as well as the timber trusses of the main rooms are one of the most precious features of the building. These elements are also a major concern for preservation, significantly contributing to the overall vulnerability of the palace.

The diagnosis was based on a detailed inspection that involved more than two professionals to reduce the time-requirement. The staff was interviewed providing a series of relevant information to complement the main findings of the inspection. Degradation and alteration likely due to moisture and water penetration, fostered by the inadequate and poorly maintained drainage system, were identified. Loss of material in the walls is also present, likely due to the incompatibility between the granite blocks and the mortar used in recent repointing works. Finally, superficial cracks are found at the ground floor in load bearing walls. The building features several integrated movable assets, including hundreds of art pieces dating back to the 17th and 18th centuries. The inventory and condition survey of the most significant ones were carried out. Few pieces showing an unsatisfactory conservation state were closely inspected by the HeritageCare team.

Based on the criticalities emerged from this condition survey and according to SL2 protocols, an ad-hoc monitoring system was installed in October 2018 to track structural and environmental parameters. The goals and demands of the monitoring system were defined upon agreement with the directions of the DRCN (Northern Regional Directorate of Culture), aiming at a trade-off between costs, visual impact of the sensors and quality of the collected information for the conservation purpose. The network, still operating, is composed of: (i) 12 temperature and relative humidity sensors (7 surface and 5 ambient sensors) and 5 combined sensors measuring surface temperature, relative humidity and luminosity; (ii) 3 xylophagous sensors at the timber roofing of main room and chapel; (iii) 1 carbon dioxide sensor; (iv) 2 biaxial clinometers on the outer wall; (v) 1 external meteo station recording air temperature, humidity, barometric pressure, wind direction and velocity, precipitations, rain duration, hail, solar radiation and carbon dioxide.

SL2 protocols adopted included a laser scanner survey, carried out, using a Leica ScanStation P20 (Figure 6). Four full working days on-site were necessary for the survey. 360° panoramic views ware taken, concurrently. All these sources of information contributed to the generation of the digital twin of the palace hosted on the HeritageCare platform. This is composed, at SL2, of the 360° panoramic views based virtual tour, enriched by a set of hotspots. Asset hotspots identify the significant movable assets inspected and assessed with more detail, providing the results of the on-site survey (Figure 7).

Damage hotspots locate the anomalies found during previous inspection on-site. Sensor hotspots allow reading the most recent instrumental acquisitions (Figure 8). Here, samples are updated hourly. Based on pre-set threshold values, the acquisition presents a colour-based warning so that the manager can easily identify parameters that are deviating from the acceptable condition. The results of the laser scanner survey are also navigable on the platform.

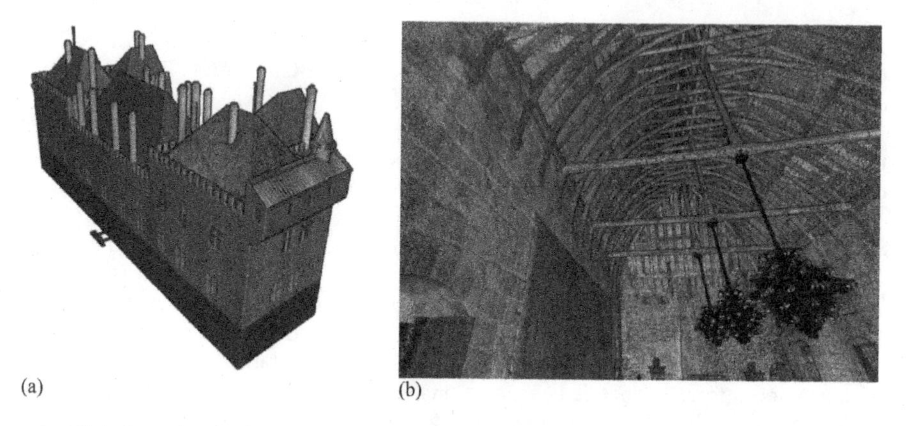

(a) (b)

Figure 6. Guimãraes ducal palace point cloud: (a) external view; (b) detail of the interior.

Figure 7. Virtual tour enriched with asset hotspots, linked to condition survey report.

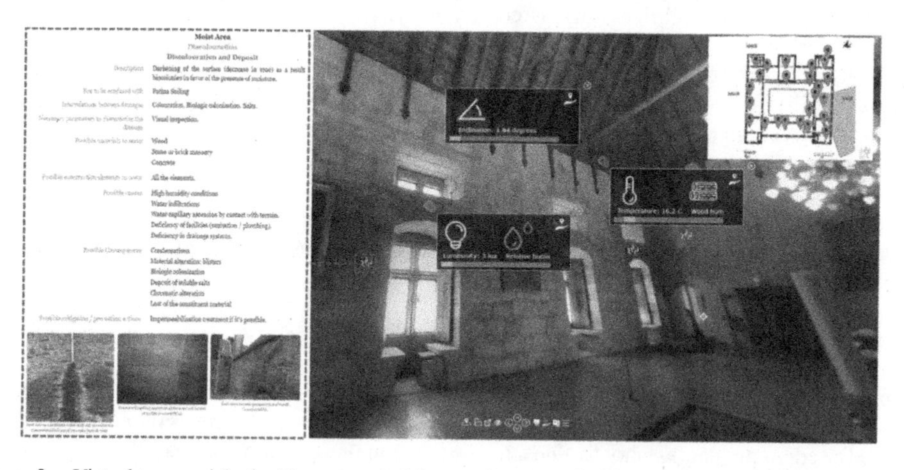

Figure 8. Virtual tour enriched with sensors and damage hotspots, linked to recent acquisition and damage report, respectively.

Finally, SL3 protocol was applied, by building a BIM model of the palace, enriched with all the aforementioned data (Figure 9). The model was generated first resorting to traditional survey techniques and existing documentation, then it was validated through the laser scanner acquisition. The purpose of the model, namely supporting an effective exchange of information between asset manager and surveyors regarding assets condition and ongoing or emerging anomalies is fulfilled by a wise trade-off between graphical and non-graphical information. This allows a reduction in time and costs to generate the model, ensuring a sufficient level of geometrical detail to correctly localise, within the building and its components, damage, movable assets or sensors. An augmented reality aided inspection was carried out by means of a pair of HoloLens and a smart glass inspection app. More information on this case study are provided in (Masciotta et al. 2019).

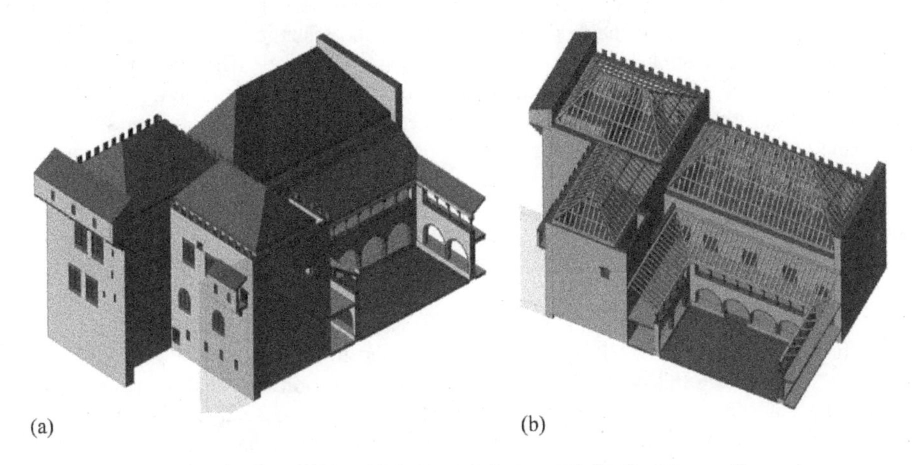

Figure 9. Guimarães ducal palace BIM model: (a) global view; (b) detail of the roofing system.

5 NEW TECHNIQUES FOR THE INSPECTION AND PRESENTATION OF THE NON-VISIBLE PARTS OF THE BUILT HERITAGE: HWITHIN METHOLODOGY

The Heritage Within (HWITHN) European Research Project (Ortega Heras et al. 2021) aimed to develop new technologies to produce an innovative visualization of the cultural heritage by showing nonvisible features of buildings and archaeological assets. 3D surveying and modelling techniques (e.g. photogrammetry and laser scanning) have greatly evolved in the recent years but they can only reconstruct the exterior surfaces of the elements.

The project aimed to go beyond this barrier to image also relevant information of the interior of its constructive elements and other non-visible data. To this end, the project not only resorted to existing techniques (Ground Penetrating Radar) but also developed new ones, namely a system to perform on-site ultrasonic acoustic tomography of complex architectural elements in an automatized way. As a result, the inner hidden morphology of several columns could be reconstructed, on an almost stone-by-stone basis, and the inner damage and state of conservation of the material could be evaluated.

The final 3D model and associated information was implemented into a Virtual Reality (VR) application to offer an interactive visualization of our heritage, on-site and remotely. The approach proposes to relate the visible with the invisible, looking beyond the surface of the object, which facilitates the identification of inner morphology, cavities, hidden objects or damage. The action proposed as a pilot case study the Archaeological Museum of Carmo, in Lisbon (Portugal), which occupies the ruins of the old church of the Carmo Convent, destroyed during the 1755 Lisbon earthquake.

The Carmo Convent was commissioned in 1389. After two attempts, failed due to local subsoil conditions, the construction works started in the last decade of 14th century and continued until

1423, with alternate vicissitudes, including structural problems still related to subsoil capacity. In the following centuries, further addition and works embellished the church that was also populated with various pieces of art, until 1755 when a catastrophic earthquake caused the collapse of most of the structure. Although in ruin, the building was preserved due to its emblematic value and today is a museum.

All the activities of the Heritage Within project relied on the synergistic contribution of a multidisciplinary team of architects, civil engineers, telecommunication engineers, archaeologists, art historians, and geophysicists, driven by a collaborative definition of goals and demands. The project took advantage of the wide expertise of this team and generated layers of specialized information of the case study (e.g. virtual reconstruction of the original aspect of the church, structural analysis results, thermography studies or location of old art pieces), inserted within the VR platform with two main objectives: (i) make the public more aware of the essential role of professionals in the field of conservation (from art historians to engineers and architects), showing the importance of surveying, diagnostic and analysis activities for the conservation of built heritage; (ii) help specialists in the interpretation of their own results related to other specialists outcomes, by means of novel visualization tools and integration of the results of diagnostic investigations from different sources into a single VR platform.

The use of virtual reality for the dissemination and storytelling of complex results is expected to enhance accessibility to cultural heritage and enrich the visitor's experience. Moreover, it is likely to support technical activities, allowing an easy and informative visualisation on-site instead of at their desks on their computer.

Knowledge of the monument is a key aspect of conservation activities. The HWITHIN project essentially explores the use of VR as a work environment to read and visualize multiparametric information, facilitating the interpretation of technical inspection and analysis results (obtained from non-invasive inspections or structural analysis). The use of such platforms to interrelate heterogeneous data can help to understand cause-effect mechanisms between constructive characteristics, damage and structural behaviour. Nevertheless, beside the primary preservation purpose, the project set as primary objective an enhancement of the visitors' engagement with the building, for example, through the virtual reconstruction of the original aspect of the church. To this end, the multi-layered digital twin created was enriched with technical information, but also made suitable for the implementation of a virtual reality visit of the church in the actual condition and in the reconstructed aspect before its collapse. Based on the goal and the intended user, the complexity of the model can range from a virtual tour based on interlinked 360° panoramas to an advanced 3D photorealistic restitution (Figure 10).

Figure 10. Virtual tour and linked layers: photorealistic 3D restitution, ultrasonic acoustic tomography (above) and structural analysis results (below).

The proposed approach constitutes an effective system for storing and analysing heterogeneous data. The future challenging goal is to integrate the virtual scenarios proposed with an Internet of Things (IoT) system to be used with a digital twin perspective. The Carmo Convent can be equipped with sensors measuring in real time environmental and structural health monitoring parameters. Results can be evaluated on the digital twin and possible interventions can be assessed and managed remotely. The association between physical object and virtual reality makes it possible to activate data analysis and monitoring of the monuments in such a way that it is possible to operate in predictive mode, identifying problems even before they occur. A digital model continuously connected with its physical counterpart and capable to be managed in an interactive form can highly optimize conservation activities.

6 CONCLUSIONS

Although preventive conservation is recognised as the potentially most cost-effective strategy for cultural heritage preservation, its widespread application is complicated by several factors, as its multidisciplinary nature, the sparsity and case-specificity of the available information, the lack of effective guidelines or standards, the limited expertise of the professionals and the lack of awareness of users, owners and managers. Here, a list of essential requirements in the field of preventive conservation is presented, within an ongoing process of rethinking this discipline towards the definition of a comprehensive and cost-effective framework. Moreover, innovative practices developed and validated, with an active involvement of the authors within two European Projects, are presented.

ACKNOWLEDGEMENTS

This work was financed by ERDF funds through the V Interreg Sudoe program within the framework of the project "HeritageCare - Monitoring and preventive conservation of historic and cultural heritage" (Ref. SOE1/P5/P0258). The Heritage Within research project was co-funded by the Creative Europe Programme of the European Union, under the Cross-sectorial subprogramme, Bridging culture and audiovisual content through digital action (Project 614719-CREA-1-2019-1-PT-CROSS-SECT-INNOVLAB). Moreover, this work was partly financed by FCT/MCTES through national funds (PIDDAC) under the R&D Unit Institute for Sustainability and Innovation in Structural Engineering (ISISE), under reference UIDB/04029/2020.

REFERENCES

Abbott, G.R., McDuling, J.J., Parsons, S.A., Schoeman, J.C., 2007. Building condition assessment: a performance evaluation tool towards sustainable asset management.

Balen, K.V., 2015. Preventive Conservation of Historic Buildings. Restoration of Buildings and Monuments 21, 99–104. https://doi.org/10.1515/rbm-2015-0008

Barontini, A., Alarcon, C., Sousa, H.S., Oliveira, D.V., Masciotta, M.G., Azenha, M., 2021. Development and Demonstration of an HBIM Framework for the Preventive Conservation of Cultural Heritage. International Journal of Architectural Heritage 0, 1–23. https://doi.org/10.1080/15583058.2021.1894502

BS EN, 2020. 17412-1: Building Information Modelling. Level of Information Need. Concepts and principles.

CEN, 2010. Maintenance — Maintenance terminology.

Della Torre, S., 2010. Critical reflection document on the draft European standard CEN/TC 346 WI 346013 conservation of cultural property-condition survey of immovable heritage (unpublished discussion document), in: Proceedings of the Seminar on Condition Reporting Systems for the Built Cultural Heritage, Monumentenwacht, Vlaanderen. pp. 22–24.

EN, 2012. 16096: Conservation of cultural property - Condition survey and report of built cultural heritage.

EN 15898, 2019. Conservation of cultural heritage - Main general terms and definitions.

Giovanna Masciotta, M., Sánchez Aparicio, L.J., Bushara, S., V. Oliveira, D., González Aguilera, D., García Alvarez, J., 2021. Digitization of cultural heritage buildings for preventive conservation purposes, in: 12th international conference on structural analysis of historical construction | 29-Septiembre a 1 Octubre 2021 | Barcelona, pp. 1559–1570.
Gonçalves, J., Mateus, R., Silvestre, J.D., 2018. Comparative Analysis of Inspection and Diagnosis Tools for Ancient Buildings, in: Ioannides, M., Fink, E., Brumana, R., Patias, P., Doulamis, A., Martins, J., Wallace, M. (Eds.), Digital Heritage. Progress in Cultural Heritage. Springer International Publishing, Cham, pp. 289–298. https://doi.org/10.1007/978-3-030-01765-1_32
Gonçalves, J., Mateus, R., Silvestre, J.D., Vasconcelos, G., 2017. Survey to architects: challenges to inspection and diagnosis in historical residential buildings. Green Lines Institute for Sustainable Development.
ICOMOS, 2003. ICOMOS Charter – Principles fora the Analysis, Conservation and Structural Restoration of Architectural Heritage.
ISO, 2010. 13822: Bases for design of structures — Assessment of existing structures.
Jagodzińska, K., Sanetra-Szeliga, J., Purchla, J., Van Balen, K., Thys, C., Vandesande, A., Van der Auwera, S., 2015. Cultural Heritage Counts for Europe: full report.
Kioussi, A., Labropoulos, K., Karoglou, M., Moropoulou, A., Zarnic, R., 2011. Recommendations and Strategies for the Establishment of a Guideline for Monument Documentation Harmonized with the Existing European Standards and Codes. Geoinformatics FCE CTU 6, 178–184. https://doi.org/10.14311/gi.6.23
Marjanović , V., 2014. Europe's endangered heritage (No. Doc. 13428). Report, , to Committee on Culture, Science, Education and Media.
Masciotta, M.G., Morais, M.J., Ramos, L.F., Oliveira, D.V., Sánchez-Aparicio, L.J., González-Aguilera, D., 2019. A Digital-based Integrated Methodology for the Preventive Conservation of Cultural Heritage: The Experience of HeritageCare Project. International Journal of Architectural Heritage 1–20. https://doi.org/10.1080/15583058.2019.1668985
Masciotta, M.-G., Ramos, L.F., Lourenço, P.B., 2017. The importance of structural monitoring as a diagnosis and control tool in the restoration process of heritage structures: A case study in Portugal. Journal of Cultural Heritage 27, 36–47. https://doi.org/10.1016/j.culher.2017.04.003
Ortega Heras, J., Gonzalez, M., Izquierdo, M., Masini, N., Vasconcelos, G., Pereira, C., Navarro, R., Gabellone, F., Vitale, V., Abate, N., Liébana, J., Anaya, G., Secanellas, S., Anaya, J., Leucci, G., Sileo, M., Borghardt, J., Ferreira, T., 2021. Heritage Within. European Research Project.
Pereira, C., de Brito, J., Silvestre, J.D., 2021. Harmonized Classification of Repair Techniques in a Global Inspection System: Proposed Methodology and Analysis of Fieldwork Data. Journal of Performance of Constructed Facilities 35, 04020122. https://doi.org/10.1061/(ASCE)CF.1943-5509.0001529
Ramos, L.F., Aguilar, R., Lourenço, P.B., Moreira, S., 2013. Dynamic structural health monitoring of Saint Torcato church. Mechanical Systems and Signal Processing 35, 1–15. https://doi.org/10.1016/j.ymssp.2012.09.007
Taylor, J., 2005. An Integrated Approach to Risk Assessments and Condition Surveys. Journal of the American Institute for Conservation 44, 127–141. https://doi.org/10.1179/019713605806082365
Wagg, D.J., Worden, K., Barthorpe, R.J., Gardner, P., 2020. Digital Twins: State-of-the-Art and Future Directions for Modeling and Simulation in Engineering Dynamics Applications. ASCE-ASME J Risk and Uncert in Engrg Sys Part B Mech Engrg 6. https://doi.org/10.1115/1.4046739

Geotechnical Engineering for the Preservation of Monuments and Historic Sites III – Lancellotta, Viggiani, Flora, de Silva & Mele (Eds)

Protecting the Sagrada Familia temple from railway tunnel construction

E.E. Alonso & A. Ledesma
Department of Civil and Environmental Engineering, Barcelona School of Civil Engineering, UPC, Barcelona, Spain

ABSTRACT: The high-speed railway line crossing downtown Barcelona was immediate to the main façade of Gaudi's modernist temple of Sagrada Familia. The paper describes the precautions adopted to avoid any damage to the church. The façade was protected by a large diameter pile wall. The performance of this wall against volume losses induced by the Earth Pressure Balance Shield construction procedure is analysed by two computational methods which rely on some fundamental solutions for the elastic half-space. The first 2D plane strain method allows a dimensionless formulation which facilitates a sensitivity analysis and helps to adopt main design decisions. The more general 3D approach described may reproduce the spatial distribution of volume loss, it solves the interaction between general pile foundations and tunnel-induced deformations and it is useful to follow the effect of advancing tunnel excavation on pile foundations (pile walls in particular). The very small settlements observed at the street level (less than 2 mm) are compatible with an overall volume loss of 0.1%. The paper describes also the organization and precautions taken to ensure a very small volume loss. Of particular interest is the procedure followed to maintain in perfect conditions the machine cutting head. It required the construction of deep shafts and auxiliary enclosures to permit the maintenance works under atmospheric conditions.

1 THE SAGRADA FAMILIA TEMPLE

The temple, an "expiatory" church to be financed by public donations, was initially designed (not by Gaudí) as a neo-gothic structure and the construction started in 1882. It was designed as a classic, Latin Christian basilica, having a central nave, two aisles, a transept, a circular apse and a tower, 85 m high, above the nave-transept crossing. Gaudí (1852–1926), at the time a 32-year-old architect, was appointed director of construction works in 1883. He started soon to introduce changes in the original design. By 1891 he had already re-designed one of the façades of the church (the Nativity façade) in a modernist (art nouveau) style. The construction of this facade, that became later an iconic landmark of Gaudí's modernist style (Figure 1), began in 1892 when the gothic apse of the church was already well advanced. Gaudí faced successfully the challenge of integrating the apse into the modernist style that he had in mind. The Nativity facade, at one end of the transept, with its four slender and very characteristic geometry and texture was a reference for the two additional portals of the temple: The Passion portal, at the other extreme of the transept, and the main Glory facade, at the entrance of the church.

The towers of the Nativity had a height of 98–107 m and a diameter of no more than 7.2 m. In 1902, when Modernism was widely adopted in Barcelona, Gaudí fully redefined the architecture and dimensions of the temple. In addition to the 12 towers defining the three portals of the church, Gaudí designed six more towers of increasing height located at the nave transept intersection: four named after the four evangelists, one for the Virgin Mary and the tallest one (170 m) to honor Jesus Christ. Gaudí paid special attention to building plaster models of parts of the structure. He was

DOI 10.1201/9781003329756-5

Figure 1. The Nativity façade (darker colour), the first to be built, in a recent photograph, showing the iconic modernist towers designed by Gaudí.

also aware of the increasing loads of the new design, but was confident that new developments in construction materials and procedures would be available in the future to complete the works. He was right. Portland cement and reinforced concrete was used in the pinnacle of the first tower completed of the Nativity portal before the date of Gaudí's death.

Modernism received strong criticisms as an anarchic, decadent and "romantic" style during the first decades of the 20th century and it was replaced in the history of architecture by a rational, ordered, "modern" style. Gaudí designed columns for the nave, inspired in nature, of great beauty. He was an accomplished geometer and his favorite shapes for vaults, column-vault connections and walls were quadric surfaces (hyperboloids, paraboloids, ellipsoids). He integrated often sculptures into his design, but he never forgot the fundamental resisting concepts of his structures. His style is unique and the Sagrada Familia temple is his masterpiece.

After Gaudí's death, his close collaborators continued the construction of the temple. In 1936, the beginning of Spanish civil war, a deliberate fire destroyed many architectural plans and drawings. The plaster models survived and they perpetuated Gaudí's original design. The construction advanced slowly because its financing depended on public donations.

Natural stone (sandstone, granite, basalt and porfyr, in order of increasing strength) was used by Gaudí to build the pillars. Later, as the temple grew in size, reinforced concrete was necessary to guarantee stability. The stone covered the inner concrete structure in a sort of "lost formwork".

Construction advanced slowly until 1980's. Then, tourist revenues allowed a much faster construction pace. Today, the nave and aisles, transept, transept facades and apse are completed. The central high towers are under construction. The main Glory facade will complete the temple. A tentative date for its completion is 2026. Figure 2 is a view of the arborescent columns, vaults and stained-glass windows of the nave. Figure 3 is a recreation of the temple, once finished. The facade in the first place is the Glory façade; on the right, the original modernist Nativity facade completed in 1930.

The Sagrada Familia is currently the second most visited European church after St. Peter's in the Vatican. The initial modernist Nativity facade and the crypt are protected UNESCO's world heritage sites, as well as six additional sites built by Gaudí in or near Barcelona.

Figure 2. A view of the columns and vaults of Sagrada Familia's nave.

Figure 3. Recreation of the finalized temple. Courtesy of Sagrada Familia Foundation.

2 THE TUNNEL PROJECT

Crossing Barcelona, a densely populated city with relatively narrow streets and a well-developed underground metro network, was a real challenge for the designers of a new high-speed railway line. The alternative selected involved the construction of a deep tunnel under Mallorca street (Figure 4), which is at the southeast limit of the Sagrada Familia temple. In fact, the access to the Gloria portal of the church will probably "fly" over Mallorca street. The Foundation in charge of building the temple and other citizen associations expressed their deep concern for the perceived risks of damaging the temple. UNESCO shared these concerns, based on the special care which required any intervention menacing this world heritage site. Therefore, the protection of the temple was a major concern for the Spanish Administration of Railway Infrastructures (ADIF).

The soil profile in the area is given in Figure 5. The substratum is a stiff tertiary (Pliocene) irregular sequence of sandy clays and clays, rather stiff and overconsolidated. The tunnel, whose centerline is 30 m below the street, remains in this Pliocene level. Overlying the Pliocene soils is a quaternary deposit of alluvial fans of clayey gravels and, on top of them, yellow silts, lightly cemented and calcareous crusts. These quaternary deposits are unsaturated and they offer also a good support for shallow foundations of moderate loading. The phreatic level is 15 m below the street level.

The type of foundations of the temple evolved during the long construction period. Figure 6 is a plan view of the different foundation typologies, which correspond to the historic construction phases of the church. The Nativity façade (Figure 1), including its four towers, was founded on a thick slab of "cyclopean concrete" (large pieces of rock in a concrete mass) resting on the ancient quaternary stiff soil layer. The Passion facade, at the other end of the transept, is supported by a 2.50 m thick reinforced concrete slab, lying over the quaternary soil. The main nave and aisles, whose construction began in 1983, the central towers and the towers of the Glory facade, that faces Mallorca street, are supported by large diameter excavated piles, 20 m long, in several configurations, capped by stiff slabs. The original slab foundation of the crypt under the transept

Figure 4. Work in progress and Mallorca street at the time of the construction of the tunnel.

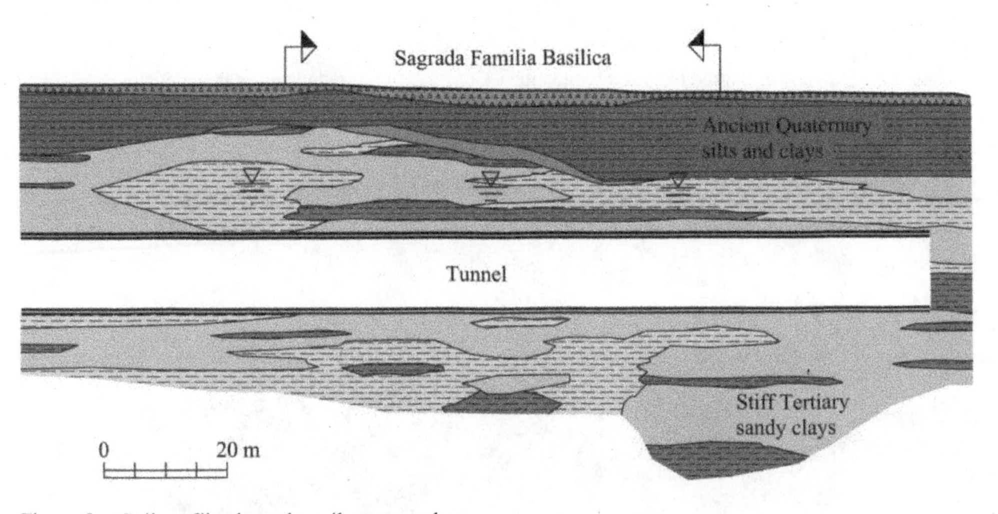

Figure 5. Soil profile along the railway tunnel.

and apse dating back to the beginning of construction in the late 19th century, was reinforced by micropiles to provide an adequate support of the central towers currently under construction. No pathologies associated with foundations have been noticed.

The initial proposal to protect the temple from displacements induced by tunnel construction was to improve the tertiary sands and clays by means of a jet grouting massive treatment shown in Figure 7. The idea was to create a massive arch of stiff treated soil around the tunnel, which would be excavated later on. This solution was discarded, among other foreseen difficulties, because of the risk of deforming the piled foundations of the Gloria facade by the high-pressure jet injection.

The accepted solution was to build a pile wall, represented in Figure 8. The 1.50-meter diameter bored piles are 41 m long and their tips are 5 m deeper than the invert of the tunnel. Pile centers

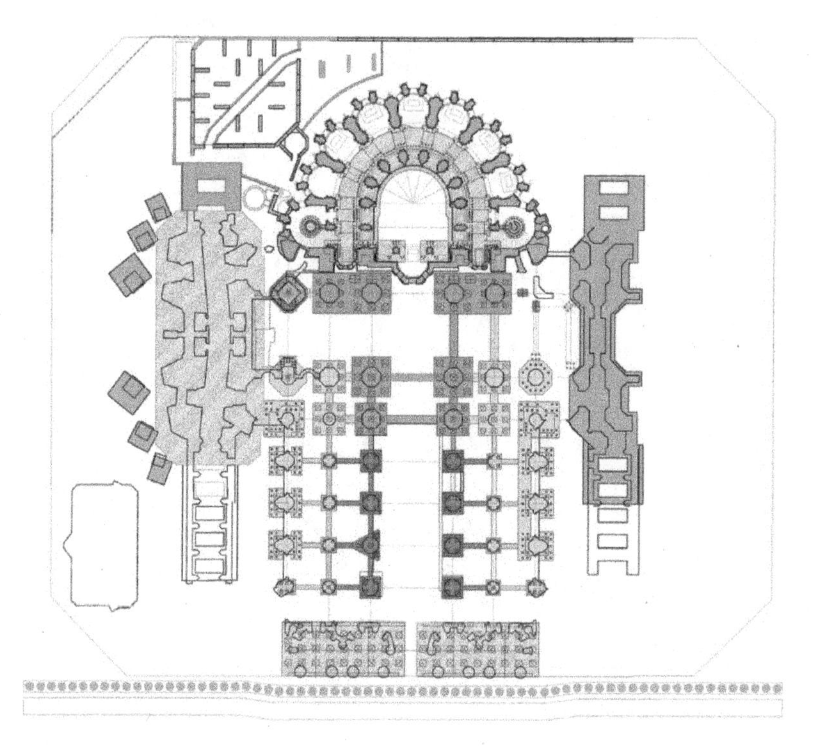

Figure 6. Plan view of Sagrada Familia foundations. Courtesy of Sagrada Familia Foundation.

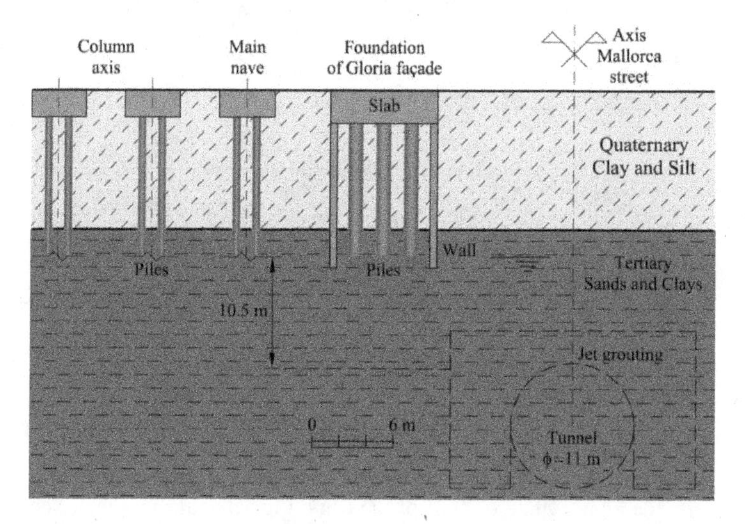

Figure 7. Pile foundations of the Glory façade and nave. Also indicated is a preliminary proposal to protect the church foundations by means of a jet grouting thick shield around the tunnel.

are 2 m apart to allow for the natural seepage flow in the area that goes in SE direction towards the sea.

Figure 6, in its lower part, shows the position and length of the pile wall built to protect the Glory portal and its four towers.

Pile walls are a well-known procedure to protect buildings against tunnelling-induced damage. Figure 9 shows an early example (Peck, 1969). The construction of the high-speed railway tunnel profited from accumulated experience in metro lines in Barcelona and, in particular in the most recent long metro line 9, which provided useful information of virtually all the geotechnical environments of Barcelona and neighboring municipalities: delta areas, Palaeozoic rocks, ancient quaternary soils, tertiary substratum and the transitions among them. Also, the layout of the metro line and the topography of the city required deep tunneling excavations.

Figure 10 (Di Mariano, 2017) shows the relationship between maximum measured surface settlements and the cover ratio H/D (H: depth of the center line; D: tunnel diameter). Two classes of excavated soils are shown: "soft soil" (typically deltaic deposits) and stiff soils (substratum). Note that maximum settlements for alluvial soils in Besós delta are quite significant (20–100 mm).

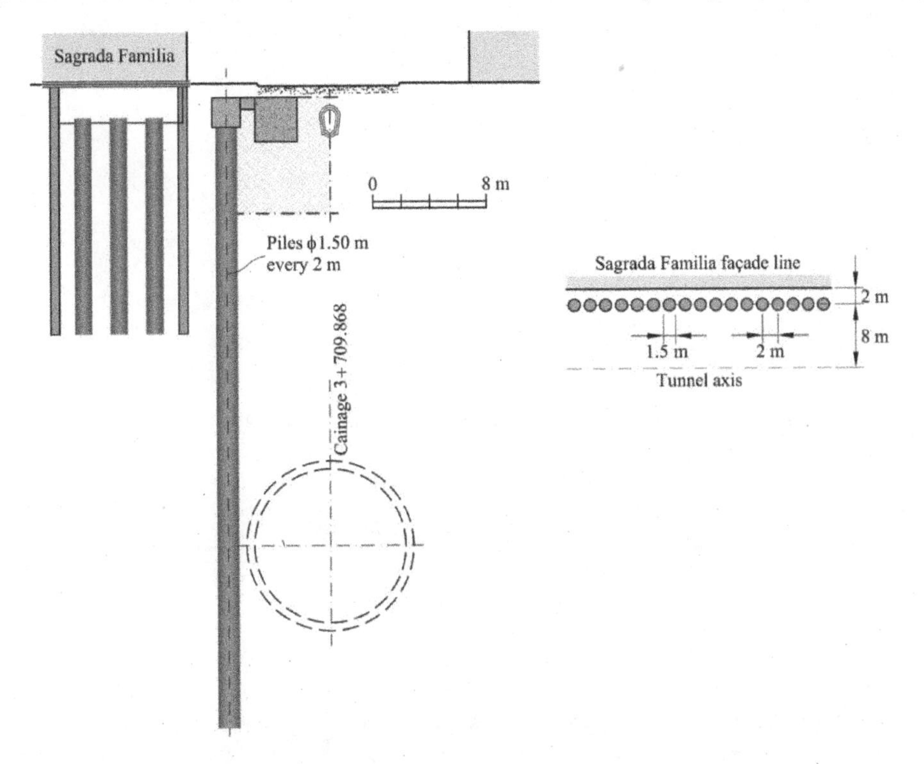

Figure 8. Protection pile wall.

The range of cover ratios of the Mallorca AVE tunnel is also given in the left part of the figure. For $H/D = 3$, which is the cover ratio in the vicinity of the Glory portal, expected settlements are in the range of 3 to 12 mm and an average value close to a settlement of 5 mm.

Figure 11 (Di Mariano et al., 2007) shows a pile wall protection of a building close to metro line 9. The tunnel was excavated in saturated alluvial sand and gravel. Figure 12 shows the measured surface settlements and the results of some FE calculations. The presence of the wall drastically reduced the settlements in the protected part of the wall.

Shown in the figure are two numerical analyses (Plaxis 2D) under greenfield conditions (no wall) and in the presence of the wall, which follows the actual measurements. The figure also suggests that the volume loss was not affected by the presence of the wall. Therefore, the protection of the wall is made "at the expense" of increasing the settlements of the non-protected side (if compared with greenfield conditions).

Walls are also efficient to reduce horizontal displacements, which also contribute to induce damage (Figure 13). The condition of conservation of volume loss may be used to "predict" the

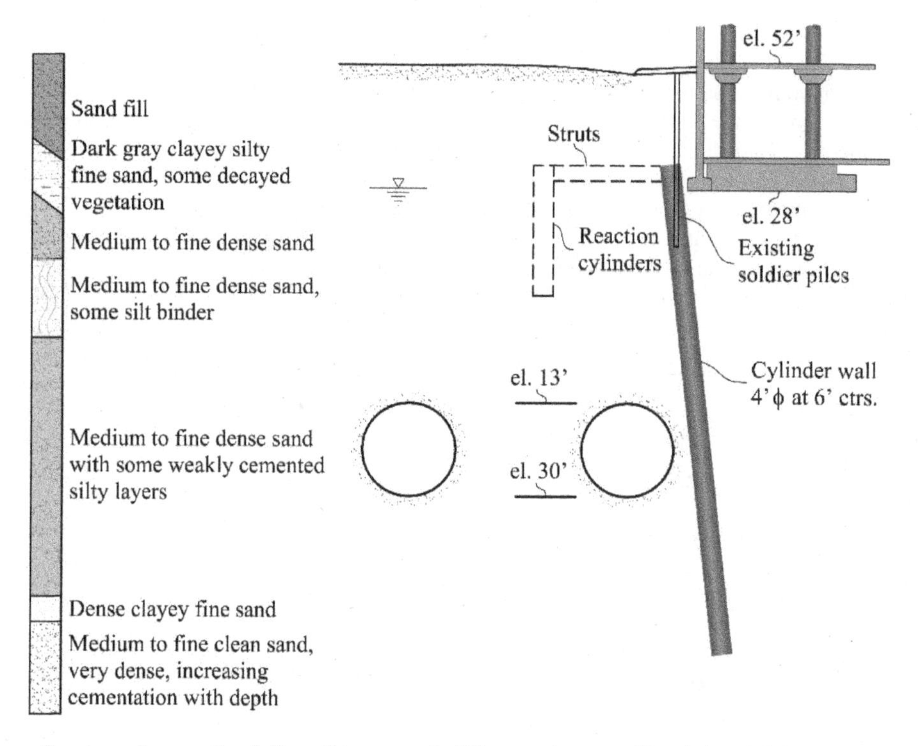

Figure 9. An early example of pile wall to protect buildings against tunnelling induced deformations (Peck, 1969).

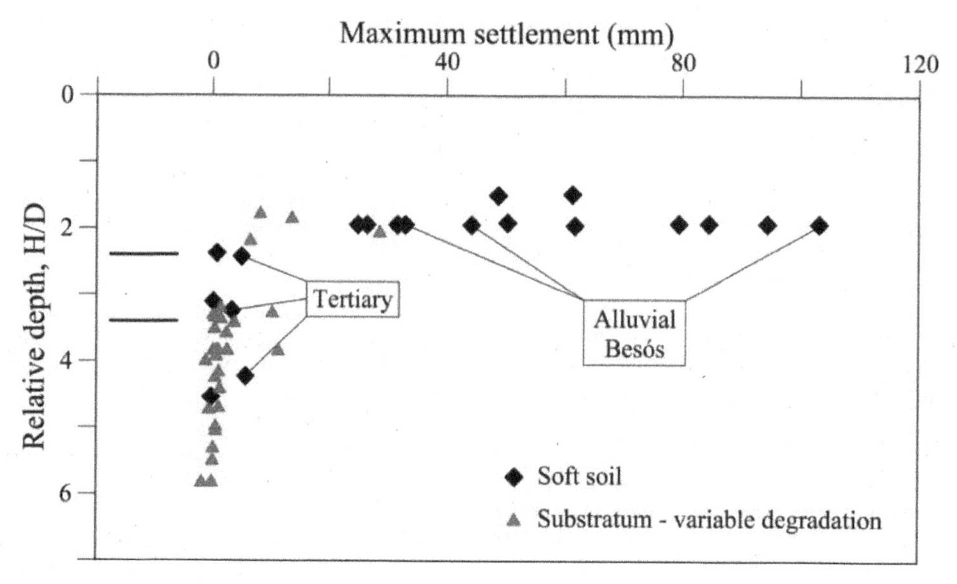

Figure 10. Maximum surface settlement measured in Metro Line 9 of Barcelona plotted against relative depth of tunnels for Alluvial and Tertiary soils of Barcelona. The range of depth ratios of AVE railway crossing of the city are shown on the left of the plot (Di Mariano, 2017).

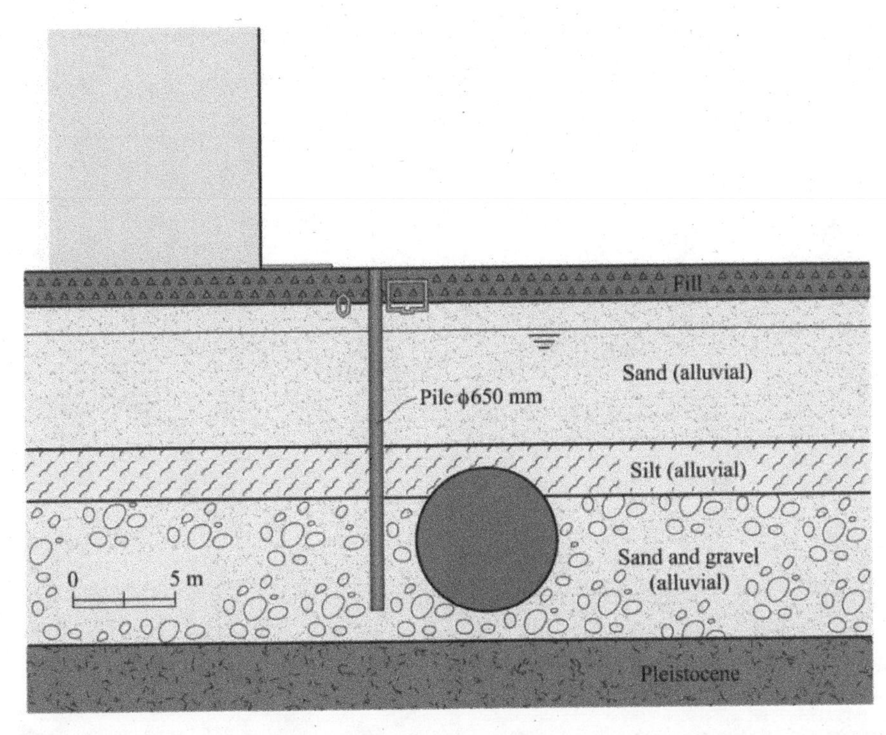

Figure 11. Pile wall protection against a metro tunnel in alluvial saturated soils in Barcelona. Di Mariano et al, (2007).

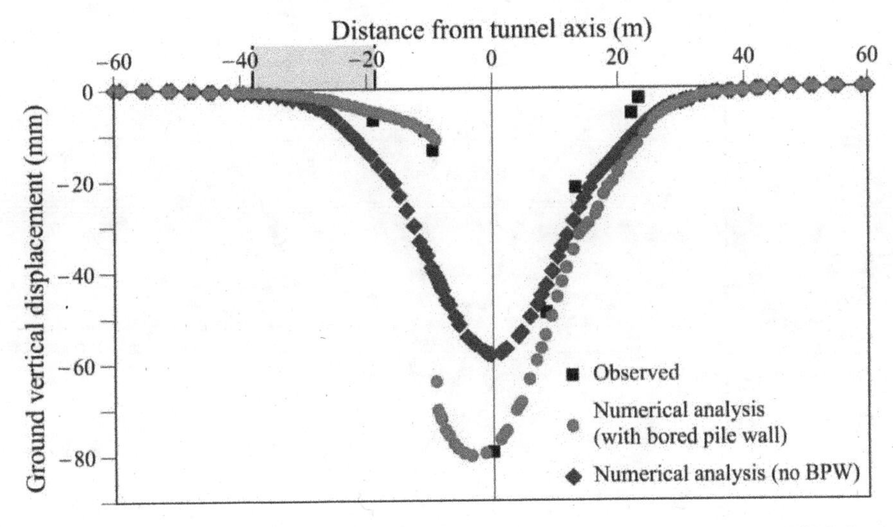

Figure 12. Measured and calculated settlement trough for the case represented in Figure 11. Di Mariano et al, (2007).

effect of a protecting wall without resorting to finite element calculations in view of previous considerations. Figure 14 presents a preliminary exercise of "predicting" the maximum settlement response at street level for an assumed volume loss of 0.5%. The greenfield settlement trough was approximated by the Loganathan & Poulos (1998) semi-analytical formula and the effect of the proposed pile wall relied on the condition of constant volume loss, irrespective of the presence of the wall.

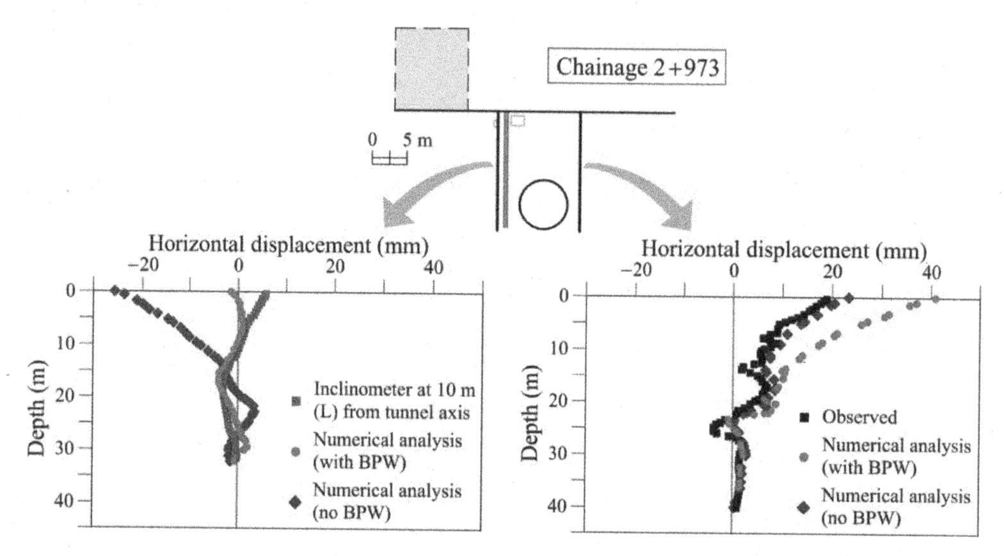

Figure 13. Horizontal displacements measured and calculated, in two vertical profiles on both sides of the tunnel. Di Mariano et al., (2007).

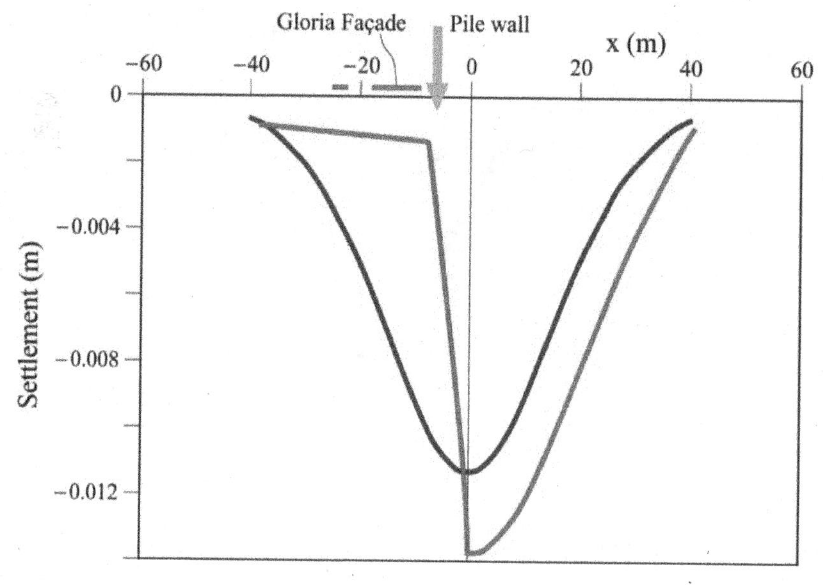

Figure 14. A sketch showing the effect of a protecting wall on settlement trough.

The protection wall designed for the Sagrada Familia maximizes the internal forces developed in the piles subjected to the tunnel-induced displacements by fixing the pile cap, attaching it to a stiff reaction beam (cross section 3×3 m), shown in Figure 8.

3 ANALYTIC SOLUTIONS

Finite element analyses are often performed to quantify the interaction between tunnel construction and building foundations. However, reproducing the set of operations involved in an Earth Pressure Balance Shield (EPBS) construction sequence is a difficult task. In practice, the success of a EPBS construction is the result of a careful planning, past experience, real time interpretation of machine performance and monitoring data, etc. A successful construction minimizes soil volume lost, a critical information difficult to estimate a priori and very difficult to reproduce by a numerical representation of machine operations.

It turns out that volume loss and soil stiffness provide the most significant information to analyze tunnel-foundation interaction. This simplicity opens the possibility of building predictive models by incorporating analytical solutions into relatively simple calculation procedures. One of the advantages of analytical solutions is the possibility of building dimensionless formulations, which offer a powerful and simple procedure to perform sensitivity analyses of controlling variables, a useful tool at the design stage.

Two approaches are described below: A 2D formulation for the tunnel-wall interaction and a more general 3D analysis of tunnel and piled foundations, which can be used also to analyze a piled wall as a particular case of a pile group.

3.1 *A 2D plane strain analysis of protection wall*

Consider in Figure 15a the geometry of the problem. The purpose of the analysis is to calculate the surface settlement induced by a tunnel construction in presence of a wall, as indicated (Ledesma & Alonso, 2018). Displacements are the result of the tunnel construction and the influence of forces developing along the wall shaft. The tunnel effect will be approximated by the solution given by Loganathan & Poulos (1998). Their analysis starts in the previous publication by Verruijt & Booker (1996), which was modified by the authors to better approximate some field data of surface settlements. They argue that the actual volume loss in EPBS tunneling reaches a maximum on the tunnel crown and a minimum at the invert (Figure 16). This is a consequence of the settlement of the tunnel lining where it abandons the protection of the shield. The vertical displacement for $\nu = 0.5$ (no volume change of elastic soil) is given by

$$u_z(x,z) = R^2\left(-\frac{Z-H}{x^2+(Z-H)^2}+\frac{Z+H}{x^2+(Z+H)^2}-\frac{Z\left[x^2-(Z+H)^2\right]}{\left[x^2+(Z+H)^2\right]^2}\right)\varepsilon \tag{1}$$

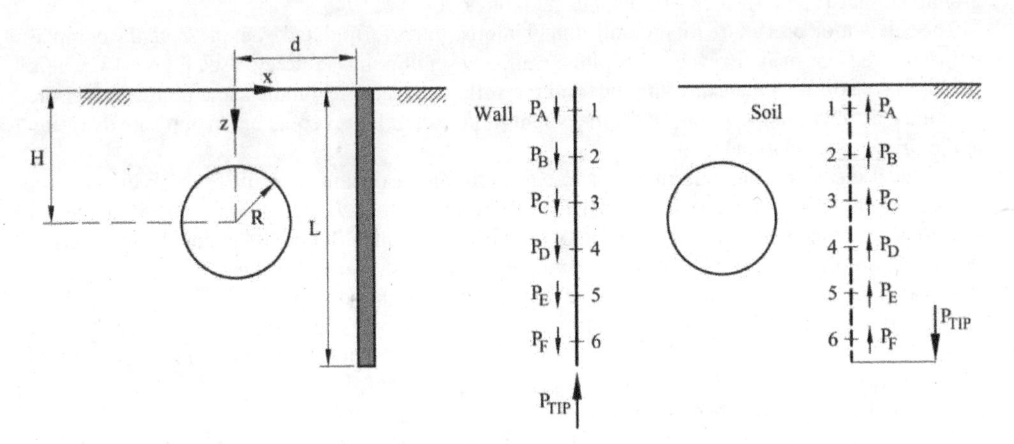

Figure 15. Geometry of the problem and interaction forces (Ledesma & Alonso, 2018).

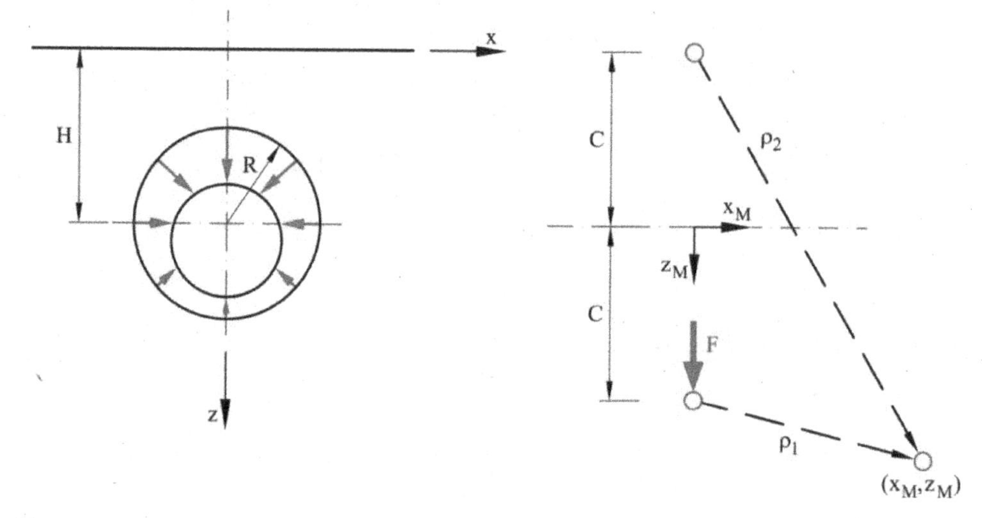

Figure 16. Sketch of volume loss simulated in Loganathan & Poulos (1998) solution and geometry of the 2D Melan' theoretical line loading problem.

where ε is a modified ground loss (illustrated in Figure 16):

$$\varepsilon = \varepsilon_o \exp\left[-\left(\frac{1.38x^2}{(H+R)^2} + \frac{0.69Z^2}{H^2}\right)\right]\varepsilon \tag{2}$$

where H is the depth of the tunnel axis and ε_o is the normalized ground loss: area of the surface settlement trough divided by the tunnel cross sectional area. Note that displacements do not depend on the elastic soil modulus in these expressions.

An additional fundamental solution required, concerns the line load, F, per unit length, applied inside an elastic half-space (Melan, 1932). We use here Sneddon (1951) and Verruijt & Booker (2000) formulae for displacements. The vertical displacements on the wall axis ($x_M = 0$) are

$$u_Z = \frac{3F}{2\pi E_s}\left[\ln\frac{\sqrt{c^2+w^2}}{c+Z_M} - \frac{1}{2}ln\frac{|Z_M - c|}{Z_M + c} + 1 - \frac{c^2}{c^2+w^2} + cZ_M\left(\frac{1}{(c+Z_M)^2}\right)\right] \tag{3}$$

(see Figure 16) where E_s is the soil modulus, w is a horizontal distance from the wall on the ground surface ($z_M = 0$) where the displacement is negligible.

The calculation procedure for the wall-tunnel interaction is formulated by expressing the compatibility between vertical displacements of the wall, at a set of points (1, 2, etc., see Figure 15, where the interface soil-wall loads are applied) and the soil displacements on the same points, under the combined action of tunnel-induced displacements and a set of forces equal but in opposite direction to the forces applied to the wall.

The wall displacements due to forces P_A, P_B, etc. and a tip force, result from an elementary calculation in terms of unit cross sectional area, A_W, and modulus, E_w, of the wall. The fundamental solutions mentioned allow the calculation of soil displacements. Making them equal and ensuring vertical equilibrium, lead to a set of algebraic equations that allow the calculation of interaction forces and displacements at any point. Note that the validity of a superposition principle is applied.

The problem may be formulated in terms of four dimensionless coefficients which are helpful to perform sensitivity analysis: $\Pi_1 = H/R$ (tunnel cover ratio: 2 to 5); $\Pi_2 = d/R$ (wall distance to tunnel: 1 to 3); $\Pi_3 = L/R$ (wall length: 1 to 10) and$\Pi_4 = E_sR/E_wA_w$ (stiffness ratio: 2.5 E–5 to 2.5 E–1). Usual ranges in practice are indicated.

It is of interest to examine some results for a reasonable case ("base case" in the following, characterized by the following data $\Pi_1 = H/R = 3$; $\Pi_2 = d/R = 2$; $\Pi_4 = 0.0025$and a volume loss of 1%. Figure 17 shows the effect of the length of the protection wall on the expected surface settlement. The greenfield case is a symmetric gaussian-type curve. A useful way of describing the protection of a given wall design is to define an efficiency coefficient (Bilotta & Russo, 2011) as follows:

$$\eta = \frac{S_{greenfield} - S_{behindthewall}}{S_{greenfield}} \tag{4}$$

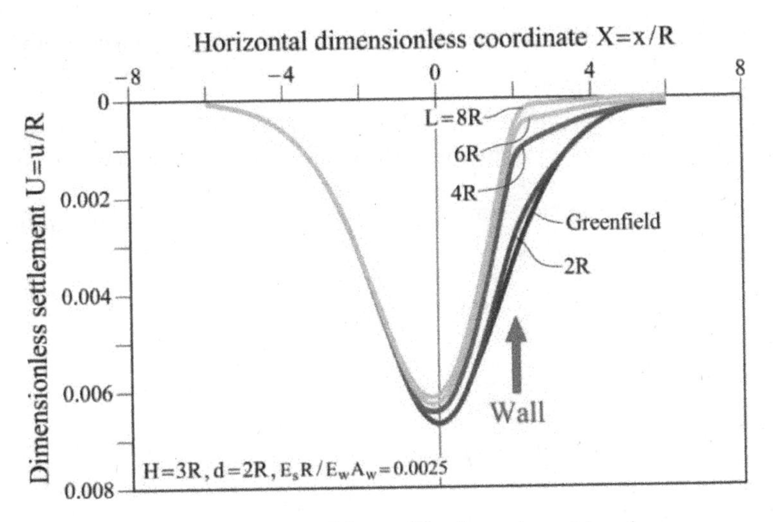

Figure 17. Base case. Effect of length of protection wall on the settlement trough.

where settlements are computed in a point immediate to the wall position, on the protected side. η varies between 0 (no effect) and 1 (full effect).

Figure 18 (curve A) provides a plot of η in terms of the wall length (L/R) for the base case just defined. Also plotted (curve B) is the result of a Finite Element analysis (2D, elastic, plane strain) which reproduces the base case. The calculated efficiency is lower in this case and the

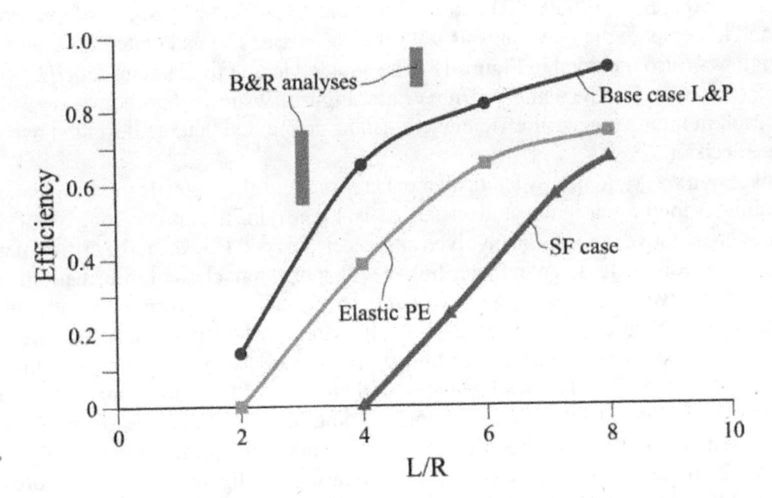

Figure 18. Efficiency of the protection wall in terms of dimensionless wall lengths. The different cases represented are explained in the text.

reason is that the Loganathan & Poulos expression for tunnel-induced displacements provides a surface settlement through narrower than a "standard" elastic solution. Loganathan & Poulos (1998) expression is expected to be closer to displacements associated with EPBS operation. There are also two estimations of efficiency (letters D, E), which were taken from Bilotta & Russo (2011) (B&R) for volume losses ranging from 0.5% to 2.5%, $L/R = 3$ (curve D) and $L/R = 5$ (curve E). Curve A (this paper) and the finite element results (D, E) are quite consistent despite the widely different calculation procedures (Bilotta & Russo introduced also a soil stiffness increasing with depth).

A third efficiency curve (C) corresponds to the Sagrada Familia protection wall discussed below. In all cases, increasing the wall length looks like an efficient procedure to increase the wall efficiency.

Figure 19a shows the effect of stiffness ratio $\Pi_4$on wall efficiency. For a given soil modulus, if$\Pi_4$decreases the efficiency increases. This is shown also in Figure 19b. However, below a given ratio ($\Pi_4 \cong 0.02$) reducing $\Pi_4$does not help to increase efficiency.

The pile wall dimensions for the Sagrada Familia (Figure 8) lead to the following dimensionless coefficients defining the wall: $\Pi_1 = H/R = 5.4; \Pi_2 = d/R = 1.4$ and $\Pi_3 = L/R = 7.1$.

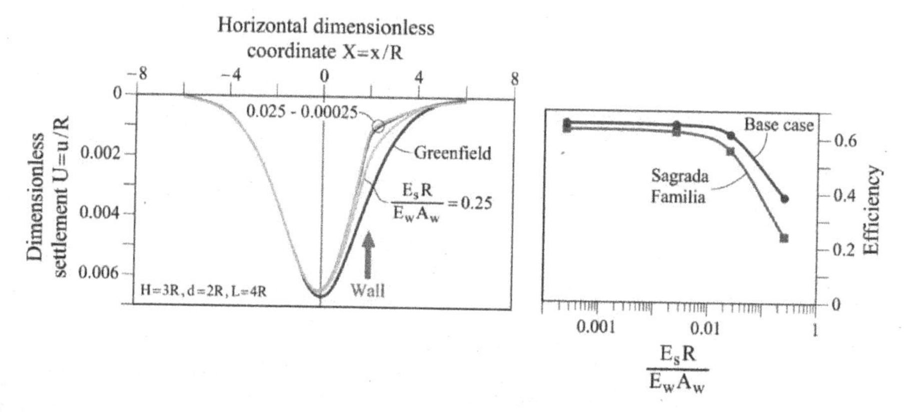

Figure 19. Effect of stiffness ratio on surface settlement (left) and variation of wall efficiency with stiffness ratio.

For $E_s \cong 30$ MPa, $E_w = 30000$ MPa and $A_w = 1$ m^2/m, $\Pi_4 \cong 0.006$. Also, a volume loss $\varepsilon_o = 1\%$ was selected to compare the new analysis with the base case. The calculated efficiency, in terms of pile length was also included in Figure 18. The adopted length to diameter ratio ($L/R = 8$) leads to $\eta = 0.65$ (curve C). For the actual stiffness ratio, Figure 19 shows that increasing wall stiffness does not result in improvement of efficiency. It can be concluded that the designed barrier was an appropriate decision.

The wall design was validated by finite element analysis and elastoplastic models, featuring small strain capabilities for the quaternary and tertiary soils. Identifying a relatively long list of parameters of a complex constitutive model is an involved and uncertain task. However, the critical parameter in these cases is the volume loss, a parameter linked with a different class of information: experience in EPBS handling, workmanship, control in real time of machine performance and interpretation of monitoring data. Figure 20 indicates the FE calculated settlements in a section perpendicular to Mallorca street axis for a conservative value of volume loss (0.5%). The maximum settlement under the street was 6 mm. The wall reduces settlements to less than 3 mm. Two calculations are represented, in terms of introducing or not, building loads. Differences are small. The figure also shows the actual settlement measured: No settlements in the protected zone and a maximum settlement of 2 mm at the street side of the wall. Measured settlements along Mallorca street, at locations away from the Sagrada Familia block, remained also very low.

The next section presents a more advanced 3D calculation procedure, using fundamental solutions for loading and local volume change in an elastic half space.

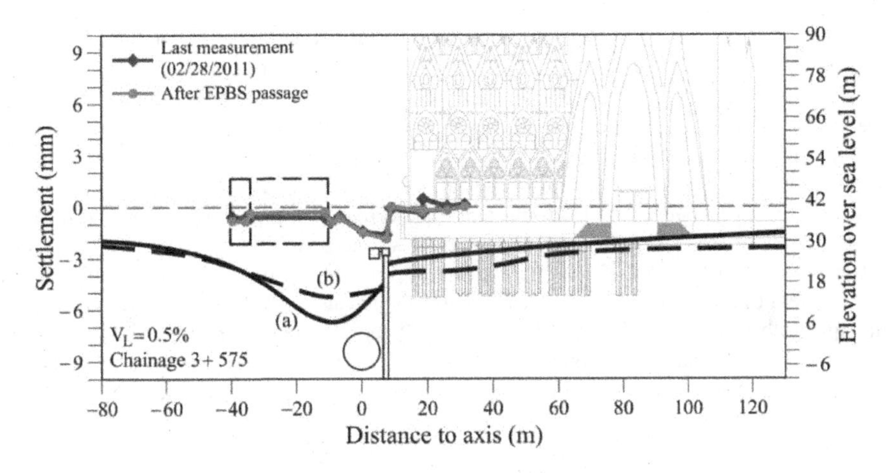

Figure 20. Settlement results of a FE analysis for a volume loss of 0.5% and actual measurements.

3.2 *A 3D analysis of interaction between piled foundations and TBM tunneling*

Volume loss of TBM tunnelling distributes around the tunnel in a non-homogeneous pattern, as mentioned before. This is sketched in Figure 21. If the objective is to reproduce the effect of the advancing tunnel in an underground structure, such as a piled foundation, the actual geometry of volume losses may be "discretized" by adopting a length interval (ΔL) to account for the advancement of the tunnel construction and a suitable subdivision of the ground loss in the cross section. Then, the volume loss can be represented by a set of elementary volumes. Sagaseta (1987) found an analytical solution for the distribution of undrained deformations of a half-space when a spherical volume is lost in an arbitrary position. This solution is useful to analyze the effect of tunnel-induced volume loss in foundations.

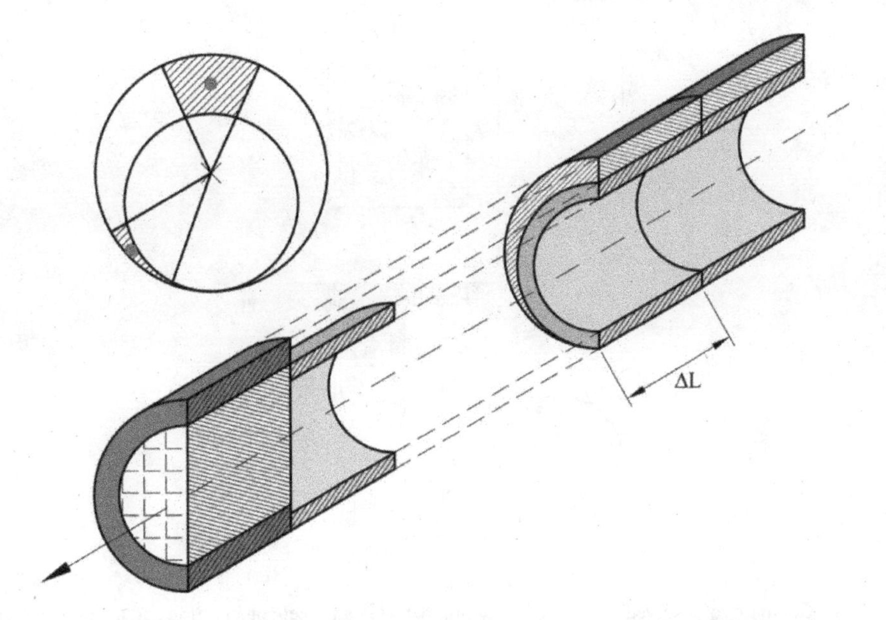

Figure 21. Definition of volume loss and sketch showing the advancement of tunnel construction.

Consider (Figure 22) a piled foundation in the vicinity of the tunnel. The figure represents (left) the capped pile structure subjected to external, as well as "internal", forces and stresses. On the right, the soil is represented. The soil will deform because of the interaction stresses received from the structure and the volume loss associated with tunnelling. Under two hypotheses: superposition of loads applies and there is a compatibility of displacements (vertical and horizontal) of structure and soil, a procedure can be devised to solve the interaction problem. If the analysis is performed in a homogeneous, isotropic and elastic half-space, the well-known fundamental solutions of Boussinesq (1885) for a point load on the surface, Mindlin (1936) for the point load in the half-space and Sagaseta (1987) for the volume loss, compatibility and equilibrium conditions lead to a system of algebraic equations that provide a solution.

This approach was followed to build a computer program, fairly general, which is useful to analyze the effect of an advancing tunnel on an existing piled foundation. Consider one example in Figure 23. A 3×3 capped group of nine 25m long, 1 m diameter piled foundation will experience

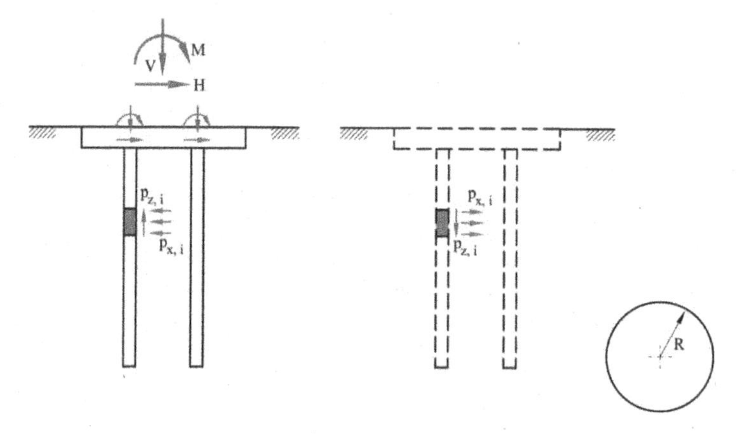

Figure 22. The analysis of a pile foundation in presence of a tunnel construction. Left: the capped pile structure. Right: Soil half space and tunnel.

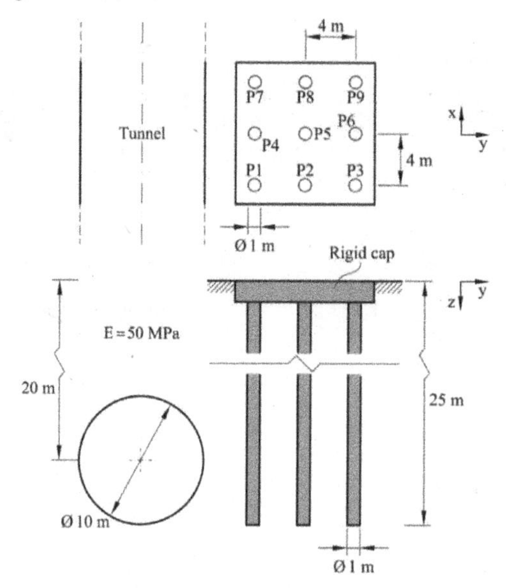

Figure 23. Geometry of a solved example: Interaction between a tunnel construction and a 3x3 capped pile foundation.

the effect of a 10 m diameter tunnel construction in the vicinity. Table 1 summarizes the main parameters of the case.

Table 1. Parameters for the tunnel-piled foundation interaction

Piles (rigid cap; 3×3)		Soil		Tunnel	
Length	25 m	E	50 MPa	Volume loss	1%
Diameter	1 m	v	0.5	Diameter	10 m
E_C	30,000 MPa			Centerline depth	20 m
v_C	0.3				

Figure 24 provides a 3D view of the example solved and the length of tunnel excavation, which was divided in 100 steps, 2 m long each.

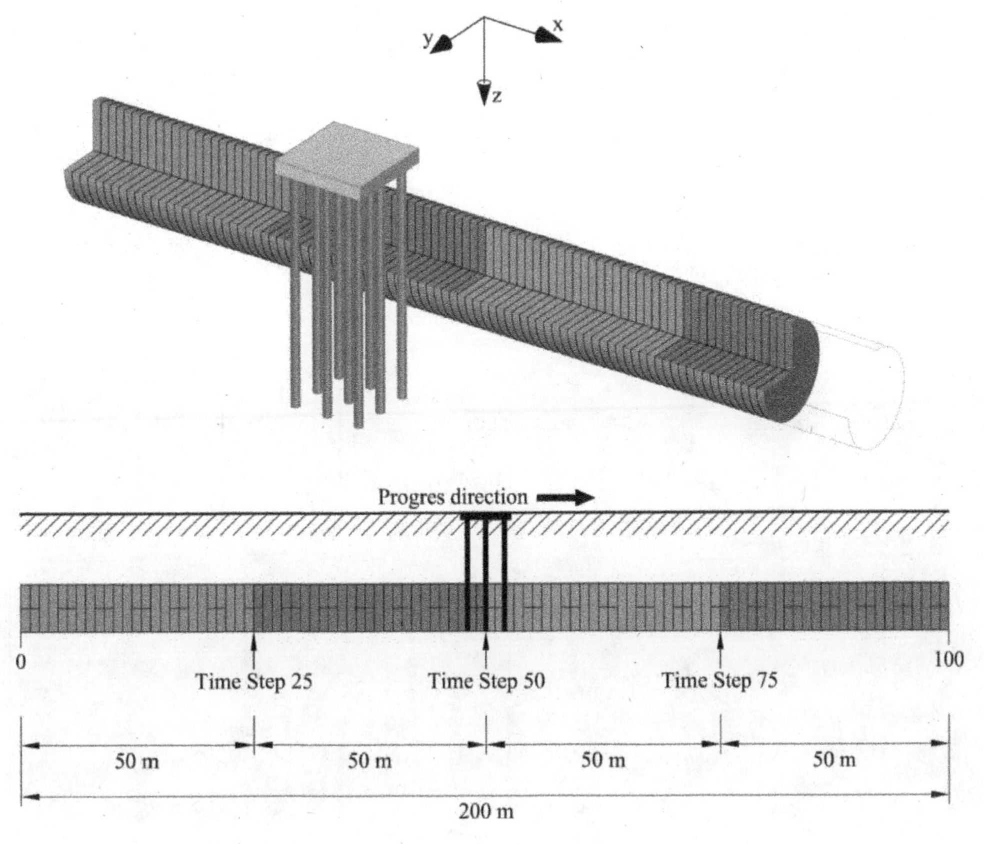

Figure 24. 3D view of the pile foundation and the excavated tunnel. The construction of a 200 m long tunnel is divided in 100 increments 2 m long each one of them.

One pile (number 4, in Figure 25) is selected to provide some calculated results to illustrate the capabilities of the analysis. The solution is general and provides forces, bending moments and displacements of the structure. Soil stresses and displacements are also known.

Figure 25 shows the horizontal x displacements of pile 4. The x direction is the tunnel axis. Displacements are zero when the tunnel front starts operating 100 m apart from the center of the pile group and accumulate progressively as the tunnel head crosses the center position of the pile

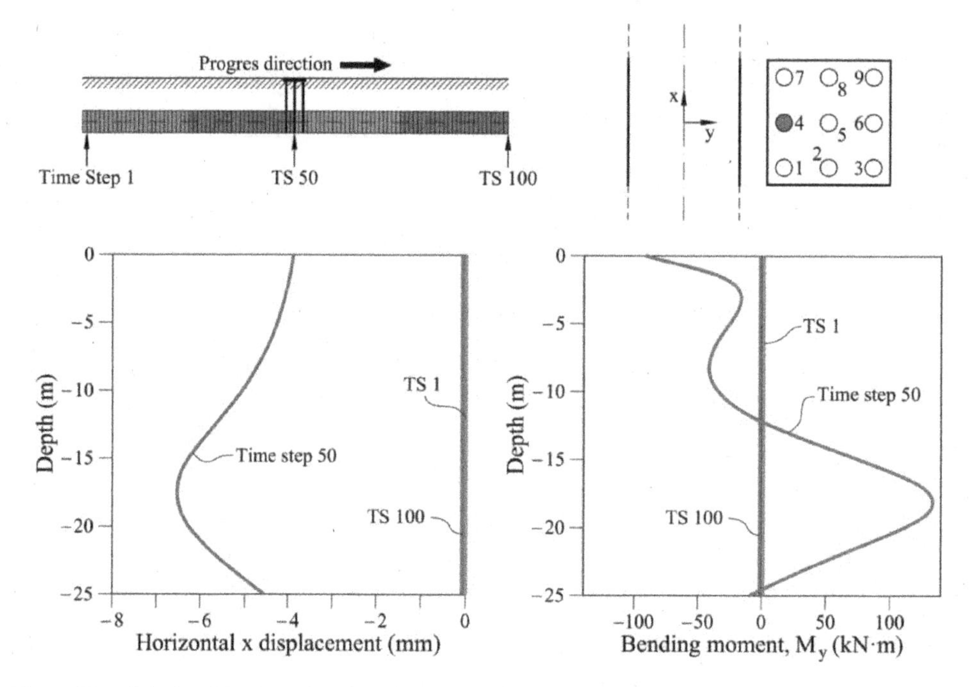

Figure 25. Calculated horizontal x displacements and associated bending moments developing along the shaft of pile 4 during tunnel construction.

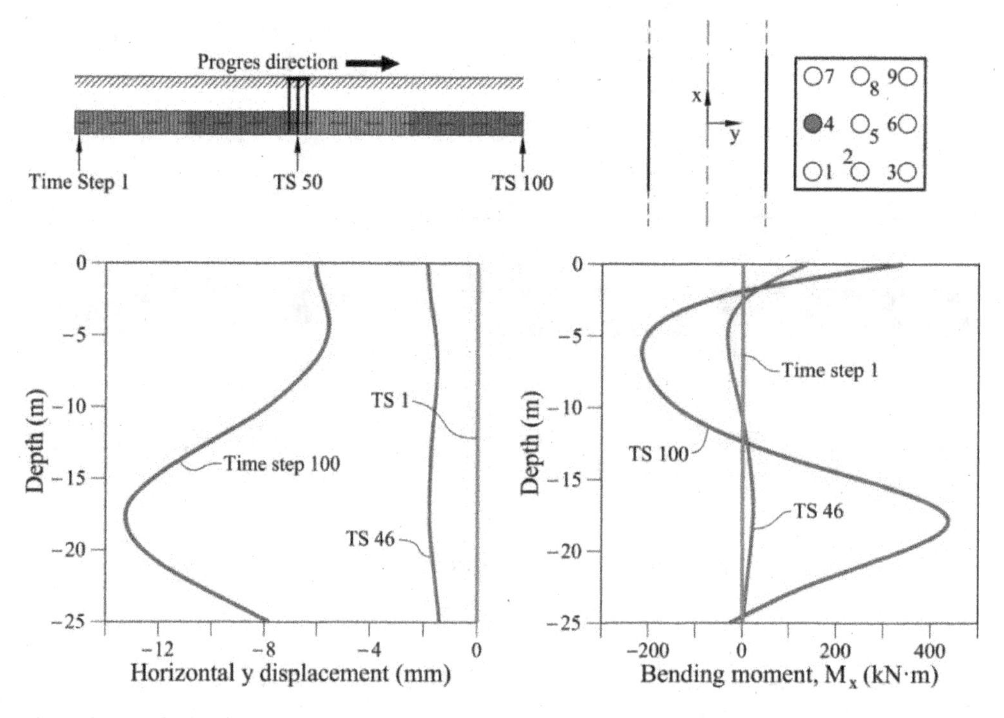

Figure 26. Calculated horizontal y displacements and associated bending moments developing along the shaft of pile 4 during tunnel construction.

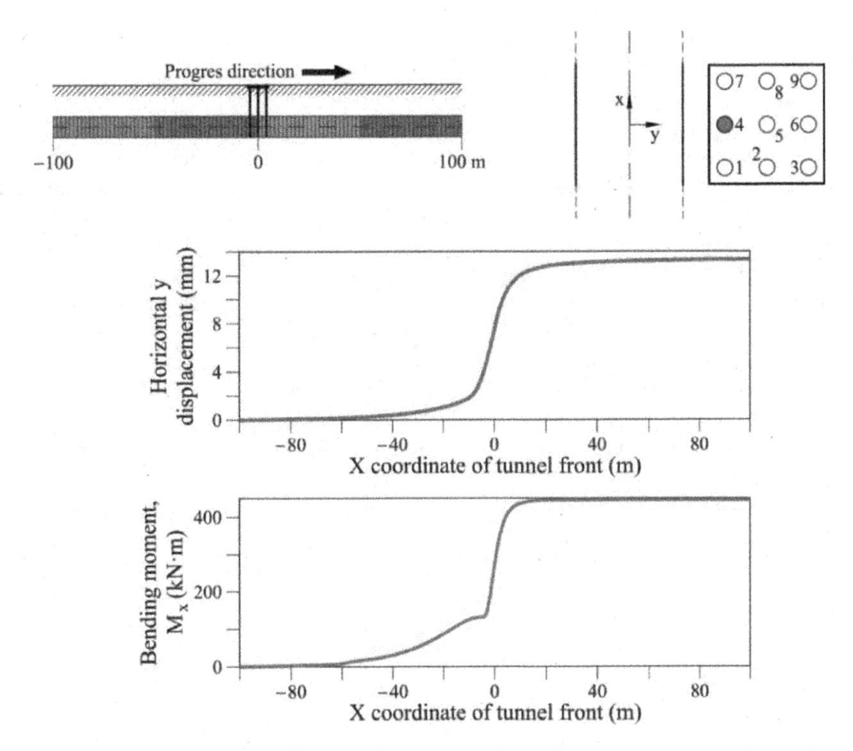

Figure 27. Accumulation of maximum horizontal y displacements and maximum M_x moments in pile 4 during tunnel construction

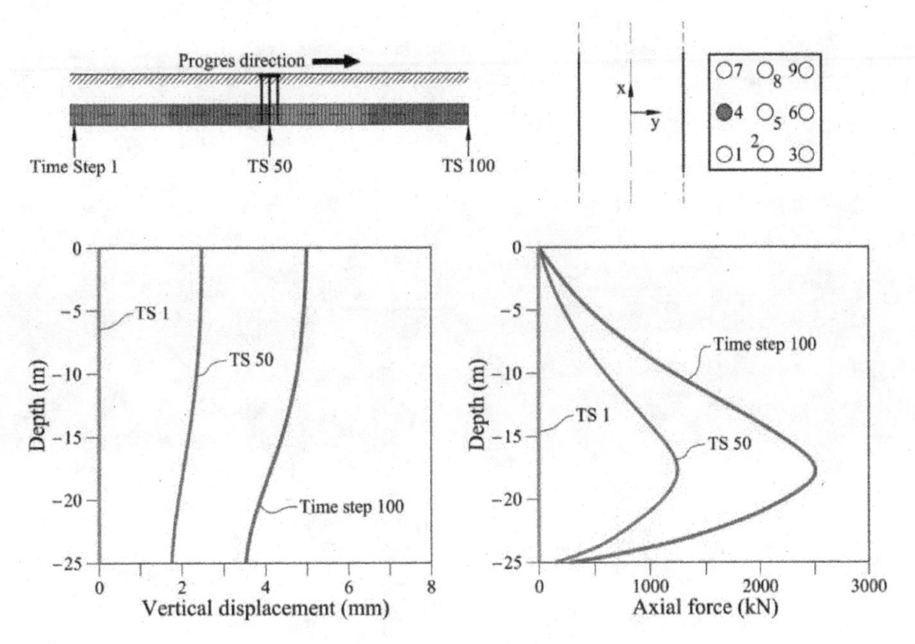

Figure 28. Calculated vertical displacements and axial forces developing along the shaft of pile 4 during tunnel construction.

group (time step 50). These are displacements towards the tunnel front. Further tunnel excavation induces x displacements equal but of opposite sign to the previous excavation stage. At the end of the simulated length of construction (step 100) the x displacements of pile 4 vanish. Note also that the position of the tunnel and the distribution of the volume loss leads to a maximum x displacement of the pile at a depth of 17 m, which is 3 m above the position of the tunnel centerline.

The M_y bending moments follow the distribution of x displacements along the pile and the fixity condition of the cap. They also vanish when the tunnel is already built and the front is 100 m away (step 100) of the pile group centerline.

In contrast, y-displacements accumulated constantly during tunnel passage. This is shown for pile 4 in Figure 26. Maximum y-displacements double the maximum x-displacements and maximum M_x bending moments multiply by 3 the maximum M_y ones.

The development of maximum displacements towards the tunnel, in terms of the position of the tunnel front, indicate that the tunnel effect on the foundations mobilizes when a 20 m long stretch of the tunnel, centred on the location of pile 4 is built (Figure 27). Maximum bending moments follows a similar pattern. It is also interesting to realize that tunnel excavation, because of the particular distribution of volume losses, leads to vertical settlements of the pile and may compress them as shown in Figure 28. This is also a case of continuous accumulation of vertical deformations and loads during the tunnel construction.

A protection pile wall is a particular case of a pile group. Consider in Figure 29 the pile wall protecting the Glory facade of the temple. Given the length of the wall and the large number of piles, let us pay attention to a single representative pile (P7) centred on a short pile group (9 piles

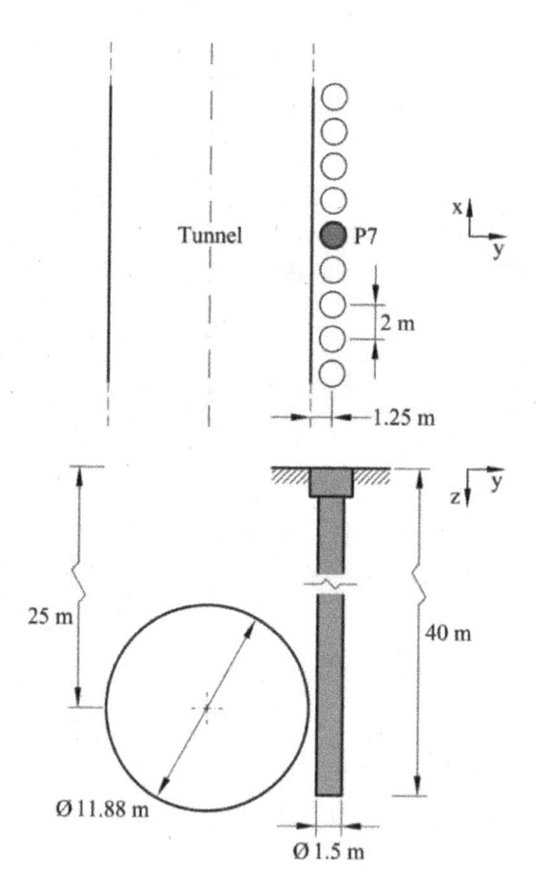

Figure 29. Geometry of the analysed protection pile wall of the Glory façade of Sagrada Familia.

and a rigid cap) at a sufficient distance of the end of the barrier. The excavation length of the wall was also divided into 100 steps. Adopted soil and pile stiffness are $E_s = 30$ MPa; $\nu_s = 0.5$; $E_w = 30{,}000$ MPa; $\nu_w = 0.3$. A volume loss of 0.10% reproduces the measured settlements. It was found by fitting measured surface settlements with the model.

Y-displacements of the pile accumulate during the passage of the tunnel and reach a calculated maximum horizontal displacement of 2.6 mm at a depth of 23 m, seven meters above the level of the tunnel centerline. The calculated maximum bending moment (375 kN m) is a very small value for the reinforced 1.50 m diameter pile (Figure 30). Maximum calculated pile settlements are also very small (0.33 mm) (Figure 31). An axial compressive force develops and reaches a maximum (670 kN) when the tunnel front is away from the location of the pile wall. The maximum vertical stress in piles is also a very small value for a reinforced concrete (380 kN/m^2).

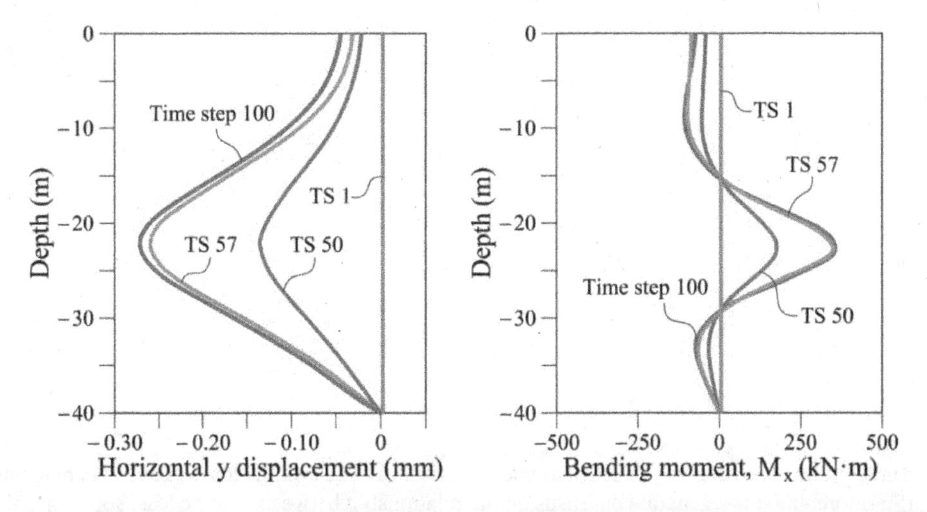

Figure 30. Sagrada Familia protection pile barrier. Calculated y displacements and M_x bending moments of a single pile of the group during tunnel construction.

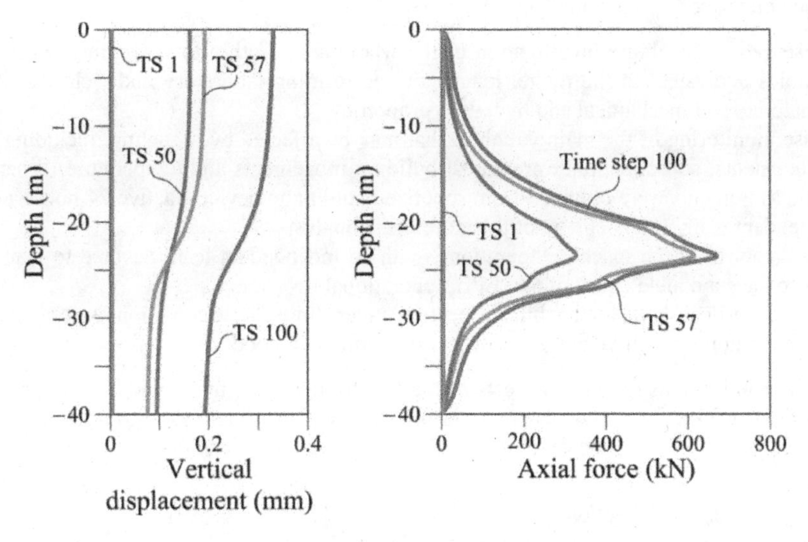

Figure 31. Sagrada Familia protection pile barrier. Calculated vertical displacements and axial forces of a single pile of the group during tunnel construction.

The comparison of measured and calculated settlements in the reference cross section is shown in Figure 32. The agreement is good for a volume loss of 0.1%.

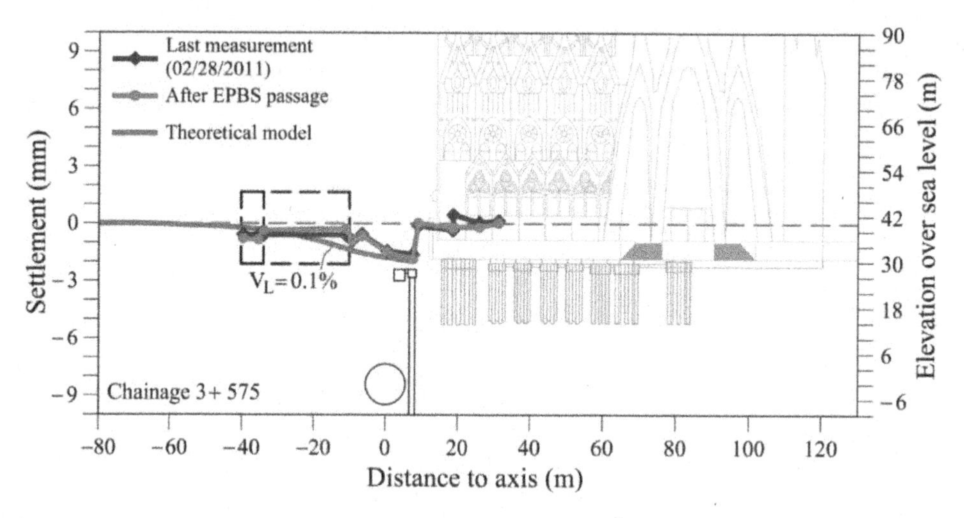

Figure 32. Sagrada Familia. Calculated and measured surface settlement for a volume loss of 0.1%.

4 TUNNELING CHALLENGES

4.1 *General aspects*

The magnitude of the tunnel induced settlements depends on the volume loss of the EPBS machine. Furthermore, there is a general consensus on the relationship between volume loss, soil properties and machine operation. Therefore, settlements are related to the adequate control of the pressurized excavation chamber, the conditioning of the excavated spoil, the screw conveyor operation and both the annulus and tail void grouting (Mair, 2008). Keeping this in mind, the construction of the tunnel was based on the following principles:

- Intense geotechnical site investigation of the whole area, with a total length of geotechnical boreholes equivalent to the tunnel length and performing laboratory and field tests to fully characterize soil mechanical and hydraulic properties.
- Intense monitoring of the main variables that may be affected by tunneling, including ground displacements, soil pore water pressures, building movements and temperature. When possible, measurements were obtained from robotized automatic devices, active 24 hours a day. An independent team was in charge of the monitoring analysis.
- Intense control of the machine operations, with an independent team devoted to that task, in order to have a double check of any EPBS operational decision.
- Design of vertical shafts for machine inspection under atmospheric conditions before each tunnel section including sensitive buildings or involving difficult conditions.

This section presents the main aspects of the EPBS operations that could have impact on soil and building displacements, with emphasis on the last two principles abovementioned. Some hydrogeological issues are also described.

4.2 *Control of EPBS operations*

An independent contract was set up by the Spanish Railway Administration (ADIF) to control all tunnels under construction during that period, including the tunnel next to the Sagrada Familia

Temple. An independent technical team reporting to the Administration was organized, having a replica of the pilot cabin monitors presenting all the control parameters of the EPBS machine and the online measurements from field instrumentation. Any decision considered by the pilot out of the programmed operations had a double check by that team. This was a key factor in the reduction of any risk, allowing a consensual action in case of unexpected warnings. The control was based on several numerical indicators evaluated every 200 mm of advance for which preventive thresholds were defined. In this way, any discrepancy with respect to the theoretical values could be analyzed in an almost continuous manner. The indicators involved:

- Face pressure measured through the sensors in the EPB chamber. Warning is issued if more than a 10% variation was measured and an alarm sign if a 20% variation was detected. An average value of 2 bars was used in the vicinity of Sagrada Familia Temple.
- Apparent density of the material in the chamber. A minimum value of 1.4 g/cm^3 was defined. Lower values could indicate the formation of an air bubble at the top of the chamber. This should be analyzed considering the pressures measured in the chamber as well.
- Weight of excavated material, based on estimations from site investigation. Warning when more than 10% variation was observed.
- Volume of bentonite injected around the shield and values of the injection pressure.
- Volume of mortar injected in the gaps between ground and concrete segments and injection pressure used. Generally, warning was activated if actual mortar volume was less than 90% or more than 120% of the theoretical one. An alarm was turned on when the actual value was below 70% or above 150% of the expected value.
- Measured displacements of specific targets on buildings and on extensometers located just above and close to the tunnel crown. Warning of settlements in buildings in the Sagrada Familia area was fixed in 5 mm, tilting limit was defined as 1/3000 and the warning limit for the ground vertical displacement close to the tunnel crown was 20 mm. Separating displacements due to temperature from movements caused by tunneling was not obvious and became an important task.
- Additional EPBS variables as total thrust, applied torque and velocity.

All these indicators were controlled in a continuous manner, so that any potential incident could be anticipated at early stages. A conventional color code was adopted: green for normal operation, orange for warning and red for alarm. The tunnel length was divided into sectors and for each one a plan with the expected values for the indicators was carried out. Using that code, a protocol to follow, depending on the colour of the indicators, was prepared in advance.

In general, warning situations were very limited and corrected quickly: settlements were below 2 mm not only in buildings, but also in the soil, just above the tunnel crown.

This type of control was fundamental to keep all indicators within limits, avoiding fluctuations of pressure or volume that have been reported to be the source of volume loss increments, particularly in shallow tunnels (Wonsaroj et al., 2006, Mair, 2008).

4.3 *Hydrogeological issues*

An environmental concern was expressed by the Catalan Water Agency before the tunnel construction: any global change in the average water table in the city should be below 1 m. The tunnel runs parallel to the seashore and there is a regular groundwater flow from the hills to the sea that could be interrupted by the tunnel. This is called the barrier effect (Pujades et al., 2014, 2015). Some tunnels that were built in the past in the city became a barrier for the natural groundwater flow, raising the water table in the upstream side of the tunnel and wetting basements in nearby buildings. A particular attention was devoted to the pile wall designed to separate the Sagrada Familia basilica from the tunnel. A barrier effect could raise the water table below the Temple, affecting eventually its foundation or the crypt. To avoid that, a 50 cm gap between consecutive piles was designed. Both upper and lower aquifers became connected due to the construction of the piles and that effect was more important than the barrier effect of the pile wall on the upper aquifer. Nevertheless, variations of water table in the area were always below 1 m as required by the environmental Agency (Pujades et al., 2015). In fact, the variations of the water table were due to other effect: the construction of the piles connected the upper unconfined aquifer and the lower confined aquifer. The deep aquifer had a piezometric head about 4 m greater than the upper one. Overall, the barrier effect was negligible. The piezometric head of the lower aquifer had, however, some impact on the face pressure to be considered during tunnel excavation.

4.4 *Vertical shafts*

Vertical shafts were designed, intersecting the tunnel, with the aim of allowing the inspection and maintenance of the cutting head of the EPB under atmospheric conditions. After tunnel construction, the shafts were adapted as emergency exits.

Before approaching the Sagrada Familia Temple, a vertical shaft was excavated at Padilla Street (300 m away from the Glory façade), so the excavating tools could be replaced if damaged. A complete revision of the EPBS was carried out before approaching that sensitive area (Figure 33). The construction of the shaft itself was carefully planned to avoid any impact on nearby buildings, also concerned about the works. In fact, the outer diaphragm wall was excavated by means of hydromills, to guarantee a perfect contact joint between vertical panels and to reduce noise during construction in this urban environment. Dewatering was also limited to the excavation of the vertical shafts in order to minimize the potential impact on soil settlements. Thanks to the overconsolidated nature of the soil due to low water tables in the past, changes in effective stresses were always in the elastic range and displacements due to dewatering were small. The unsaturated nature of the upper geotechnical profile implying high soil stiffness had a positive influence on settlements as

Figure 33. Photograph of the Padilla shaft and the head of the EPB Shield, before approaching the Sagrada Familia Temple.

well. Note that water table was about 10 m above tunnel crown and it was reduced inside the shaft to a position below the shaft bottom slab, that is, a reduction of more than 24 m of piezometric head.

The design of the vertical shaft had to consider the risk of bottom uplift. A theoretically closed excavation was possible if the outer walls were deep enough, reaching low permeability layers (tertiary clays below the tunnel). However, there was a need to reduce water pressures at depth, to guarantee stability. The solution was to build relatively deep walls and a parallel control of water pressure under the shaft by permanent pumping from wells (Pujades et al., 2014). That is, the main design criterion for the circular diaphragm wall was not the dewatering condition, but the mechanical containment of the enclosure before and during tunnel crossing of the EPBS. An inner wall was required additionally to reinforce the lower part of the outer wall when the machine was crossing the shaft. Regarding materials, glass fiber reinforced concrete was used in the areas crossed by the shield.

An important issue with respect to vertical shafts is the operations required to enter and leave the enclosure, considering that a face pressure is used by the EPBS and there is a risk of a water inlet to the shaft from the surrounding soil. To cope with that, two additional enclosures were designed, as shown in Figure 34a. These enclosures were not excavated, but diaphragm walls were built to isolate the ground in a particular area. In those additional enclosures, a progressive dewatering was possible when reducing face pressure from 2 bars to atmospheric pressure when entering the shaft, or, when leaving the shaft, a progressive increase of face pressure and ground water pressure was adopted. In this manner, the dewatering operations were concentrated on enclosures, thus limiting the impact of potential settlements on nearby buildings (including the Sagrada Familia Temple). Moreover, when reducing EPB face pressure to atmospheric conditions, or increasing from atmospheric to 2 bars, ground should have enough strength to avoid any local collapse due to insufficient face pressure. For this reason, mortar piles where built inside the enclosures to increase ground shear strength. Figure 34 shows the basic steps followed during the operations of entering and leaving the shaft, defined with the aim of minimizing the ground disturbance:

- EPBS cutting head approaching the first enclosure (Figure 34a).
- EPBS cutting head inside the first enclosure (Figure 34b). Pumping well active. Two pumping wells were installed just in case tail grouting may clog one. Progressive reduction of face pressure.

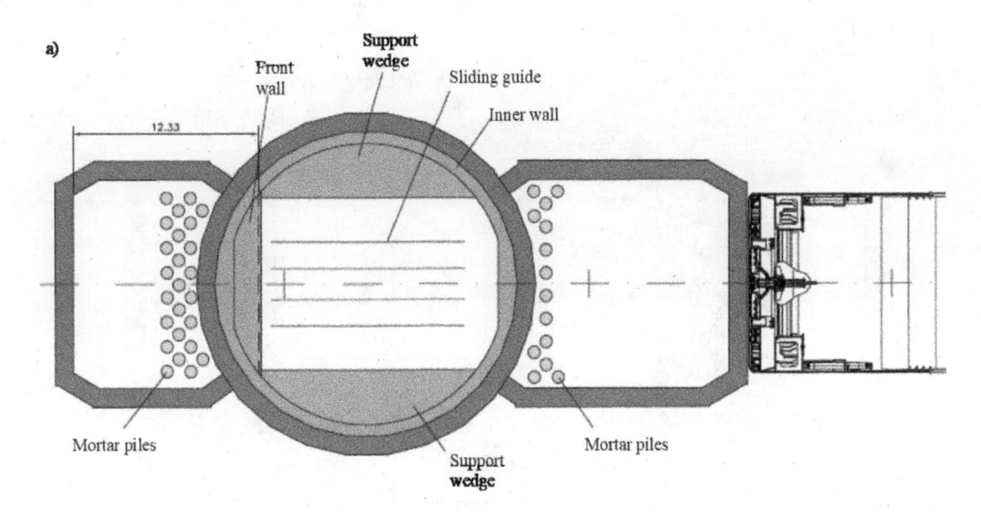

Figure 34. Operation steps when crossing the vertical shaft by the EPBS, a) before entering the enclosure, b) EPBS head in the first enclosure, pumping well active, c) EPBS head in the shaft, d) EPB head in the second enclosure, pumping well active, e) EPBS head leaving the second enclosure.

b)

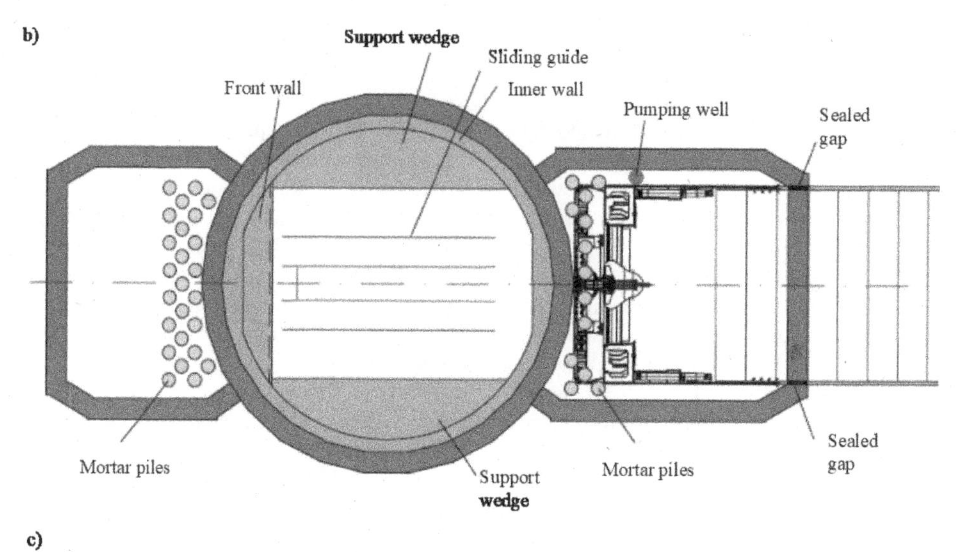

c)

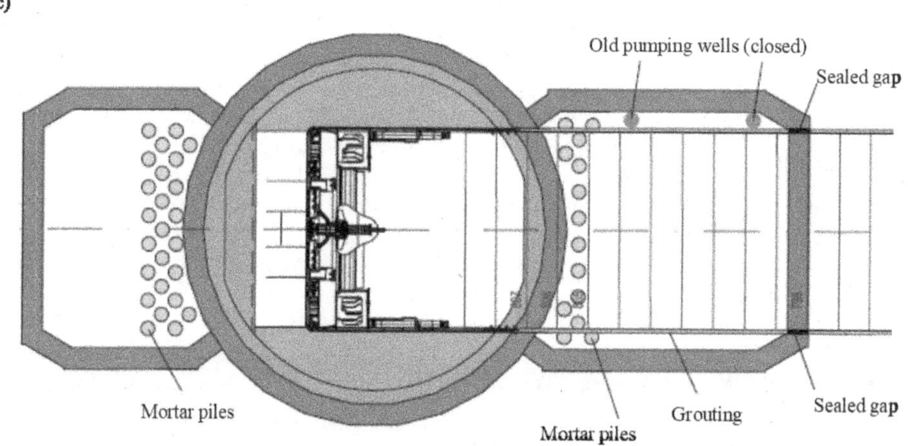

d)

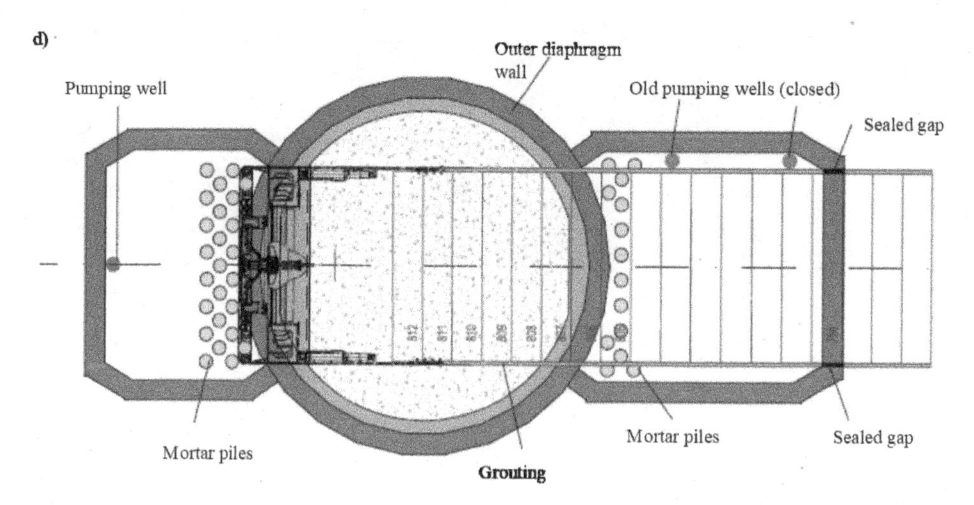

Figure 34. Continued.

e)

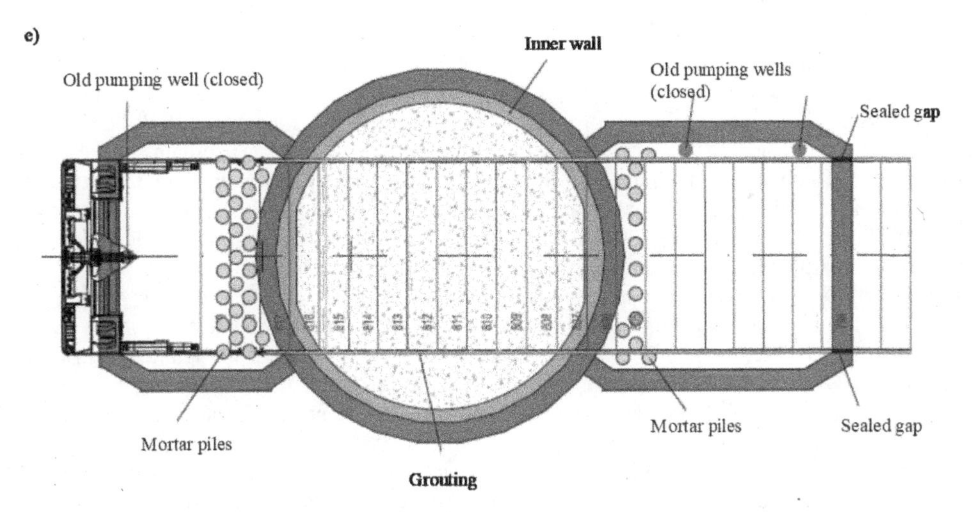

Figure 34. Continued.

- EPBS cutting head inside the vertical shaft at atmospheric pressure. Inspection of the machine and repair of damaged tools (Figure 34c).
- EPBS cutting head leaving the shaft and entering the second enclosure. Filling the lower part of the shaft with low quality mortar, using a geomembrane to protect the tunnel segments. This is to confine the tunnel to gradually increase face pressure and tail grouting. (Figure 34d). Initially a pumping well is active, but pumping is reduced and eventually stopped as face pressure increases.
- EPBS cutting head leaving the second enclosure (Figure 34e).

These operations were carried out smoothly before approaching the Temple and definitely, they had a positive influence on the success of the works and on the negligible impact on the buildings, including the Sagrada Familia Temple.

5 CONCLUSIONS

The Sagrada Familia Temple is a modernist architectural landmark of Barcelona designed by Antonio Gaudi at the transition between the 19th and 20th centuries. It is included in the list of UNESCOs protected monuments. Plans to build a railway tunnel immediate to the main façade of the temple raised a public concern about the risks involved in tunnelling operations. The paper describes briefly the Gaudi's masterpiece, which is expected to be concluded in 2026.

The chosen design by the tunnel was a deep, 11 m diameter segmented structure, having a cover of 25 m and built by an EPBS machine. The stiff Barcelona soils in the area (a surficial quaternary deposit of clay, silt and sand, cemented in some layers, partially saturated, overlying a stiffer tertiary clayey sands) was a positive geotechnical environment. However, the tunnel had to be built under a moderate water pressure (15 m of head over the crown). A deep (41 m) and stiff pile wall (1.5 diameter piles at 2 m intervals) was an added protection for the temple structure, which was founded on piles.

In addition to numerical calculations performed at the design stage, the paper highlights the value of predictive analytical solutions to analyse the behaviour of protection walls under the displacements induced by EPBS tunnelling. They are particularly useful because displacements induced by tunnelling operations have a strong kinematic nature and they are mainly explained by the total amount and distribution of volume losses. Volume losses are much dependent on a few aspects (experience of EPBS machine operators, control of a significant number of machine variables, integrated evaluation of monitoring surface structures and others) which are difficult to reproduce in numerical models.

The paper describes two analytical approaches: a 2D, plane strain analysis of the tunnel-wall interaction and a more general 3D approach. They require different fundamental solutions for the elastic half space. The 2D case shows the advantage of defining the solution in dimensionless terms, which provide a rapid sensitivity analysis. This is a nice feature to help designers to define the key variables of the project, in particular the position, length and stiffness of the protection barrier. The 3D analysis developed may account for a precise definition of the geometry of the volume loss and it can handle arbitrary pile groups. Displacements, forces and bending moments of piles as well as soil displacements and stresses are found in a rapid manner. The evolving geometry implied by tunnel construction and its accumulated effects on pile foundation behaviour is also a feature of the procedure. The protection pile wall of the Sagrada Familia main portal is a long pile group which was also analysed. It was found that the measured settlement trough at street level (maximum of 2 mm at the unprotected side of the wall and essentially no settlement at the protected side) could be found for a total volume loss of 01%.

The paper concludes by a description of the organization and precautions adopted to ensure a successful operation of the shield machine. The sequence of control operations and warning limits are given in some detail. The permeable pile protection barrier did not create any significant alteration of the natural ground flow towards the coast line. An important precaution to ensure the correct performance of the excavation machine is to guarantee a perfect state of the cutting head. The solution adopted was to inspect and repair, if necessary, the cutting wheel, under atmospheric conditions, before coming close to the temple. This was achieved by excavating a deep shaft at a street intersection 300m away from the Glory façade. The design of this shaft and auxiliary enclosures to allow the machine to enter the shaft to be repaired are described. This operation, which was repeated a few times in other points of the 5.6 km long tunnel crossing downtown Barcelona was an important feature of the tunnel construction aimed at eliminating risks.

ACKNOWLEDGEMENTS

The authors contributed as geotechnical advisors to the design and construction of the project. This important and challenging project is the result of the effort and dedication of many engineers. A special mention is made to the engineers from the Spanish Railway Administration (ADIF), the Contractor (Sacyr – Cavosa – Scrinser), the technical assistance and design team (Intecsa – Inarsa – Censa) and the online external control unit provided by Sener. Our appreciation goes also to Genis Mayoral and Anna Ramon for their important role to develop the 3D analysis described, to Cristian de Santos and Andrés Pinto for their collaboration to some of the FE analyses, all of them from the Universitat Politècnica de Catalunya. Thanks are extended to Architect Director of Sagrada Familia construction, Jordi Faulí, for his help to describe some aspects of the Temple structure.

REFERENCES

Bilotta, E. & Russo, G. 2011. Use of a Line of Piles to Prevent Damages Induced by Tunnel Excavation. *J. Geotech. Geoenviron. Eng.*, 137: 254–262.

Boussinesq, M.J. 1885. *Application des potentiels a l'eìtude de l'eìquilibre et du mouvement des solides eìlastiques, avec des notes eìtendues sur divers points de physique matheìmatique et d'analyse*. Paris: Gauthier-Villard imprimeur libraire.

Di Mariano, A., Gesto, J.M., Gens, A. & Schwarz, H. 2007. Ground deformation and mitigating measures associated with the excavation of a new Metro line. In V. Cuéllar et al. (ed.), *Geotechnical Engineering in Urban Environments; Proc. 14th Eur. Conf. Soil Mech. & Geotech. Engng., Madrid, 24-27 September, 2007,* Rotterdam: Millpress Science, Vol. 4: 1901–1906.

Di Mariano A. (2017). "Large Diameter Tunnels in Urban Environments – The Barcelona L9 Case". Invited Lecture at the Workshop for the Singapore Building Control Authority (BCA) and Land Transport Authority (LTA) organized by Infraestructures.cat, Barcelona, May 31st.

Ledesma, A. & Alonso, E. E. (2017) Protecting sensitive constructions from tunnelling: the case of world heritage buildings in Barcelona. *Géotechnique*, 67(10): 914–925

Loganathan, N. & Poulos, H.G. 1998. Analytical prediction for tunnelling-induced ground movements in clays. *J. Geotech. Geoenviron. Eng.*, 124(9): 846–856.

Mair, R.J. 2008. Tunnelling and geotechnics: new horizons. *Géotechnique,* 58(9): 695–736.

Melan, E. 1932. Der Spannungszustand der durch eine Einzelkraft in Innern beanspruchten Halbscheibe. *Z. Angew. Math. Mech.*, 12: 343–346.

Mindlin, R.D. 1936. Force at a point in the interior of a semi-infinite solid. Journal of Applied Physics, 7: 195–202.

Peck, R.B. (1969). "Advantages and Limitations of the Observational Method in Applied Soil Mechanics", 9th Rankine Lecture, *Géotechnique* 19, No. 2, 171–187.

Pujades, E., Vázquez-Suñé, E., Carrera, J., Vilarrasa, V., De Simone, S., Jurado, A., Ledesma, A., Ramos, G., Lloret, A. 2014. Deep enclosures versus pumping to reduce settlements during shaft excavations. *Engineering Geology*, 169: 100–111.

Pujades, E., Vázquez-Suñé, E., Culí, L., Carrera, J., Ledesma, A., Jurado, A. 2015. Hydrogeological impact assessment by tunnelling at sites of high sensitivity. *Engineering Geology*, 193: 421–434.

Sagaseta, C. 1987. Analysis of undrained soil deformation due to ground loss. *Géotechnique,* 37(3): 301–320.

Sneddon, I. 1951. The Mathematical Theory of Huygens' Principle. By B. B. Baker and E. T. Copson. Second edition. Pp. vi, 192. 21s. 1950. (Geoffrey Cumberlege, Oxford University Press). *The Mathematical Gazette*, 35(311): 67–67.

Verruijt, A. & Booker, J.R. 1996. Surface settlements due to deformation of a tunnel in an elastic half plane. *Géotechnique*, 46(4): 753–756.

Verruijt, A. & Booker, J.R., 2000. Complex variable analysis of Mindlin's tunnel problem. *Dev. Theor. Geomech.*: 1–20.

Wonsaroj, J., Borghi, F.X., Soga, K., Mair, R.J., Sugiyama, T., Hagiwara, T. & Bowers, K.H. 2006. Effect of TBM driving parameters on ground surface movements: Channel Tunnel Rail Link Contract 220. In Bakker et al. (ed.) *Geotechnical Aspects of Underground Construction in Soft Ground*. London: Taylor and Francis Group, 335–341.

Geotechnical Engineering for the Preservation of Monuments and Historic Sites III – Lancellotta, Viggiani, Flora, de Silva & Mele (Eds)
© 2022 Copyright the Author(s), ISBN 978-1-032-35998-4

Tunnelling under the San Francisco church in Guadalajara, Mexico

E. Ovando-Shelley, E. Botero & M. A. Díaz
Instituto de Ingeniería, Universidad Nacional Autónoma de México

ABSTRACT: The San Francisco de Asis church is a masonry building whose construction began in 1554 was finalised in 1611. A new 51 km line to expand the massive transport system in Guadalajara was built between, 2015 and 2019, includes a 5.1 km underground stretch that required a tunnel 10.6 m in diameter that was excavated with a Tunnel Boring Machine (TBM), at an average depth of 25 m.

The tunnel passes directly under the north east corner of the church. Soil strata beneath the church comprise a complex array of sandy silts and pumitic materials underlain by basalts. To reduce the risk of damage to the church related to induced settlements during the excavation process, the designers (SENERMEX) decided to use jet grouting to improve the mechanical properties of the soils directly under the church to avoid of mitigate damage induced by the excavation of the tunnel. The San Francisco church was profusely instrumented with an automatic monitoring system. Measurements showed that jet grouting induced the larger detrimental settlements, mainly along the east wall whereas settlements induced by the excavation of the tunnel and the passage of the TBM under the church were not as large. These settlements re-activated existing cracks and produced some additional cracking without compromising in anyway the security of the church. Nevertheless, a comprehensive retrofitting program was implemented to restore the structural integrity of the church.

A numerical model was used to study the dynamic response of the church with a three dimensional finite difference computer code. Despite its simplifying assumptions, the model was able to provide fairly close approximations to the overall dynamic properties of the church. The numerical analyses confirmed that jet grouting did not increase the vulnerability of the church to earthquake shaking but, on the contrary, it decreased slightly.

1 INTRODUCTION

Guadalajara, the second largest city in Mexico, is located in the western portion of the Trans-Mexican volcanic range, some 460 km north-west of Mexico City, as indicated in Figure 1.. The range traverses diagonally (NW to SE) the Mexican Republic from its westernmost portion near Guadalajara, close to the Pacific Ocean, to the Gulf of Mexico. The largest and most active volcanoes in Mexico are situated along it. The vicinity of Guadalajara to active volcanoes has had a major influence on the formation of soils and on their stratigraphical sequences.

The city lies on a paleo-valley in which ignimbrites are the uppermost rocks followed by volcanic tuffs that are then underlain by basalts. Guadalajara is notorious for the presence of pumitic silty sands known locally as jales that comprise the uppermost strata. These materials are generally saturated and in terms of compacity, they are medium-dense to dense. Jales originated from pyroclastic flows which, in turn, provide them with various degrees of cementation that can easily be obliterated by water and humidity.

Seismicity in the city is determined by the interaction of the Rivera and Pacific Plates with the North America Plate, as seen in Figure 2. Several large earthquakes have hit the city in the past

 DOI 10.1201/9781003329756-6

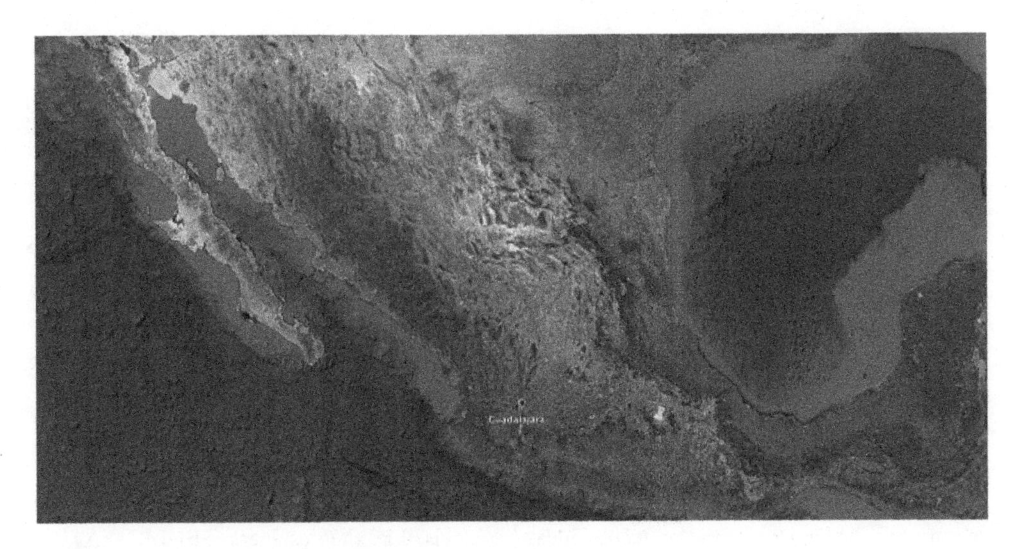

Figure 1. Location of the city of Guadalajara.

and seismicity must necessarily be taken into consideration in the design of urban infrastructure of any type.

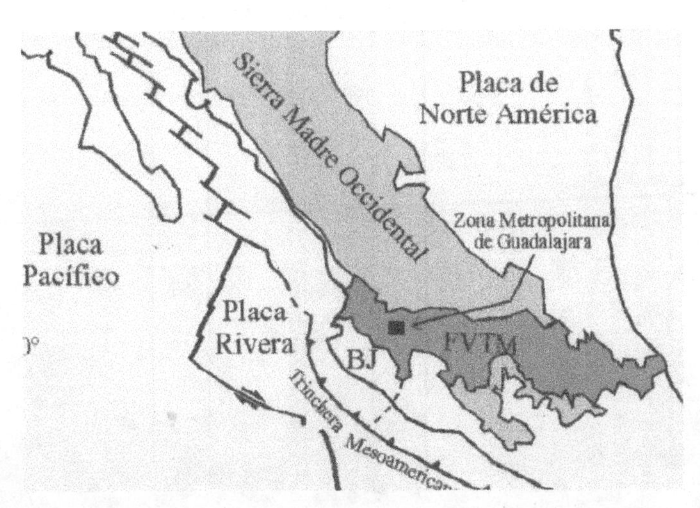

Figure 2. Tectonic environment around Guadalajara.

A new 51 km line to expand the massive transport system in Guadalajara was built between, 2015 and 2019. It includes a 5.1 km underground stretch that required a tunnel 10.6 m in diameter that was excavated with a Tunnel Boring Machine (TBM), at an average depth of 25 m. In most of the underground segment of the line, the tunnel traverses the uppermost silty sands (jales) which appear in irregular layers and with the water table at shallow depth, which makes it particularly sensitive to be compacted by vibrations.

Ever since the onset of the project it was recognised that a major challenge during the excavation process was the protection of the historical buildings located near the tunnel which runs along one of the main streets in the city and it passes only a few metres away from 20 of its most iconic buildings, a situation that prompted initially an assessment of their structural condition before the excavation of the tunnel and then their potential vulnerability to suffer damage due to the construction of the underground stretch.

In order to closely follow the structural behaviour of the buildings during the protection works and tunnel excavation, a monitoring system was installed, consisting of a set of high-precision sensors, as well as of automatic surveying equipment, with which the vertical and horizontal displacements of more than one hundred reference points on each structure could be recorded. This allowed real-time monitoring of temple behaviour, so that any signals of displacements exceeding danger thresholds could be detected in a timely manner.

The Institute of Engineering of the National Autonomous University of Mexico, (II UNAM), was called to review the project in regard to the integrity of the historical buildings that could be affected by the excavation of the tunnel and of the underground stations. The preventive and corrective actions to protect structures along the line were also evaluated in terms of settlements and tilts induced by tunnelling and of their consequences with respect to structural damage. In this paper we review, from the perspective of Geotechnical Engineering, the main aspects of the response and behaviour of one of the most important heritage buildings in Guadalajara to the excavation of the tunnel, the temple of Saint Francis (San Francisco). In doing so, we describe and provide evaluations of the protective measures devised by the designers (SENERMEX), jet grouting around and under the San Francisco temple.

2 THE TEMPLE OF SAN FRANCISCO

The San Francisco de Asis church is a masonry building whose construction began in 1554 was finalized in 1611. This structure in addition to the Aranzazu church are the remains of the former San Francisco monastery that was demolished when it was relocated elsewhere in the city. The photographs in Figure 3 show a view of the main façade and of a reconstructed archery of the former cloister.

Figure 3. Left hand side: view of the main façade; right hand side: view of the reconstructed archway of the former cloister along the East side of the church.

The church base dimensions are approximately 63 × 26.5 m, and the structure can be divided into three different sections. The sketches in Figures 4 and 5 illustrate the geometrical features of the church. Its roof vaults rest on half-point arches supported by pilasters, whose position does not match with the axis of the external abutments. In the past, the two lateral façades experienced significant differential settlements with a general rotation from north to south. Maximum differential settlements was 70 mm on the west axis of the main nave which, additionally, is leaning outwards, probably because it lost part of its counterbalance when the attached convent was demolished. These rotations have caused a large number of vertical cracks in the longitudinal wall, along with a transverse crack in the southern part of the vault.

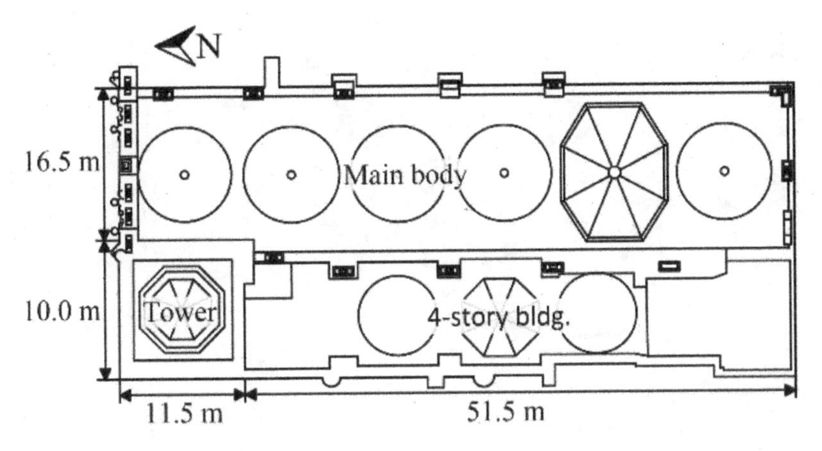

Figure 4. Plan of the San Francisco Church.

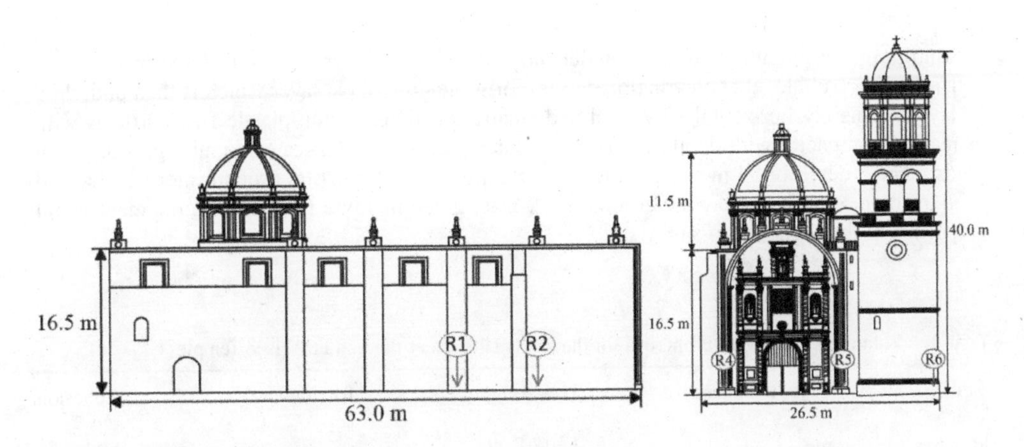

Figure 5. Architectonic details along the west wall and the façade.

Soil strata beneath the church were characterized based on the information of standard penetration tests (SPT tests) and laboratory tests executed by a geotechnical consultant during the design phase of the project. SPT tests were performed every 100 m along the entire length of the project. Laboratory testing included index tests for soil classification (particle-size analyses, plasticity index, moisture content, and mass density). Strength parameters were obtained from UU test and a limited amount of CU tests with pore water pressure measurements were used to derive effective stress strength parameters. A geotechnical profile along the tunnel axis, on the east side of the church shows a complex array of sandy silts and pumitic materials underlain by basalts and, as seen in Figure 6, there is a geologic fault that produces a discontinuity in the main soil strata.

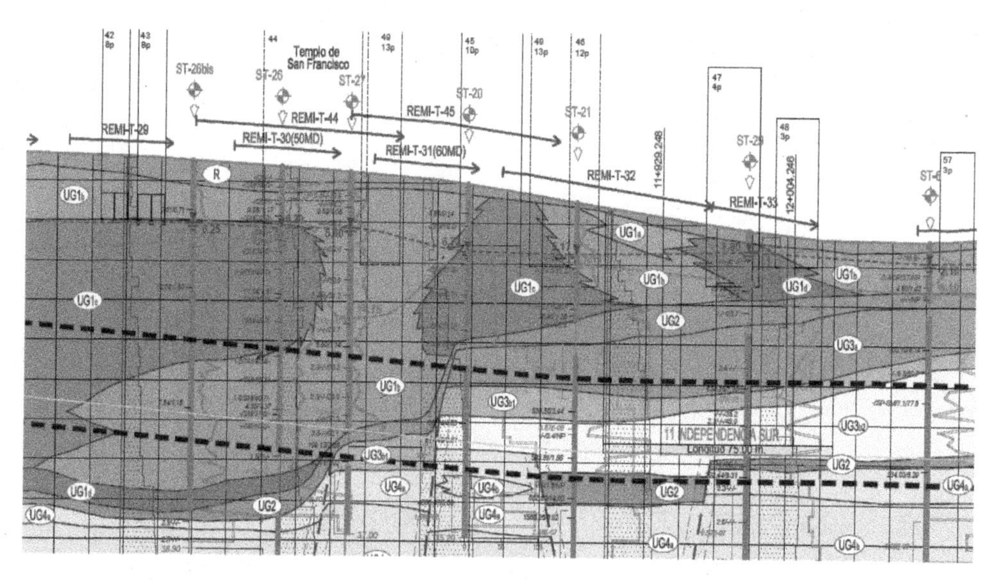

Figure 6. Geotechnical profile along the tunnel axis, on the west side of the church.

Stratigraphical conditions directly under the church are less erratic, as depicted in Figure 7. Artificial fills (rubble and construction debris) form the uppermost layer which is then underlain by layers of nearly horizontal silty sand and smaller portions of non-plastic fine particles with increasing compacity with depth, occasionally interspersed with lenses of pumitic gravels. Silty sands reach a depth of 34 m and are then underlain by basalts. On-site groundwater was located at an average of 5.6 m below the ground level. Data given in Table 1 summarise the mechanical properties of the sequence of sandy silts.

Table 1. Summary of relevant properties for the sandy silts under the San Francisco Temple

Soil layer Id	Layer thickness t m	Unit weight γ kN/m^3	friction angle ϕ' –	Cohesion c' kPa
1	2.0	17.0	29.0	5.0
2	4.0	17.5	31.0	15.0
3	10.0	18.0	33.0	20.0
4	14.0	18.0	31.0	15.0
5	2.0	18.5	36.0	25.0
6	2.0	18.5	35.0	30.0

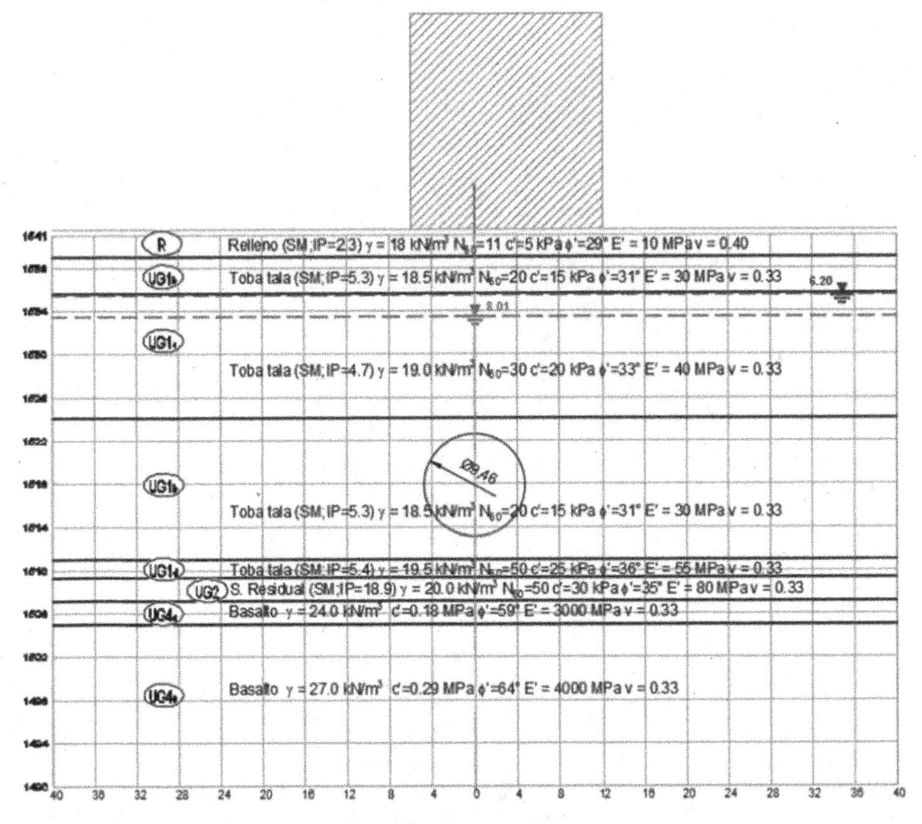

Figure 7. Simplified stratigrahical profile underneath the San Francisco church.

3 JET GROUTING

The tunnel of Line 3 passes directly underneath the north-east corner of the temple of San Francisco and that is why guaranteeing its stability during the passage of the tunnel boring machine and after it, became a source of major concern. To reduce the risk of damage to the church related to induced settlements during the excavation process, the designers (SENERMEX) decided to improve the mechanical properties of the soils directly under the church to avoid of mitigate damage induced by the excavation of the tunnel. SENERMEX chose jet grouting as the soil improvement treatment to be used. The sketch in Figure 8 shows a plan view of the location of the tunnel with respect to the church as well as the distribution of wells for grout injection. Details of the distribution of injection wells are illustrated in Figure 8.

Jet grouting was applied at a depth of 12 m and there on, the grout columns were extended 6.5 m and 18 m deeper, down to the excavation bottom mark. Vertical and inclined arrangements were used for the grouting operations. Inclined wells (as much as 32°) were used to grout areas under the church. Illustrative sketches of the grouting operations are presented in Figure 9. Examples of the mechanical properties of the grouts obtained from compression tests on control specimens are given in Table 2.

3.1 *Instrumental observations during jet grouting*

Jet grouting produces vibrations and it may eventually induce mechanical disturbance in the soils surrounding the injection wells. Consequently, it was the necessary to monitor closely the effects of

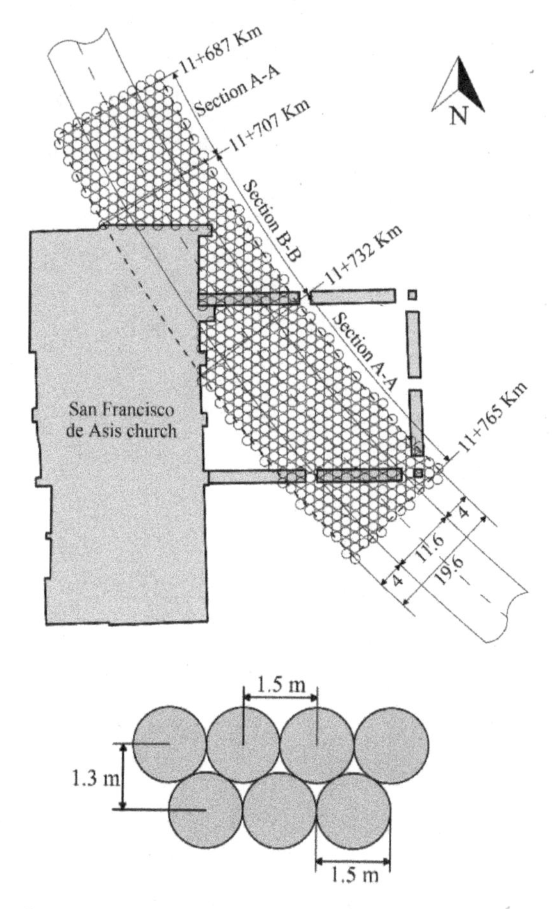

Figure 8. Array of injection wells for jet grouting along the tunnel axis near the San Francisco church.

Table 2. Mechanical properties of the grouts

Sample Id	Sample depth m	Unit weight γ kN/m^3	Compressive strength σ_c MPa	Young Modulus E_{50} MPa
1	22.03–22.35	17.2	24.43	11,647
2	22.68–29.00	17.2	24.83	10,667
3	14.70–15.05	16.3	12.76	9,071
4	18.35–18.70	16.1	15.83	7,208
5	14.25–14.75	16.1	10.44	5,373
6	17.55–18.00	16.1	11.62	3,882

the grouting operations. Topographical levellings were performed daily while grouting was under progress. An example of the results obtained with these levellings is given in Figure 10, showing that induced settlements by grouting from nine wells reached about 5 mm. Vibrations induced during grouting were measured with seismometers along the base of the west wall in terms of particle velocities. As seen in Figure 11, particle velocities fall way below the permissible values for the safe guard of historical buildings (DIN 4150-3 standard). Additional observational data will be presented later.

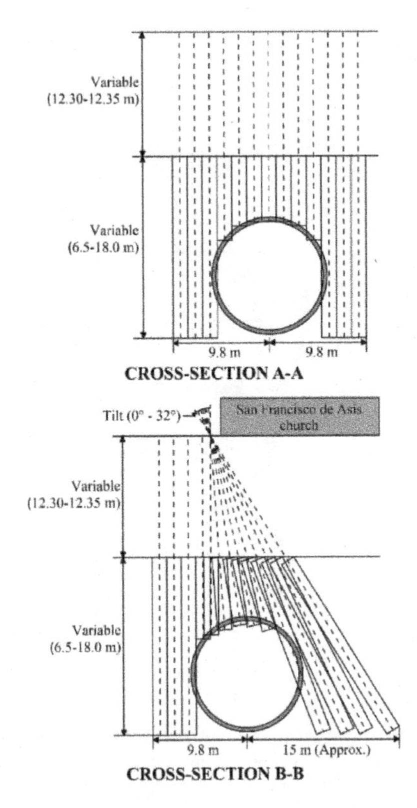

Figure 9. Vertical and inclined grouting wells used along the boundaries of the church. Cross sections are identified in Figure 8.

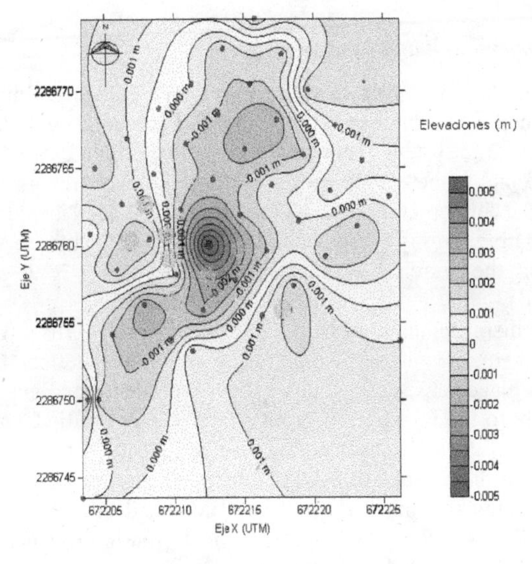

Figure 10. Differential settlements obtained from levellings performed before and after grouting in nine points (indicated by green circles). Blue circles indicate levelling points.

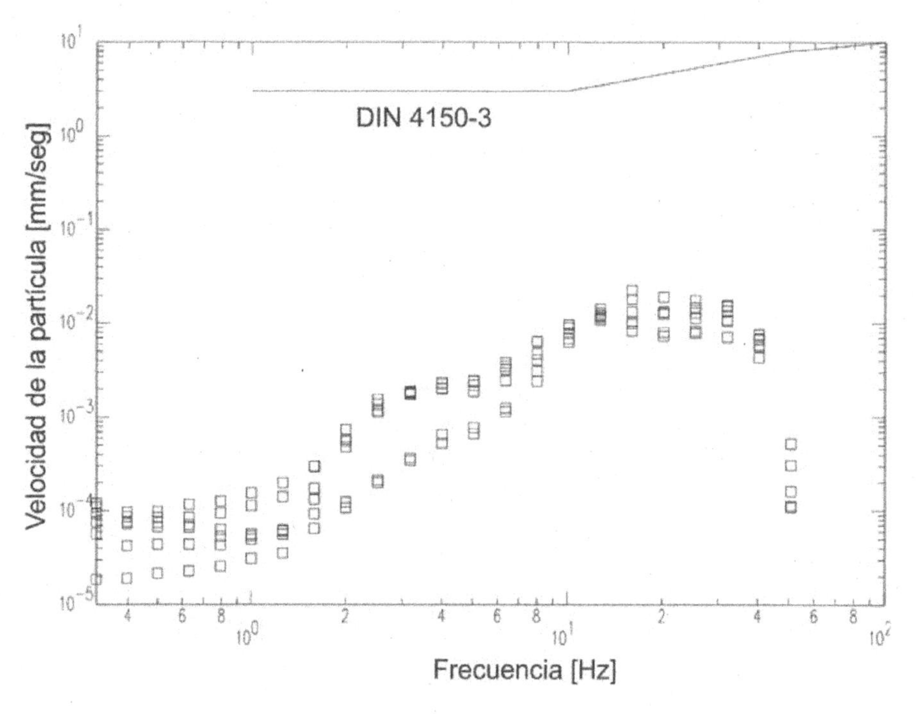

Figure 11. Values in the y axis indicate particle velocities in mm/s as a function of frequency. Measurements performed along the base of the East wall during grouting operations. Allowable values according to accepted standards are also shown.

3.2 *Additional Instrumentation and observations*

The San Francisco church as well as two other important heritage buildings were instrumented profusely with an automatic continuous monitoring system provided by SOLDATA-SIXENSE. For the San Francisco church the system comprised 36 levelling points located along the lateral walls and the main façade, 5 inclinometers were installed around the church as well as 7 vibrating wire piezometers and also 2 accelerometers; 6 tilt meters were also placed inside the church. The system operated before jet grouting, during the injection of grouts and also before and after the passage of the TBM machine along the church. It kept on working until March 2020, some four months after Line 3 was inaugurated.

Records of vertical displacements measured at several points on the west wall are presented in Figure 12. As seen there, the automatic monitoring system began operating towards the end of 2017. Jet grouting took place between 9 January and 20 April that year. Seven of the levelling points responded quite rapidly to the injection of grouts; five of them settled 1 to 2 mm and the other two, located at the base of the wall, displaced vertically almost 7 mm. The effects of jet grouting ceased in April 2017 and the passage of the TBM induced another 2 mm, also as recorded at the base of the west wall. At the beginning of May, the TBM was detained on account of uncontrolled water flowing into the face of the excavation. Additional grouting was carried out until the TBM was able to advance towards the end of August. This induced an additional settlement of about 6 mm at the base of the west wall but in the upper levelling points it only reached about half that amount. Settlement rates reduced thereafter and by the end of September, no significant additional settlements were observed after that date and when the operation of the monitoring system stopped in March 2020, settlement rates were nil.

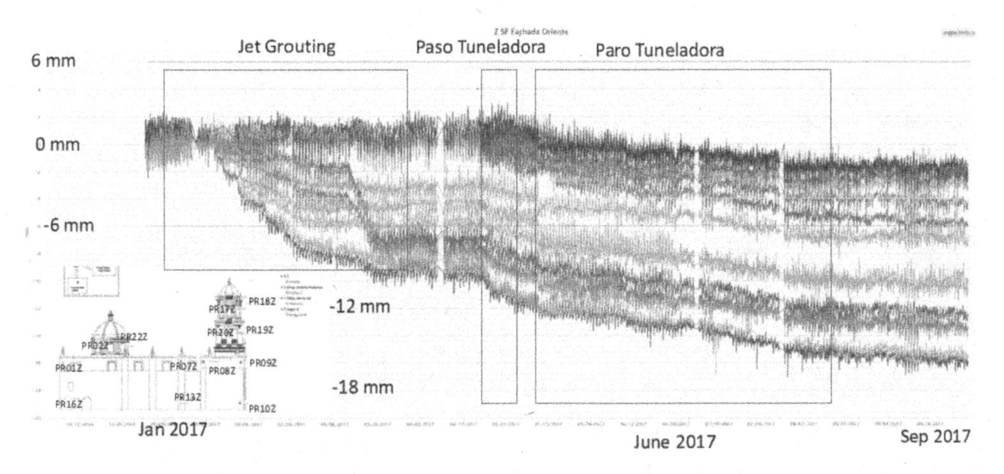

Figure 12. Evolution of vertical displacements along the west wall. Positive values indicate heave. Settlements are associated to negative values.

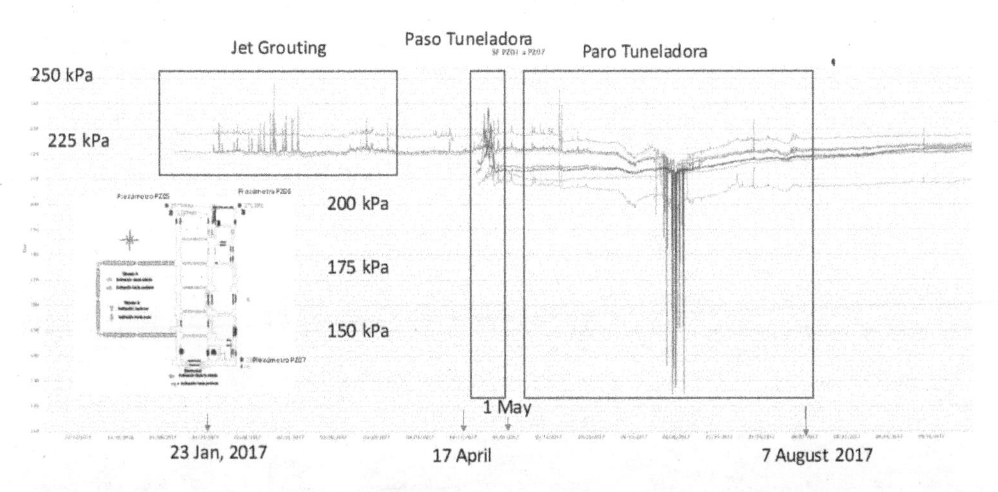

Figure 13. Pore pressure response during jet grouting, passage of the TBM and stoppage of tunnel excavation.

Jet grouting induced pore pressure increments in peaks of about 5 to 20 kPa and exceptionally of nearly 40 kPa all of which dissipated rapidly as soon as injections ceased. The passage of the TBM under the church also induced pore pressure peaks that were not larger than those recorded near the church. Pore pressure transducers placed near the tunnel axis recorded larger peaks as expected (see Figure 13). Excess pore pressure was also observed during the passage of the TBM but it rarely exceeded 20 kPa and it also dissipated in just a few days.

Shortly after the TBM passed under the church, the TBM was detained under the archway (see Figures 12 and 13) between 10 May and 7 August. As seen in the geotechnical profile of Figure 6, a geological fault passes across the axis of the tunnel. The materials before the fault (south side) are the sandy silts that have already been described but much harder volcanic tuffs and basalts were found north of the fault. Necessary adjustments to the TBM were delayed because uncontrolled seepage occurred and it was then necessary to perform additional injections of mortar. As seen in Figure 13 pore pressures decreased sharply indicating that water was flowing without control into the tunnel for a few days, at the beginning of June 2017. Pore pressures went back to normal almost immediately after the tunnel was sealed again and after adjustments to the TBM were over.

Changes in pore water pressure, i.e. in effective stress within the soil mass in the vicinity of the temple due to jet grouting, to the passage of the TBM under it and to the stoppage near it, induced settlements, mainly along the west wall, as indicated previously. Note however, that settlements and cracks occurred mainly during jet grouting, while the effects of the excavation of the tunnel were smaller and of the stoppage of the TBM were smaller. According to structural experts involved in the project, in no case was the security of the temple compromised (Meli et al. 2018).

4 EFFECTS OF JET GROUTING ON SOIL PROPERTIES

The presence of the tunnel under the temple as well as jet grouting changed soil conditions and, given the seismicity of the zone, it was necessary to assess its seismic response taking into consideration how jet grouting modified soil properties. Ambient vibration measurements (accelerations) were performed at several sites close to the temple before and after jet grouting. Spectral ratios between horizontal (H)and vertical (V) ambient vibration measurements show, as seen in Figure 14, that the dominant frequency in front of temple's main façade was 2.35 Hz before jet grouting and increased slightly to 2.4 Hz after it. In comparing ambient vibration measurements before and after jet grouting over a thickness of about 10 t0 12 m, it became evident that grouting produced a very slight stiffening of the whole soil column. Shear wave velocity profiles obtained with the combining the SPAC and MASW techniques indicate that grouting affected surficial strata (10 m or less) and it also had a slight to moderate stiffening effect over the projected grouting thickness. It should also be noted that the shape of the spectra changed and that after grouting, larger amplifications are to be expected at higher frequencies.

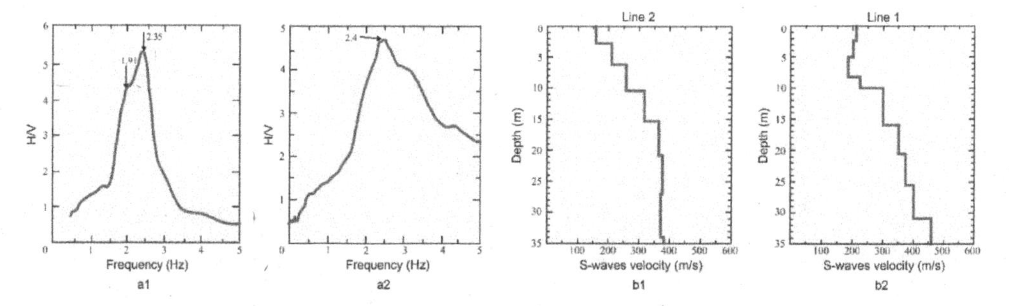

Figure 14. a) Spectral ratio between horizontal and vertical vibration of the soil in the northeastern corner of the San Francisco temple, before (a1) and after (a2), the soil treatment with jet grouting; b) Profile of S-wave velocity in the ground at different depths, measured in front of the main façade (a1) and behind the temple (a2).

Additional geophysical surveys were required to assess the effects of jet grouting. A 3-D seismic tomography was carried out, based on analyses of ambient noise vibrations recorded simultaneously in station-pairs, cross-correlating and stacking data (Bensen, et al, 2007). As an example, the graph in Figure 15 shows the distribution of shear wave velocities under and around the temple, along an E-W cross section near the main façade (CFE, GEIC, 2017). As seen there, shear wave velocities were clearly larger in the zones that were jet-grouted and, it is evident that there is a clear contrast between the surface silty sands that were not grouted.

A plan view of the shear wave distribution at a depth of 25 m, roughly at the tunnel's crown, is given in Figure 16. This graph shows that stiffening by grouting was not uniform and that it concentrated mainly near the façade, N-E corner and, also, outside the church in the N-E atrium.

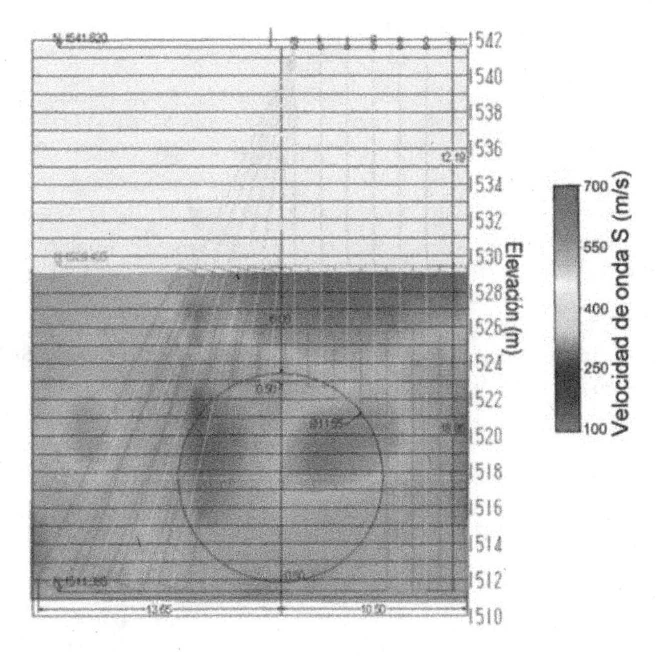

Figure 15. Distribution of shear wave velocities along an E-W cross section near the main façade.

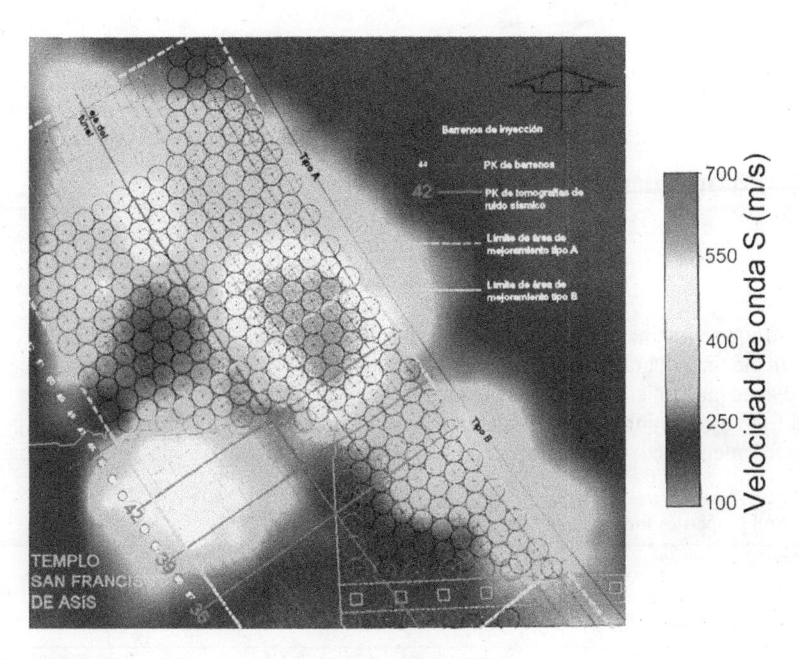

Figure 16. Distribution of shear wave velocities at a depth of 25 m.

5 CONSIDERATIONS FOR SEISMIC RESPONSE

As explained previously, the dominant frequency at the N-E corner was not greatly affected by jet grouting (see Figure 14). However, given the non-uniform distribution of shear wave velocities and also, given the seismicity of the zone, it was only natural that concerns were raised in regard to the capacity of temple to resist strong earthquake motions. Consequently, a numerical model was used to perform an analysis of dynamic soil-structure interaction.

Acceleration histories applied at the base were obtained by using the time histories recorded during the 9 October, 1995, earthquake (Mb=6.5, Ms=7.3 and Mc=7.5) at a site having similar geotechnical some 1500 m north of the church. This earthquake originated off the Pacific Coast, 260 km from Guadalajara and it is the largest recorded event recorded in the city. In applying this procedure, the recorded acceleration history was adjusted so that it would comply with the design spectrum specified for the city by local recommendations for seismic design (Abramson 1993 & Lilhanand & Tseng, 1988). Recorded movements were then transported to the site and were then deconvolved using a one dimensional wave propagation program to obtain incoming accelerations at the base of the relevant strata at the church (Schnabel et al., 1972). Finally, uniform hazard spectra for different return periods, shown in Figure 17, were created using PRODISIS, a procedure specified by Mexican regulatory agencies (CFE, 2015). Note that the peak spectral ordinates occur at periods that closely resemble dominant periods at the site. Our analyses were made for a return period of 250 years using the component that displayed the largest energy.

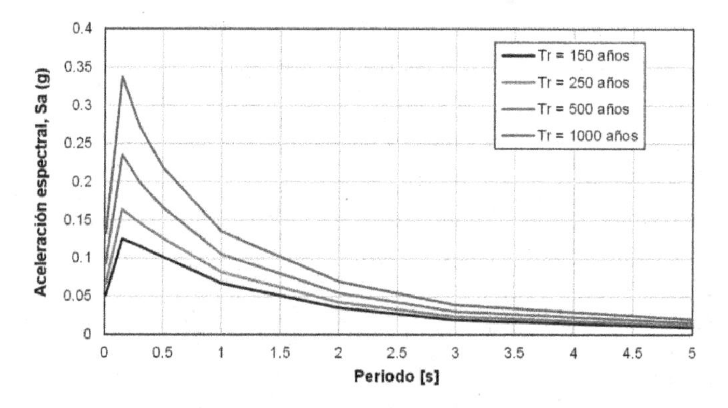

Figure 17. Uniform hazard spectra for different return periods.

5.1 *Analysis of dynamic soil-structure interaction*

The soil structure interaction analysis was performed with a three-dimensional finite difference program in the time domain (Itasca Consulting Group 2003). Soils were modelled as free draining viscous-elastic materials with degrading strain-dependent dynamic stiffnesses and damping ratios obtained from a calibration of the model made with respect to a site response analysis performed with the aforementioned one dimensional wave propagation computer code (Shake, Schnabel et al.,

Table 3. Soil properties for initial soil-structure interaction analysis.

Depth m From to	Geotechnical unit	E kPa	Reference G kPa	Reference internal damping %	Shear train %
0.0–2.00	Fill (rubble)	2044	730	8.2	0.058
2.0–5.40	1b (SM)	6414	2411	8.4	0.065
5.40–16.90	1c (SM)	8275	3111	8.9	0.113
16.90–30.20	1b (SM)	6082	2287	9.3	0.180
30.20–32.00	1d (SM)	11404	4287	8.9	0.109
32.00–34.10	2 (SM)	16963	6377	8.5	0.075
34.10–36.30	4 (Basalt)				

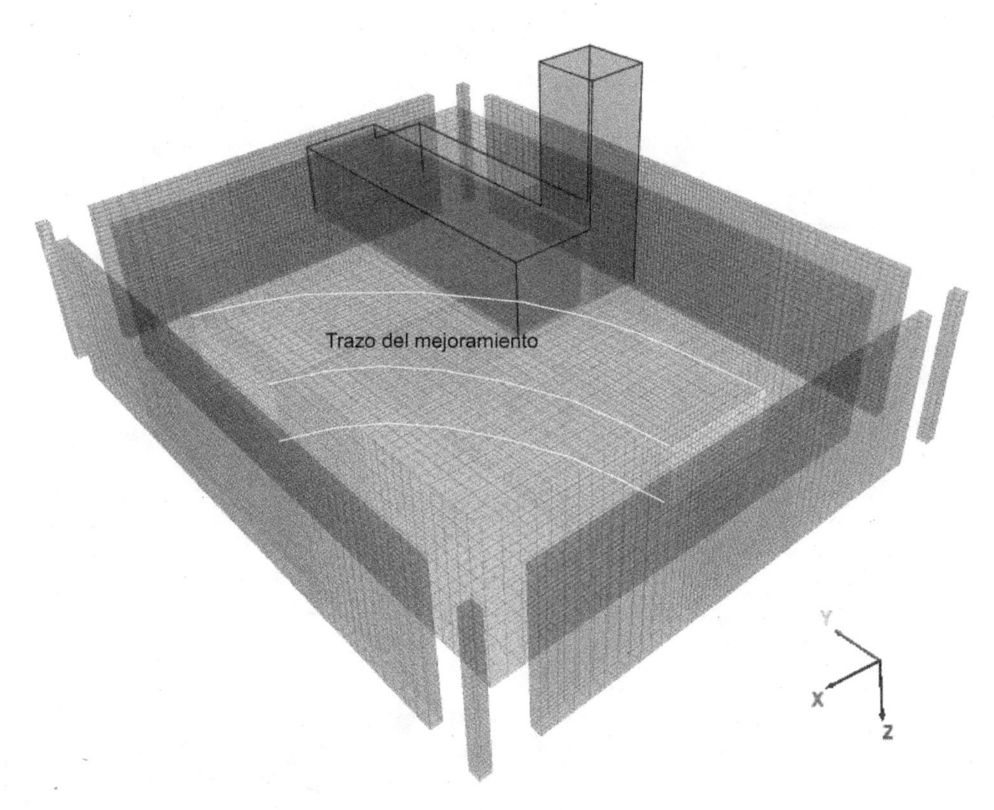

Figure 18. Three-dimensional finite difference model for dynamic soil-structure interaction analyses.

Table 4. Properties of the structural elements used in the dynamic soil-structure interaction analyses.

Structural element	Density kN/m^3	G MPa
Foundation	22.0	4082.5
Walls	22.0	3689.2
Ceiling	18.0	3102.5
Attached building	22.0	2467.5
Tower	22.0	3102.5
Grout columns	16.0	400.0

Note: The value of Poisson's ratio for all elements was 0.3 and 0.4 for the grout columns.

1972). Table 3 shows the soil parameters used in the initial calculations, before taking into account the effects of jet grouting. Properties for the structural elements are summarised in Table 4.

In the analysis the structure was modelled with linearly elastic solid elements with a base raft, perimeter walls and a flat cover, all of which resemble a hollow box. Despite its simplicity, the model still retains the prototype's main dynamic characteristics. The three-dimensional model is depicted in Figure 18. Lateral and base boundaries dissipate incoming shear waves. The seismic excitation was applied at the base of the model at contact of the soil-basalt interface in a direction parallel to the temple's shortest plan dimension.

Analyses were made using the soil properties estimated before considering the effects of jet grouting and after it, modifying shear wave velocities and stiffnesses, as a result of it. The results of the seismic tomograhies were used to this end albeit in a simplified manner. An example of the modified shear wave velocities along a cross section perpendicular to the tunnel axis in the numerical model is shown in Figure 19.

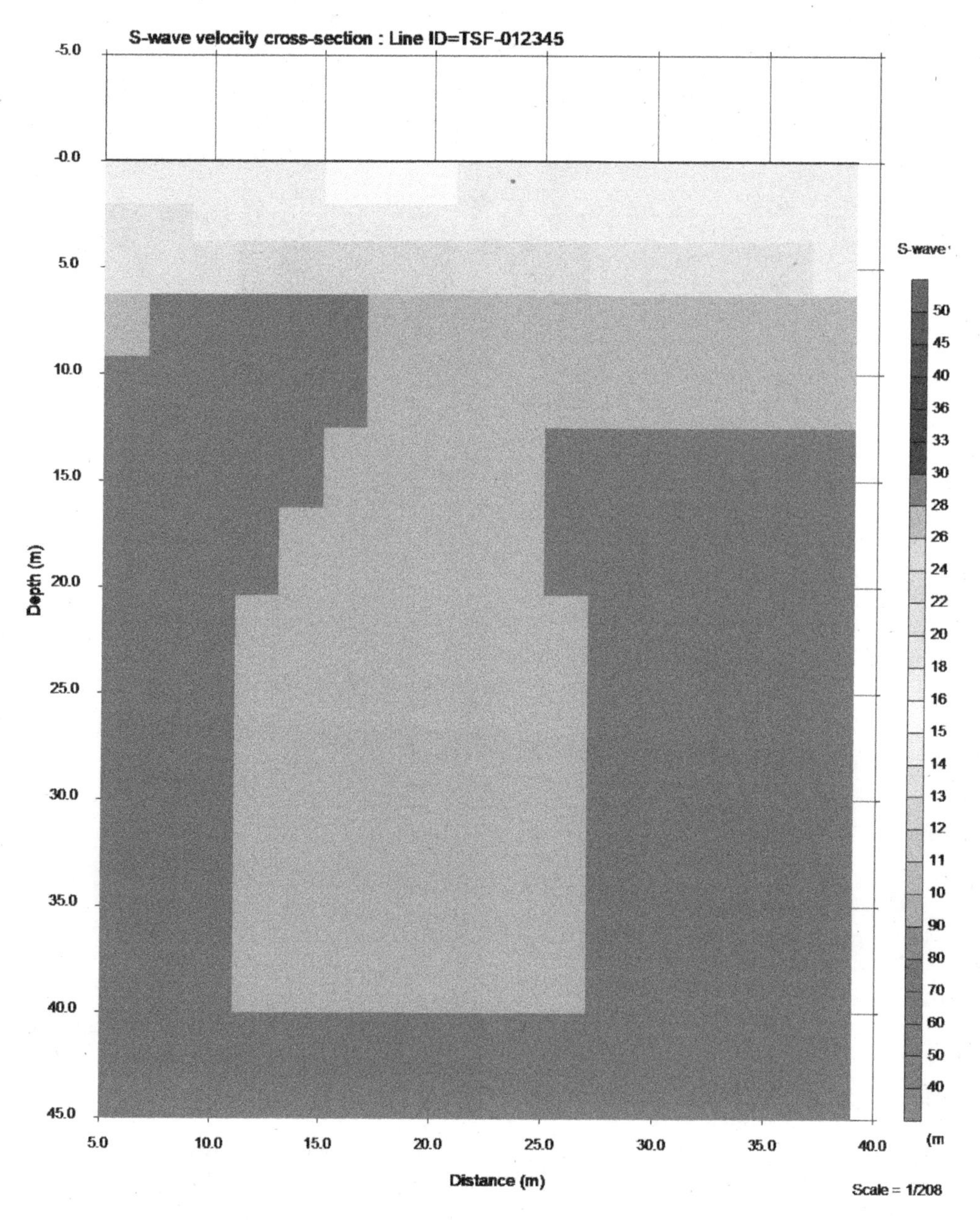

Figure 19. Example of shear wave velocity distribution along a cross section of the numerical model. In the case shown, shear wave velocities were modified to account for the stiffening effect induced by jet grouting.

5.2 *Results of the dynamic soil-structure interaction analysis*

These results are presented by means of pseudo-acceleration response spectra obtained from acceleration time histories calculated during the analyses. Six control points located in different location along different points in the ceiling as well as another one outside the church, in the free field, were used to analyse the results (see Figure 20).

Response spectra from acceleration time histories calculated at the ceiling and located in Figure 20 are shown in Figure 21. These spectra were calculated without considering modified soil properties by jet grouting. A summary of these results is given in Table 5, including data from spectra obtained with modified soil parameters after jet grouting. In Figure 20, peaks that appear at short periods (about 0.28 s, 3.5 Hz) have been associated to the lateral fundamental vibration mode and their amplitude is also influenced by energy concentrating around those shorter periods in the input excitation (see Figure 17). Peaks for the FI and FD points (that appear in periods close to 0.58 s, 1.72 Hz) in the northern portion of the church are also influenced by the vibration of the tower which was estimated to be around that same value and varies about 15% from values estimated from ambient vibration measurements (1.5 Hz). Other researchers have argued that this mode is related to the vibration of the bell tower, and is also detected in the longitudinal direction, thus showing that the vibration is symmetrical in both directions (Meli and Sánchez, 2018). Ambient vibration measurements were made at points FI CD and PI and the frequencies at which peak vibration amplitudes were obtained were, respectively, 1.6, 2.0 and 2.0 Hz; the frequencies for maximum pseudo acceleration from the calculated response spectra of Figure 21 at these same points were 1.72, 2.17 and 2.17, which overestimate the measured values in about 8 %.

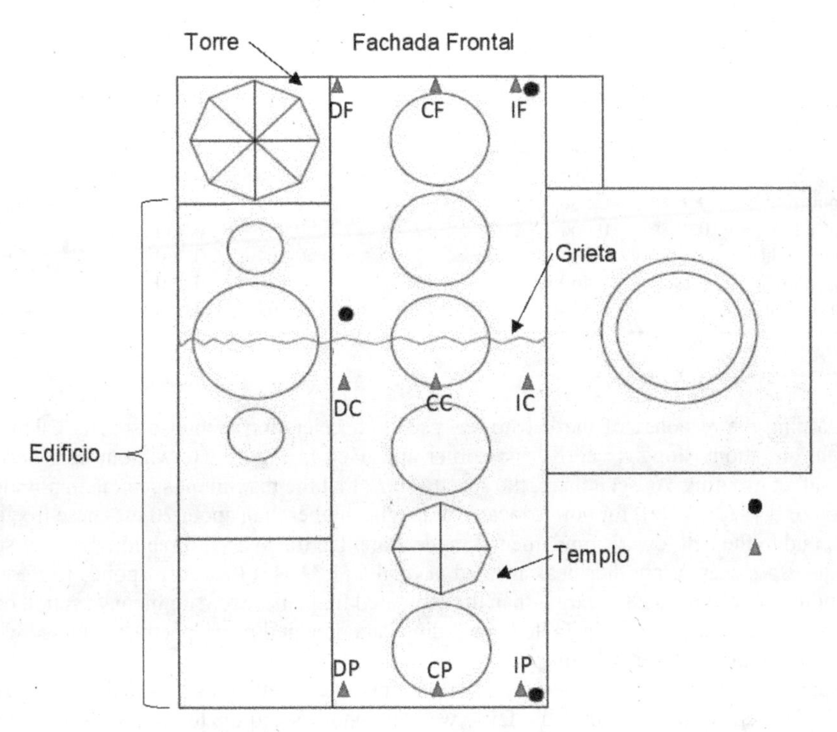

Figure 20. Control points to study the results of the dynamic soil-structure interaction analysis. Dots indicate sites where ambient vibration measurements were made; triangles indicate points where time histories were calculated with the numerical model.

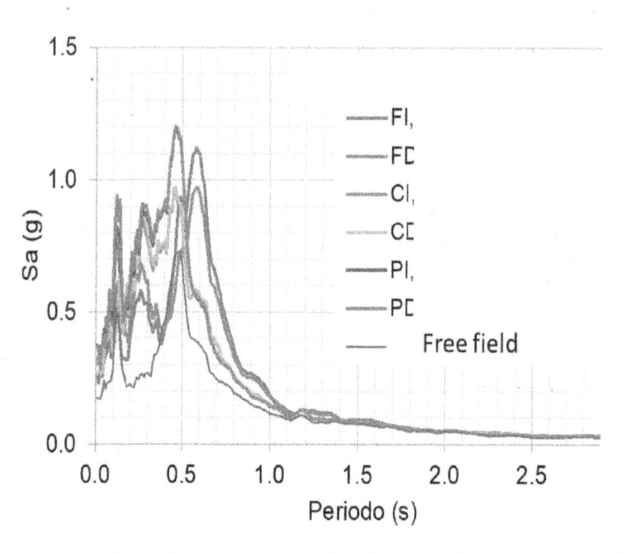

Figure 21. Response spectra obtained at the control points in the ceiling without modifying soil properties due to jet grouting.

Table 5. Ceiling: Peak spectral pseudo accelerations and associated periods and frequencies obtained from the numerical analysis.

Control point	Sa peak g	No grouting Period, s	Frequency, Hz	Sa peak g	With grouting Period, s	Frequency, Hz
FI, front left	1.21	0.580	1.72	1.086	0.590	1.69
FD, front right	0.971	0.580	1.72	0.939	0.590	1.69
CI, centre left	0.961	0.460	2.17	0.957	0.460	2.17
CI, centre right	0.972	0.460	2.17	0.964	0.460	2.17
PD, Back left	1.192	0.460	2.17	1.198	0.460	2.17
PD, Back right	1.201	0.460	2.17	1.208	0.460	2.17

Regarding the response of the bell tower, pseudo acceleration response spectra calculated at different elevations along its north east corner are given in Figure 21, without considering the effects of jet grouting. As seen there, the spectra show that the maximum spectral amplitude lie at a period of 0.49 (2.04 Hz), for points located at heights higher than about 20 m. These frequencies correspond to the bell tower's fundamental mode along the transversal direction. Smaller spectral amplitudes are seen at another peak located at 0.58 s (1.72 Hz) that corresponds to the tower's longitudinal mode and is 23 % larger than the estimated frequency from ambient vibration records. Spectral data are summarised in Table 6, including data obtained from spectra calculated with soil parameters modified by jet grouting.

Response spectra obtained at several control points in the ceiling and the bell tower displayed peaks at frequencies that approximate fairly well the temple's and the tower's fundamental modes identified from the analysis of ambient vibration records. Data given in Tables 5 and 6 confirm that jet grouting had a minor effect on the temple's fundamental frequencies and, also, that the spectral amplitudes did not change significantly. These findings clearly indicate that jet grouting did not increase the vulnerability of the church to earthquake shaking. On the contrary, maximum peak pseudo accelerations turned out to be slightly lower.

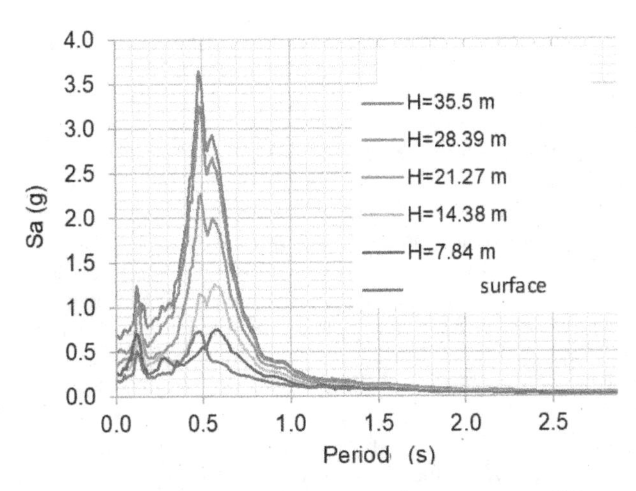

Figure 22. Response spectra obtained at the bell tower's north east corner, at different elevations from the surface.

Table 6. Bell tower: Peak spectral pseudo accelerations and associated periods and frequencies obtained from the numerical analysis.

Height along tower's front right corner	No grouting			With grouting		
m	Sa peak g	Period s	Frequency Hz	Sa peak g	Period, s	Frequency, Hz
35.5	3.641	0.480	2.08	3.532	0.490	2.04
28.39	3.254	0.480	2.08	3.171	0.490	2.04
21.27	2.264	0.490	2.04	2.209	0.490	2.04
14.38	1.251	0.570	1.75	1.214	0.570	1.75
7.84	0.753	0.590	1.69	0.732	0.580	1.72
Surface	0.729	0.480	2.08	0.738	0.490	2.04

6 CONCLUSIONS

Geotechnical studies carried out before the excavation of the tunnel suggested that preventive measures were needed to avoid or mitigate damage induced by tunnelling on the San Francisco church, given the nature and characteristics of the silty sands under it. The authors of this paper formed part of the technical staff called to analyse and assess the effects of jet grouting, the preventive measure adopted by the designers, of the excavation of the tunnel directly under the north east corner of the church, of the stoppage of the TBM right in front of the temple and, also, of its behaviour during the initial months of the operation of Line 3 of the Guadalajara Metro system.

The San Francisco church was profusely instrumented with an automatic monitoring system with which its response to the execution of the preventive works and the excavation of the tunnel were followed closely in terms of induced settlements, tilts, and pore pressure. In the San Francisco church, as well as in other historical constructions of the area, settlements occurred mainly during preventive protection works. Measurements indicate that jet grouting induced the larger detrimental settlements, mainly along the east wall and that settlements induced by the excavation of the tunnel and the passage of the TBM under the church were not as large; further settlements were induced during the stoppage of the TBM in front of the east wall. After that, settlement rates reduced continually and for all practical purposes, settlement rates were practically nil by mid 2020 when the automatic monitoring system ceased to operate. These settlements re-activated pre-existing cracks and also induced some additional cracking. Nevertheless, according to structural experts the security of the temple was not compromised (Meli & Sánchez 2018). In any case a retrofitting project was initiated towards the end of 2017.

Religious and civil authorities were concerned about the seismic vulnerability of the church since it was thought that jet grouting could have had a detrimental effect on its capacity to resist strong earthquake motions. Ambient vibration measurements as well as seismic tomographies were performed in and around the temple to obtain the dynamic properties of the underlying silty sands. These studies showed that soils were not stiffened uniformly by jet grouting. A numerical model was used to study the dynamic response of the church with a three dimensional finite difference computer code. Despite its simplifying assumptions, the model was able to provide fairly close approximations to the overall dynamic properties of the church. The numerical analyses confirmed that jet grouting did not increase the vulnerability of the church to earthquake shaking but that it decreased slightly.

The automatic monitoring system was a fundamental tool to understand the effects of the protection works, the excavation of the tunnel, and of any other non-accounted for events. It is the opinion of the authors of this paper that these systems should be considered as essential in cases as the one described here, i.e. tunnels excavated in densely urbanized areas

REFERENCES

Abrahamson N. (1993). "Non-Stationary spectral matching program RSPMATCH". Obtenidode: http://www.civil.utah.edu/~bartlett/NEUP/Rspmatch/RSPmatchdocument.pdf

Bensen, G.D., Ritzwoller, M.H., Barmin, M.P., Levshin, A.L., Lin, F., Moschetti, M.P., Shapiro, N.M., Yang, Y., (2007). Processing seismic ambient noise data to obtain reliable broad-band surface wave dispersion measurements. Geophys J. Int., 169, 3, 1239-1260.

CFE (2015). PRODISIS, A Programme for seismic design. In: Manual de Obras Civiles, Mexico, Comisión Federal de Electricidad (Mexican Electricity Board).

Itasca Consulting Group Inc. (2017) "FLAC 3D: Fast lagrangian analysis of continua in 3 dimensions, User's Guide". Minneapolis, Minnesota, USA.

Lilhanand K. y Tseng W. S. (1988). "Development and application of realistic earthquake time histories compatible with multiple damping response spectra", Proceedings of the 9th World Conference on Earthquake Engineering, Tokyo, Japan, Vol. II, pp 819-824

R. Meli and A. R. Sánchez Ramírez (2018). Protection and monitoring of three temples close to the excavation of a tunnel in Guadalajara, Mexico. Paper 42, Proceedings 11th Int Conference on structural analysis of historical constructions, Cusco, Perú, 11-13 Sept.2018

SHAKE Schnabel, P.B., Lysmer, J., and Seed, H.B. 1972. SHAKE – A computer program for earthquake response analysis of horizontally layered soils. Report No. EERC-72/12, University of California, Berkeley.

Panel Lectures

Geotechnical Engineering for the Preservation of Monuments and Historic Sites III – Lancellotta, Viggiani, Flora, de Silva & Mele (Eds)

Under the skin

P. Smars
National Yunlin University of Science & Technology, Douliu, Taiwan

ABSTRACT: Should we care about what lies behind the surface of the cultural objects we value? If a feature is invisible to our senses, does its existence really deserves our attention and concern? Engineers like doctors obviously needs to know about anatomy, materials... bones and skin diseases. Their job is to preserve what society cherish and their mean of action requires this knowledge. But should said society be concerned only by failures of plastic surgery and façade cleaning, overlook the viscera and leaving their cure to the experts? And should these last understand their duty as not only caring for what people see but also caring for what they, themselves, have the privilege to see?

These are the questions that the paper tries to develop and discuss.

1 HERITAGE

How do we conceive time?

> "Architecture is to be made historical and preserved as such" Ruskin 1849, "The Lamp of Memory", aphorism 27

In matters of heritage, what is the target of society, professionals, deciders? (Figure 1). And how does it relate to their conception of time?

In English — since Ruskin (1849), Morris and the SPAB — *restoration* became a bad word, describing an impossible and destructive operation. "Restoration (*Restaurationwesen*), the illegitimate son of the historicism of the 19th c." was writing Dehio (1905). The emergence of an awareness of the value of our built heritage was sadly also creating new threats.

This is a recurrent observation: not only nature and vandals threaten what we cherish, good intentions too, often lead to unexpected consequences and, later, repentance. The ambivalence of the UNESCO label is an example (Choay 2009; Settis 2017). Engineers too have their share in the history of what can be seen as "good" or "bad" decisions.

Terms like *conservation* and *preservation* are milder but also somehow presumptuous and, to pragmatists — looking at heritage as a useful mean (and not simply as helpful guide like Ruskin did) to create a better society — they may be delusive and backward-looking: the past is out of reach but, hopefully, the future can be shaped.

It is likely to be a personal bias or misunderstanding but *revitalisation* brings to my mind not *life* but *money* and, often, *denaturation*. *Reconstruction*, common after the two World Wars and in archaeological sites is also gaining new supporters, surprisingly or not (Stanley-Price 2009; Houbart 2020).

The "zorg" in Dutch or "Pflege" in German (*monumentenzorg, Denkmalpflege*) are more neutral, they do not inform about the mean, just about an intention: caring.

Leaving for a second what can be seen as questionable, it can be said that each term denotes a historical, cultural, and professional context and aspirations which, if they cannot be perfectly fulfilled, do serve as beacons: "Thy Word is a lamp unto my feet." (Ruskin 1849, – citing Psalm 119:105).

But who is *Thy* and what does he actually desires?

DOI 10.1201/9781003329756-7

Figure 1. Kids in Bamiyan, Afghanistan. Ph.: P. Smars, 2003.

Turning attention to the object of interest may help identifying convergences and divergences between intentions. *Monument* has acquired a connotation of institutional, aristocratic, but its Latin origin is interesting (Choay 1996; Riegl 1903). *Monere* means *to remind*, *to warn*. The term has a foot in the past (remembrance), another in the present (consider), and one in the future (warn). And that is maybe what everyone expects: a future for our past. Visions about cultural heritage are clearly related to visions about time. Saint Augustine (1991, 11–18), while admitting he cannot really understand time — as soon as he tries to explain it — writes about the presence or image of its three moments "wherever they are, they are not there as future or past, but as present": past as "memory", present as "awareness" and future as "expectation" (1991, 11–20).

Each attitude outlined above seems to weight differently the images of the three loci of time drawn by Augustine. And where the balance settled is reflected in how ideas are implemented.

But words in themselves can be manipulated. How they are translated in actions, revealing their nature, is a partly unrelated matter. Riegl (1903) dissected the reasons impeding an agreement about actions. It is a society matter and society is made of individuals which see and value differently. Decisions is then the result of a game of power and persuasion.

Assuming that individuals desire to act *well*, the question remains of what is the meaning of *well*. It is clearly a problem of ethics, carrying along all its attached complications. Moreover, if "all practical laws are the exponents of moral ones" (Ruskin 1849), "morality represents the way we would like the world to work, and economics represents how it actually does work" (Levitt and Dubner, cited by Sandel 2012, p. 88).

Let us dream for a while! The term *monument* itself also leads to ethics: we are reminded and warned! In what follows, the position advocated is that, to fare well, we need memories, attention and expectations…and too act accordingly: "Being, knowing, willing" (Saint Augustine 1991, 13–11).

As someone interested in the material and technical dimension of *monuments*, observing that this position — for a long time shared by many — is not anymore predominant, while recognising the reasons behind this evolution — in part first hand as Figures below will illustrate — I nevertheless sense a danger.

Section 2 describes it; section 3 indicates an escape route: *seeing*; section 4 advocates engineers to follow the route; section 5 discusses one of the practical complication: *uncertainty*; section 6 looks at the decision process and outlines some recommendations.

The main goal of the paper is to shed a bit of light in a dark area: the inside of the fabric.

2 SUPERFICIALITY

What is the context?

> "Le divorce accablant de la connaissance et de la mythologie" Barthes 1957, "Bichon chez les nègres"
> (The overwhelming divorce between knowledge and mythology)

To the dismay of some, movies are now full of super heroes! And for countries and patriots, World Heritage sites (*property* is the official denomination, UNESCO 2021) are also super heroes…thankfully, often in better oufit! Writing about music, Adorno (1991, p. 36) refers to the construction of a "pantheon of bestsellers", there to entertain us. "To like it (music) is almost the same as to recognise it." (Adorno 1991, p. 30), "high" and "voluminous" voices are favoured, and old instruments or conductors are becoming fetishes (Adorno 1991, p. 37). For him, the same can be said of sport and films. Built heritage is not immune to the trend: it has to transmit a clear and simple message. Architecture is particularly vulnerable because it "(…) has always represented the prototype of a work of art the reception of which is consummated by a collectivity in a state of distraction." (Benjamin 1980). People are interested by painting, music, sculpture and literature but not by architecture (Zevi 1948).

Is this a remain of bourgeois thinking? For Benjamin (1980), the advent of means of mechanical reproduction was a way to democratise access to Art. This is certainly a noble intention: another *light beacon*. He saw in particular cinema as a revolutionary and promising medium, liberated of the fetishist *hic and nunc* of the original and unique object. Aura may be lost but reproductions allow access to a greater number (Mozart brought to your home), letting them perceive details or perspectives otherwise left in the dark (approach to closest detail of the Arnolfini Portrait). Translated in the context of built heritage, loosing material authenticity sometimes let visitors see what otherwise would have disappeared. But Benjamin seems to always assume that copies are perfect reproduction of an original. But is it the case? Liberated from the aura, its cult or later merchant value, both residing in the mind of the observer, said observer can access a message. But is it possible to reproduce perfectly the fabric, to preserve the material support, the data for future interpretation and not one which is imposed, pre-digested, ("purified" Stanley-Price 2009)? "Auf der einen Seite die vielleicht verkürzte, verblaßte Wirklichkeit, aber immer Wirklichkeit — auf der andern die Fiktion." (On the one hand, the perhaps diminished, faded reality, but always reality — on the other, fiction) (Dehio 1905). Is following Benjamin ideas in the realm of visual arts, not transforming them in arts of representation, in some form of theater? And does that lead to a long-sighted look on heritage? To the production of a synthetic food, leaving no place to our invention, forcing us to think (Barthes 1957, "Photos chocs").

This is not easy to settle but one cannot help observing that history also followed a path of money, propaganda and delusion. Cinema was rarely used by *communists* and *capitalists* alike with citizen empowerment in mind. Adorno saw Walt Disney as the most dangerous man in America. And the evolution does not only concern only cinema. The "pervasive" and "mundane" role of markets in todays life described by Sandel (2012, p. 15) is steadily growing and influencing the evolution of built heritage management (Montanari 2015). In particular, the influence of tourism on the interest for reconstructions is not new (Giovannoni 1931b). Cities, architecture, archaeology are all concerned. Houbart (2020) cites Roberto Pane who already complained about the excessive

Figure 2. View from the Big Wild Goose Pagoda. Xī''ān, Shā'n-xī province, China. Ph.: P. Smars, 2005.

reconstruction to satisfy the appetite of the tourists. Stanley-Price (2009) lists it among the reasons behind reconstructions in archaeological sites. Money is a universal value and cultural heritage is there to be exploited like petrol (Choay 2009, quoting a communication othe ministry of culture in 2012: "gérable et exploitable", p. XL) (Montanari 2015, quoting Napolitano in 2012: "sfruttare fino in fondo", p. 7). Choay (2009) describes the "great void" created by this evolution: the disappearance of the *memorial* dimension and of the *historical monument* as such. The case of Xī''ān in China (Figure 2) is particularly emblematic (Dunne 2018) and also an opportunity to make a transition to the theme of *identity*. "Historic cities have often been redesigned or undergo a process of renewal in order to promote their ancient historic credentials." (Zhu 2017). Stanley Price also mentioned these political reasons, giving the example of the Parthenon after the independence of Greece or Babylon during the Iran-Irak war. In some cases, it may even lead to real crisis when significance of an heritage is not interpreted and simplified in the same way by various groups (Yang 2018) (Figure 3. Similarly, in Taiwan, after WWII, the Japanese heritage was preserved only if it was of practical utility. Most Shinto shrines were destroyed, abandoned or converted. But nowadays, any trace of Japanese presence become an opportunity to justify and reinforce the country's identity. The way a society deal with its cultural heritage betrays its intentions and values.

Returning to Walter Benjamin and the *aura*, did the infinite reproductions of Mona Lisa, Maneken Pis, the Eiffel tower or the tower of Pisa actually eliminated their aura? Or did their meaning just shifted? Do the modified anatomy of a monument, liberated from the aura of having being built by someone from another time, who left minute traces only seen by a few, change anything about its accessibility, in particular if it is a reproduction? Giovanonni was describing the difference of expectations between the scholar ("érudit"), the architect and the simple citizen (Giovannoni 1931a). Who benefits from the evolution described above? The complexity and richness of signification is displaced, deformed, impoverished by what becomes a myth (Barthes 1957). At least, as long as the material object hiding behind the myth is still there, to paraphrase Barthes, there is a possibility

Figure 3. Hashima Island (Battleship Island), Japan. Ph.: P. Smars, 2016.

to look through the window and see the landscape. They coexist. But what happens when it is not there anymore? We are left with the myth!

In this context which is unlikely to change much, the growing virtualisation of the society — whose negative impact is discussed by Choay (2009) and others — also presents an opportunity (which should be taken) to fight if not defeat superficiality, an opportunity for those acting in the material world: architects, engineers and others, to *virtue-alyse* interventions on the substance. And, doing so, to engage in the positive and democratic process advocated by Benjamin (1980) and some French Revolutionaries (Choay 2009, pp. 84–101: "Félix de Vick d'Azyr"): pass a message, give access and visibility to hidden areas and features; but also clarify the limit between data (keeping it *actual*) and interpretation (using the power of the *virtual*), in a word dispel what would be otherwise deception, in our world of fake news and endangered environment.

But, how does that translate in actions that we, individuals, in the limit of our power and responsibility, can take? First step: to see.

3 SEEING

What is the challenge?

> "far scaturire colori e forme dall'allineamento di caratteri alfabetici neri su una pagina bianca."
> Calvino 1988, "Visibilità"
> (reveal colors and shapes from lines of black alphabetic characters on a white page)

Common sense tell us that what cannot be seen does not matter. But alterity is disliked by common sense (Barthes 1957, "Martiens"). And otherness is what makes the "language" of cultural heritage not a tautology. If one does see, he learns about the other and about himself (Settis 2004, p. 3). If one does not see, it may be because one does not look.

In the countryside of Taiwan, many adobe and wattle and daub farmhouses can still be seen. Few are still in good condition (Figure 4), some are still occupied but most of them are slowly

Figure 4. Adobe house in Jiŭ -ā'n, Yún-lín county, Taiwan. Ph.: P. Smars, 2019.

disappearing, weathered down (Smars 2018; Smars 2022). Asking where such building can be found, their existence is often denied. They are not perceived, they are invisible. Asking to the – usually old– owners why they are left in such a bad condition, reveal how little weight is given to their memorial value. As soon as money permits, or money incentives leads to land sales, new constructions replace them, in the complete indifference of the planning authorities. Maybe it will result in an improvement of the global seismic resilience? But soon they all will be gone.

The Scheepsdalebrug, a 1934 swing bridge designed by Arthur Vierendeel, notwithstanding its unique features, was not deemed worthy of preservation: it was destroyed in 2009: Maybe it did not fit the medieval image of the World Heritage city of Brugge? Maybe it was his condition of *industrial* object? Various specialised associations tried to save it: The Flemish Association for Industrial Archaeology (VVIA), and the International Scientific Committee on the Analysis and Restoration of Structures of Architectural Heritage (ISCARSAH). But without success, most people and the deciders did not *see* its value. This is an example, among many, of the divorce between the smooth aspiration of popular culture and the breaking aspiration of scientific culture (Settis 2004, p. 17).

Long before, in the staircase of a tower of Notre-Dame de Paris, Hugo (1832) saw, and later lamented the disappearance, of a Greek inscription 'Αναγχη. "Ainsi, hormi le fragile souvenir que lui consacre l'auteur de ce livre, il ne reste plus rien aujourd'hui du mot mistérieux gravé dans la sombre tour de Notre-Dame (...)" ("Thus, with the exception of the fragile memory which the author of this book here consecrates to it, there remains to-day nothing whatever of the mysterious word engraved within the gloomy tower of Notre-Dame" Translation by I.F. Hapgood). He saw something that the restorer, the priest or whoever decided to whitewash the wall did not see. Writing about it, somehow, well-aware Victor Hugo preserved its memory.

This is actually a good example to follow: if because of professional or personal inclination something destined to disappear is identified, if it is deemed important or just interesting, writing about it may — or may not — help but it should be a duty to, at least, document it. Even recognising that Victor Hugo himself was only partially successful in his enterprise.

It is from feeling our misery that proceeds our compassion for indigence (Saint Augustine 1991, 13). So it is likely that, initially at least, some dissatisfaction triggered the work, exhibitions

Figure 5. Temple of Baachus, Lebanon. Ph.: P. Smars, 2015.

and books of Bernd and Hilla Becher on water towers, gas tanks, and other industrial relics (Becher & Becher 1967), and the exhibition and book "architecture without architects" of Bernard Rudofsky (1964). These two examples were emphasising the formal and aesthetic dimension of something broader and richer but they are efficient means to attract the attention of the inattentive citizen (exhibitions are effective), to open eyes, to inspire further enquiries. They also betray motivations which, without any doubt, go deeper than formalism; they document objects, periods and lives (only through their material imprint) which, often, have now disappeared.

In general terms, discussions around the preservation of some built artefact often result in fights, fights between the one who see and the one who don't. If there is progress in the field, it is about raising awareness to a level at which society starts to care.

It is not an easy task. We all speak different languages and each trace of our built heritage, shaped by centuries, speaks its own. There lies its richness. Progress is greatly related to the growing understanding of what we see and then communicate. For John Scotus Erigenia, cited by Borges, the holy scriptures contains an infinite number of meanings (Borges 1980, "Poesía"). And for the Argentinian poet and essayist, it is the case of any book worthy of relecture (Borges 1939; Borges 1980). So are our monuments: our seeing and reading is source of knowledge and imagination. Often, help is needed to perceive, indicate new angles, and then promote the care for the preservation of what is the source of it all; to keep images raining (Calvino 1988, disgressing on Dante in "Visibilità"). That is the challenge: how can knowledge, emotions, about foundations, structures, materials be communicated.

Languages are also often misunderstood. Each profession uses its own, specific, which does not translates well in the language spoken by others (Barthes 1957, "Dominici ou le triomphe de la littérature"). In particular, the language spoken in the theatre where decisions are taken has its peculiarity.

4 WHAT LIES BEHIND?

What do engineers see?

> "Tandis que les grands hommes, portés au fommet de l'édifice, tracent & élèvent les étages fupérieurs, les artiftes ordinaires répandus dans les étages inférieurs, ou cachés dans l'obfcurité des fondemens, doivent feulement chercher à perfectionner ce que des mains plus habiles ont créé." Coulomb 1776

(While great men, on the top of the edifice, trace and raise the upper floors, ordinary artists, on the lower floors, or hidden in the obscurity of the foundations, must only seek to perfect what more skilled hands have created)

Only broad directions are suggested below. This is actually the part that everyone of us has to write for himself, for the benefice of all.

There is a difficulty, related to the conceptions of time discussed in section 1. Engineers are naturally encouraged to look towards the future and to see the solutions of today as better that the ones of yesterday. This is obviously a simplification; the state of the world in the 21th c. challenges everyone's conception of progress. And in the context of the present conference, it is even more likely to be the case.

That is also confirmed by documents written by engineers for engineers. The ISCARSAH charter of 2003 (ICOMOS 2003) emphasises that "the value of architectural heritage is not only in its appearance, but also in the integrity of all its components as a unique product of the specific building technology of its time", that "the distinguishing qualities of the structure and its environment, in their original or earlier states, should not be destroyed" and that "each intervention should, as far as possible, respect the concept, techniques and historical value of the original or earlier states of the structure and leaves evidence that can be recognised in the future". The charter uses the term *distinguishing qualities*, while another document, the annex on heritage structures of the ISO standard on the assessment of existing structures (ISO 13822:2010) uses *character-defining elements*.

This is not by accident that engineers working on historical structures came out with these ideas. As I understand it, they formed first as purely technical problems of compatibility kept appearing; among which one can list:

- The disasters related to the use of steel and concrete in archaeological sites introduced the idea that modern materials are not always better;
- The damages induced in original materials by too rigid, water- or vapour-proof joints introduced the idea that stronger is not necessarily better;
- The observation, after the occurrence of earthquakes, that what modern codes were recommending was not as effective as expected and that traditional mitigation measures sometimes behaved better introduced the idea that a modern frame of thought does not always fit other context (Figure 6);
- The inadequacy of some European-born concepts or measures to other contexts (material, economic) introduced the idea of a relation between engineering and culture and its value.

All these observations required retrospection. They lead to a better appreciation of the value of the *otherness*, of the specificity of the engineering of the past, as a reminder of the traditions and a warning against the dangers of over-confidence.

This story is not smooth. At the conference of Athens of 1931, Paquet (1931) confidently encourages engineers to use of hidden reinforced concrete elements (which proved to be a dangerous recommendation) but he also expressed his concern for the respect of the construction system (which is still seen as important). At the same conference, Giovannoni (1931a) was expressing a great confidence in modern materials and techniques, while also urging caution in relation to monuments far away from our habits and civilisation. To this day, arguments are colliding.

- New materials vs old materials
- Visible interventions vs invisible interventions
- Respect of the original structural system vs innovative reinforcements

In the broader context of cultural heritage protection, even if, as discussed above, the importance of the authenticity of the substance is challenged. it is still officially listed as an important criterion in international documents. Among the 10 criteria used by UNESCO to assess whether a property has an Outstanding Universal Value, criteria 1 to 4 may benefit from contributions related to technique, construction and engineering (Figure 7). In the Operational guidelines (UNESCO 2021, point 82),

Figure 6. After the 1997 Umbria and Marche earthquake. Sellano, province of Perugia, Italy. Ph.: P. Smars, 1998.

Figure 7. Extrados of the vaults of King's College Chapel. Cambridge, UK. Ph.: P. Smars, 2002.

the authenticity, necessary to achieve an Outstanding Universal Value depends among other criteria of the truthfulness and credibility of "material and substance", and "traditions, techniques and management systems".

The critical question of *credibility* is discussed below, in section 6.1.

But the first stage is identification. In the course of preliminary studies and during the works on the fabric, elements and features of technical interest are uncovered by the specialist operators. This material substance has also a cultural meaning (ISO 13822:2010). And so have the technical achievements. They are historical evidences and have a spiritual nature hiding behind and complementing what only appears as mineral to the uninitiated. "Das herrschende Recht berücksichtigt sie nur als körperliche Wesen und doch ist es die allgemeine Überzeugung, daß ihr wahres Wesen ein geistiges sei." (The prevailing law only considers them as physical beings and yet it is the general conviction that their true nature is a spiritual one) (Dehio 1905). They are potentially the *character-defining elements* of Annex I of ISO13822:2010, the *traditions* and *techniques* of the operational guidelines. They may not be perceived by people outside the trade but that does not mean that they are not present and valuable. They are constitutive of the otherness making our built heritage rare and precious.

In the framework of geotechnical engineering, the contribution of the field to the conservation of cultural heritage is clear, at least to fellow engineers (Egglezos et al. 2013; Kérisel 1975). The existence of TC301, the past and present conferences, are witnessing its importance and vitality. But it seems to me that the historical and cultural dimension is less explored. It may be because, being more of a structural engineer, not aware of the existing literature. That is certainly not by lack of interest or need: my first experiences all involved problems related to settlements, my interest for masonry often oriented me towards soil and rock mechanics. And the challenges of structural and geotechnical engineering are strongly linked (Lambe 1973). Historical information about foundations (García Gamallo 2003; Kérisel 1985; Przewłócki et al. 2005) or soil mechanics (Skempton) is still relatively rare. And discussions about the cultural value and the authenticity of foundations are to my knowledge relatively recent, on a cultural heritage time-scale (Calabresi 2003; Iwasaki 2013; Iwasaki et al. 2013; Iwasaki et al. 2021). The members of committee TC301 are on the front-line of this work of raising awareness about the *invisible*. The book collection that they started publishing about case studies are a great initiative!

That is a difficult challenge, especially for foundations. All interventions are deemed to become invisible and what hides behind the skin, as long as it does not bring back to the surface one of the technical compatibility problem mentioned above, seems also to be of little concern to many engineers. The hidden parts of a structure are like the time described by Augustine: they only exist in the memory of their construction and transformation, in the attention given to their existence when they can be seen and touched and in the expectations that we formulate for their future. Documentation is critical. It can transmit what would not be otherwise accessible, bringing art (in the medieval sense of the word), closer to the people (Benjamin 1980).

5 UNCERTAINTY

How can the concept of uncertainty be conveyed to a society in search of certitudes?

> "One would hope that in predicting the performance of constructed facilities, precise data could be combined with scientifically derived methods to obtain precise predictions. The soil engineer knows that this hope remains an unrealistic dream." Lambe 1973

Predictions are at the "very heart" of the practice of civil engineering (Lambe 1973). Engineers are constantly devising and improving techniques to predict and reduce the risk of failures. In the course of history, they encountered many, often with catastrophic consequences. Their occurrence is often the motor of progress and the techniques designed to eliminate them are slowly taming uncertainty (recurrent cost overruns indicate that it is a work in progress). The whole process, from design to construction and management, is shaped by this intention of control. Without trying to be exhaustive, this involves:

- To produce and use materials of known properties;
- To use structural elements which can be reliably modelled: 1D (beams, columns…) and 2D (walls, slabs…);
- To join them using standard and tested procedures;
- To assemble them to form structures which can be reliably modelled: statically determinate or with sufficient provision of ductility;
- To design structures according to proven scheme (even if, as pointed out by Addis (1994), progress results from creative new designs);
- At each stage, to employ trained and qualified personnel following the rule of their art;
- To have each stage controlled by independent bodies.

The challenges posed by geotechnical problems and historical fabrics offer some resistance to this smooth process. Engineers have to deal with fundamentally 3D, heterogeneous, anisotropic, non-linear, time-dependant problems, only partially understood and depending of parameters often not known, difficult to measure, often indirectly, presenting a great variability, and forming an incomplete set. "Geotechnical engineering is especially damned and blessed by the importance of predictions and the difficulty of making accurate predictions." (Lambe 1973). Kérisel (1975) stresses in particular the importance of time and slow settlements. This is also a frequent concern with superstructures; creep was found responsible of non few collapses (Binda, Anzani, & Mirabella Roberti 2000; Verstrynge & Van Gemert 2018).

In a fuzzy context, formulas are often fuzzy too. They are born from observation and are working in the corresponding situation. They are often semi-empirical. Huerta Fernández (1990) has shown that this is certainly not new and that this does not preclude their effectiveness in said context. Lambe (1973) also points out how methods, parameters and problem to solve "uniquely linked together".

And indeed context is important. The 14th c. expert discussions concerning the cathedral of Milan (Ackerman 1949) or for the cathedral of Gerona, exemplify the difficulty of communication between masters with expertises alien to each other, the use of arguments making no or little or some sense to the next generation, the boldness of the trust that one has in his rules and, surprisingly or not, the effectiveness of the final result. Facing a lack of knowledge about an historical structure, it is a temptation to believe in assumptions or apply methods not fitting the situation, devised for new constructions, or to transform the historical fabric into something which better fits the current assumptions. As Dehio (1905) puts it in relations to the attitudes of the 15th-18th c., destructions can be the result of the boundless trust in the creative power of the present "(…) die Folge überströmender Schaffenslust einer sich selbstvertrauenden Gegenwart sein" (Dehio 1905). This attitude is not dead.

Humility is necessary. As mathematical models alone cannot provide sufficient confidence, observation becomes even more necessary. Monitoring is an effective tool (incidentally, "monitoring" and "monument" have the same etymological root). The "observational method" of Terzaghi and Peck (1969), a "learn-as-you-go" approach, is a modern systematic adaptation of what was used by master builders in the past. It also founds its use in structural engineering (ISO 13822:2010). Lambe (1973) also states that sophisticate methods do not always lead to more accurate predictions. These approaches may even be as or more appropriate today, when dealing with structures belonging to another time and place. This idea may be more difficult to accept by structural engineers, educated with a higher confidence in determinism and in the power of science, than by geotechnical engineers. But Torroja Miret (2000) stresses the importance of intuition and Nervi (1997) too refers to it, and to observation and experimentation.

In search of rigour, in the presence of risk and uncertainty, it is natural to consider a probabilistic approach, in an attempt to evaluate the risk quantitatively. PEER, the Pacific Earthquake

Engineering Research Center has for instance developed a very complete framework for a scientific evaluation of seismic performance of buildings: performance-based earthquake engineering (Porter 2003, 6,7). A requirement of this approach is to be able to characterise precisely hazard, vulnerability and losses (which in the framework discussed here require to consider the potential loss of cultural value). This is an interesting direction to follow but of limited applicability for historical structures (Smars et al. 2010). Quantitative evaluations are difficult: both for random uncertainties (which can be dealt by probability methods and PBEE) and for epistemic uncertainties (related to insufficient knowledge on the process involved, to the choice of models…). In many cases, data is too scarce to characterise random uncertainties, especially to estimate vulnerability and losses (precisely enough to lead to useful estimates). And there are numerous epistemic uncertainties (models used to evaluate the vulnerability; methodology to assess the values).

In an attempt to draw a conceptual framework of general applicability, Klinke & Rena (2002) categorise risks, assigning them to six specific classes characterised by the (a) the expected extend of damage, (b) the probability of occurrence and (c) the level of knowledge (the uncertainties on a and b). For each class, they recommend a specific strategy. They argue for instance that a *risk-based* approach (as the PBEE) is not the most suitable when the level of knowledge is low. A decision tree is given to assist risk managers in making their choices.

And for many historical constructions, because of their complex structure, because of their *otherness*, important constituent of their value but requiring flexibility and ingenuity, because of the numerous stakeholders with values difficult to conciliate, the level of knowledge is indeed low. In such cases, the importance of an extensive program of preliminary studies is recognised. It can provide better estimates of the random uncertainties. Research on the construction techniques, on the anatomy of the structures can reduce epistemic uncertainties and provide elements completing the set of values to protect. Some debate is then necessary to eliminate ambiguities on the level of knowledge and on the expectations of the stakeholders.

6 ETHICS AND DECISIONS

6.1 *Ethics*

How hard should we fight?

> "Wir konservieren ein Denkmal nicht, weil wir es für schön halten, sonderrn weil es ein Stück unseres nationalen Daseins ist. Denkmäler schützen heißt nicht Genuß suchen, sondern Pietät üben."
> Dehio 1905
> (We do not conserve a monument because we think it is beautiful but, rather, because it is part of our national existence. Protecting monuments does not mean seeking pleasure, but exercising piety)

Architects, engineers, conservators have a key role in the definition of interventions on cultural heritage. They are the one which define the technical and material aspect of the intervention. As such they have large responsibilities in front of the population (Settis 2017).

They are firstly contracted to do a job with technical duties, according to a professional ethic. This is not what is discussed here. They have in particular to satisfy the requirements fixed by their sponsors and other deciders which, hopefully, reflects what society judge as *good*. The problem arise from the fact that good as an adjective is not the same as red: its meaning depends of the noun to which it is attached (MacIntyre 1998) and of who makes the statement.

As discussed above, in the course of their work, engineers are uncovering features of technico-historico-cultural interest, features which are not always seen or perceived as important by others. The argument defended in the present paper is that they should care about them and act accordingly. This is (or should be seen as) a responsibility in front of the society and of their pairs, past, present and future. They see them through their professional eye, but also because they have the privilege to look behind the skin! They access places normally inaccessible, they use instruments offering them glimpses of the anatomy of the fabric: materials, construction systems, revealing the ingenuity, the mistakes and the resources of their predecessors.

In many circumstances, this does not lead to difficulties. The chance of most technical endeavours is that, as their implications and the means of their implementation are not clearly understood, if they do not interfere with what matters to others, technicians can decide how to proceed and, if they do care about certain material features, try to preserve them to the best of their ability.

In other circumstances, these material evidences are threatened of alteration or elimination. It may be because of some technical reason (internal) or because of difference of opinions (external). The process of decision is discussed in section 6.2.

The first way to deal with such a situation is passivity: either by being satisfied to be the hand providing what the majority wants, either by seeing it as a lost battle and concentrate strictly on the terms of a contract, either by internalising the discourse which section 2 tried to expose and denounce. The alternative way to deal with the situation is pro-activity:

- Actively search and identify features of technico-historico-cultural interest, the structural *character-defining elements* of the fabric;
- If their protection seems impossible for technical reasons, check whether a change of the terms of the equation would help. Are some non-structural mitigation measures possible (change of program, restriction of access…)? Would more detailed studies, budget modifications or a clarification of the objectives of the whole operation improve the chance of preservation?
- Inform stakeholders and deciders about values, options and consequences.
- Plead for their protection. This is a natural and automatic step if what is mentionned in the first point is done.
- Document and share what will disappear, either because it will be again hidden behind the surface of the fabric, either because it cannot be preserved.
- Ensure that the intervention and the way it is presented is not deceptive. For Ruskin (1849), the ethical dimension of our approach to monuments is fundamental. He thinks that a distinction must be made between malicious deceits and the works of imagination which require, as we could phrase it today, "suspension of disbelieve". He gives the example of gilding which is a deceit in jewellery but not in architecture. Stanley-Price (2009), proposing some criteria for reconstructions in archaeological sites (which are happening whether or not one believes it is a good idea), relates most of them to the ethical principle of not lying.
- Because, these features are often difficult to apprehend by non-specialists, it is also important to disseminate a translated version of the documentation work to the non-specialist. This is laying a stone which will contribute to the education role of heritage (which at the time of the French revolution was one of the most important value) and which eventually will facilitate de defense of such features.
- Consider that this work of vulgarisation also contributes to the credibility of the interventions on our built environment.

Settis (2017, p. 6) argues that the ethics of the architect should not only be to offer services to a client but also to the community of citizens. This is the idea that is extended above to include the engineer. There is a complication: the community of citizens is not homogeneous and neither are its expectations. Saying that "conservation means the greatest good to the greatest number for the longest time" (Settis 2017, citing Gifford Pinchot, p. 33), is a noble target. It is also the pursuit of democracy, but the recent evolution shows that it does not proceed as smoothly and harmoniously as expected.

How can minority opinions be heard? When you care about something and you don't have the leverage to convince, a strategy has to be devised. Since its origin, monument conservation as always required fights. Awareness does not come easily.

6.2 *Decisions*

What belongs to science and what belongs to politics?

Figure 8. Wall repair. Bamiyan, Afghanistan. Ph.: P. Smars, 2003.

Decisions are often difficult, conflictual processes. It is not possible to accommodate everyone. The expectations of the various groups in the population, the requirements of the stakeholders, the opinion of the experts, rarely align perfectly. To decide basically comes down to weight values (Smars & Patrício 2016). The current COVID-19 pandemic exemplify the problem in a topical manner: economists emphasise the importance of not hindering business activities; epidemiologists urge to confine the population; prophets prophesy; and finally, politicians weight factors and decide. Who is right? In a democracy one can just expect that:

- Decisions are taken by people which represent the population;
- Stakeholders and experts can voice and defend their opinions and advices;
- Deciders are accountable;
- The press inform.

Sadly, the process is often more opaque (Settis 2004). Even when the object whose fate is discussed is recognised as cultural heritage, it does not mean that it is the only, decisive, factor. It is not always about protecting what is in danger of disappearance. It can be also about affirming an identity, helping a community, insure a re-election…or most of the time a mixture of noble and less noble motives. Development, safety, well-being of the locals are legitimate aspirations. The transmission of a message to the future generation is only one variable of the equation.

Engineers (acting as experts or designers) are advising, proposing and for the most technical matters, they are trusted and decide.

But, in difficult and/or conflictual circumstances, if there is a fear that something may go wrong, everyone gets ready to transfer responsibility to others. This is a natural tendency. Deciders justify taking decisions based on technical ground and expert advises. And they impute hard decisions to the technicians; who themselves minimise their role: they are just operators following orders. And some press castigates the impotence of experts to form a coherent response.

The trial made to the scientists who met the population of L'Aquila the week before the earthquake of 2009 accusing them of unreasonable reassurance was a shock for the community. It illustrates the problem of vocabulary discussed in section 3. Conveying uncertainty is challenging (Scolobig et al. 2014). The danger will be that anyone possibly at risk of being misjudged will too carefully choose his/her words, make them unintelligible, open to multiple interpretations. Judges and lawyers are better prepared. They are not trialled when someone wrongly accused is found innocent.

But, hopefully, misadventures can help clarifying tasks. That was one of the outcome of the introspection of various scientific associations after the misadventure of L'Aquila. Experts and technicians study, honestly inform, propose and implement in the realm of their specific knowledge. They may also decide on matters which do not interfere with other considerations. But for conflicting situations, they are not the main deciders, the one who have to weight all the variables, all the values, in terms of resources and timing and following their conviction and/or policy. In that world, there is no best solution. At the highest level, this is the job of the politicians, elected by the population for this high and heavy responsibility: represent them. Others chose to bear this responsibility too and face the consequences of misjudgments.

How these considerations fit with the responsibility of the technicians discussed in section 6.1? The main answer is information, the duty to inform about the consequences of choices. And, ultimately, the refusal to participate to operations irreconcilable with professional ethics.

6.3 *Priorities*

As in many circumstances related to conservation, it is impossible to generalise and state principles which, because they are inapplicable or just not going to be followed, are deemed to have only a paper life. This is the difficulty or inadequacy of applying European ideas about conservation, born in a context of stone architecture, with a big A, not "bicycle sheds" (Pevsner 1943), to vernacular architecture, wooden or earth buildings, Shinto temples and other foreign traditions, which lead to recognising that there are other values besides *material authenticity* of the fabric. But that does not mean that this was just a mistake, a fetishist and morbid attachment to old flesh which can be forgotten. It seems sometimes just convenient to use it as a pretext to achieve goals which are better left hidden. Invoking the importance of skills, perpetuation of traditions but use modern materials and tools, or apply it to context where tradition has disappeared borders on hypocrisy.

Presumptively, archaeological sites should be the place where most respect is given to the material dimension, to the authenticity of the evidence. Even Lassus & Viollet-le Duc (1843) were pleading for this cause! But this is not the case, far from it. Archaeological sites are not the place where *authenticity* can be expected. This was observed at multiple occasion by the author and noted with equal bewilderment by Houbart (2020). The observer sometimes doubt about the reality of what he sees. Cinema prepared us to enter this world of suspended disbelief. But up to a point: some elements instill suspicion of dishonesty. Sometimes, the law of mechanics seem to have vanished. Sometimes, it is the absence of the traces of age which is suspicious. Sometimes, no clear evidence of corruption is perceived but it is too late, the doubt is there.

The same kind of reasoning can be made for architecture: the feeling of being told a biased story.

Fixing priorities is important. But the infinite variability of situations is rebel to generalisations. A most important point can nevertheless be put in the first position: *transparency*: it attests an intention of honesty and opens the door of accountability (it is also likely to be among the less respected principles in practice). The main point discussed in the paper is then the identification a *character-defining elements*. Not accepting the cultural value of the fabric would make the whole process insignificant for engineers. Beyond these two points, an ordered list of priorities should be drawn on a case by case basis (Figure 8). Possible other points: compatibility, reversibility, retreatability, distinguishability, preservation of the principle of construction, of the material type, of the original material, documentation, dissemination, identification, vulgarisation…

7 CONCLUSION

Engineers have first and foremost the task to understand the structural condition of the structure, evaluate the consequences in terms of risk and, if necessary, to propose action(s) to mitigate the risks. As discussed above, this usually requires a cautious approach, studies, experience.

In the last one hundred years or so, for various reasons: geographical (eurocentric), material (the first buildings for which a cultural importance were often built in stone), sociological (conservation was mostly in the hands of specialists), historical (reaction against the *restorations* of the 19th c.), great importance was given to material authenticity.

The progressive expansion of the concept of heritage, lead to the observation that this vision, outside of the context in which it formed and evolved, was difficult to apply.

For Michael Petzet (Larsen 1995, Michael Petzet ("In the full richness of their authenticity"), former president of ICOMOS,

> "It was certainly a necessary process for us to take heed not only of beautiful outer surfaces or of the appearance of a monument, but rather than to become concerned with material and structure, with the inner fabric that perhaps only the scientist or the civil engineer can explain to us (...). However, we should still be interested in the front as well, although certain exercises in our modern preservation cult seem to have forgotten this."

In many ways, this is a progress; and a logical development of the idea that in matters of care of monuments, it is not possible to have rules of general applicability. Accepting that contexts have their specificity which has to be respected is recognising that each case has its specificity and requires a tailored approach.

But shifting the problem from a geographical to a professional level, it has to be reminded that differences are not only a matter of country, religion, traditions but that various groups in the population have also their own particular relation with heritage. They see and value different features. Architects, engineers, conservators and other *modifiers* of our built environment do not apprehend in the same way. And for many of them and for various reasons, *material authenticity* has still some meaning and value.

That does not mean that all decisions have to be left in their hands and that, only their values should be protected (even if it was possible, groups are not homogeneous). But, conversely, that does not mean that they just have to be silent implementers. They are also stakeholders. Heritage also matters to them, in a very direct and acute way, and not only because it happens to feed them and their family.

Following this line of reasoning, besides the technical duties to which they are contractually binded and besides the ethical duty of designing and implementing solutions coherent with what the democracy decides, they should also consider the duty of transmitting the message of their predecessors in the full richness of its authenticity.

REFERENCES

Ackerman, J. S. (1949). "Ars Sine Scientia Nihil Est" Gothic Theory of Architecture at the Cathedral of Milan. *The Art Bulletin 31*(2), 84–111.

Addis, B. (1994). *The art of the structural engineer*. London: Artemis.

Adorno, T. W. (1991). *The culture industry*. London & New York: Routledge.

Barthes, R. (1957). *Mythologies*. Paris: Seuil.

Becher, B. & H. Becher (1967). *Industriebauten 1830 - 1930. Eine fotografische Dokumentation*. München: Die neue Sammlung.

Benjamin, W. (1980). Das Kunstwerk im Zeitalter seiner technischen Reproduzierbarkeit. In R. Tiedemann and H. Schweppenhäuser (Eds.), *Walter Benjamin – Gesammelte Schriften Band I, Teil 2*. Frankfurt am Main: Suhrkamp. (Translation: Harry Zohn (1968). "The Work of Art in the Age of Mechanical Reproduction". New York:Schocken/Random House).

Binda, L., A. Anzani, & G. Mirabella Roberti (2000). Tall and massive ancient masonry buildings: Long term effects of loading. In *G. Penelis Int. Symp. On Concrete and Masonry Structures*, pp. 273–284.

Borges, J. L. (1939). Pierre menard, author del quijote. *Sur 56*, 7–16.

Borges, J. L. (1980). *Siete Noches (Seven nights)*. Argentina: Fondo de Cultura Económica.

Calabresi, G. (2003). Problemi geotecnici nel consolidamento delle costruzioni di interesse storico. *Quaderni di scienza della conservazione* (3), 19–34.

Calvino, I. (1988). *Lezioni americane. Sei proposte per il prossimo millennio*. Sagi blu. Milano: Garzanti.

Choay, F. (1996). *L'allégorie du patrimoine*. Paris: Editions du seuil.

Choay, F. (2009). *Le patrimoine en questions - Anthologie pour un combat*. Paris: Editions du seuil.

Coulomb (1776). Essai sur une application des règles de maximis & minimis à quelques problèmes de statique, relatifs à l'architecture. *Mémoires de mathématque et de physique, présentés à l'académie royale des sciences, par divers savans, & lÃ»s dans fes Affemblées 7*, 343–382.

Dehio, G. (1905). *Denkmalschutz und Denkmalpflege im neunzehnten Jahrhundert: Rede zur Feier des Geburtstages Sr. Majestät des Kaisers gehalten in der Aula der Kaiser-Wilhelms-Universität Strassburg am 27. Januar 1905*. Strassburg: J.H. ED. Heitz.

Dunne, E. (2018). How Xi'an's past became a blueprint for its future. Public memory as a tool of development in Xi'an. *The Diplomat*.

Egglezos, D., D. Moullou, & M. Ioannidou (2013). The role of the geotechnical engineer in archaeological work: The greek experience. In Bilotta, Flora, Lirer, and Viggiani (Eds.), *Geotechnical Engineering for the Preservation of Monuments and Historic Sites*, pp. 359–365.

García Gamallo, A. M. (2003). The evolution of traditional types of building foundation prior to the first industrial revolution. In S. Huerta (Ed.), *Proceedings of the First International Congress on Construction History*, Madrid, pp. 943–956.

Giovannoni, G. (1931a). La restauration des monuments en Italie. In *Conférence d'Athènes sur la conservation artistique et historique des monuments*.

Giovannoni, G. (1931b). Les moyens modernes de construction appliqués à la restauration des monuments. In *Conférence d'Athènes sur la conservation artistique et historique des monuments*.

Houbart, C. (2020). "Reconstruction as a creative act": on anastylosis and restoration around the Venice Congress. *Conversaciones 9*, 39–58.

Huerta Fernández, S. (1990). *Diseño estructural de arcos, bóvedas y cúpolas en España ca. 1500, ca.1800*. Ph. D. thesis, ETSAM, Madrid.

Hugo, V. (1832). *Notre-Dame de Paris. 1482* (2d, definitive ed.). Paris: Eugène Renduel.

ICOMOS (2003). Principles for the analysis, conservation and structural restoration of architectural heritage.

ISO 13822:2010. *Bases for design of structures – Assessment of existing structures*. ISO, Geneva, Switzerland.

Iwasaki, Y. (2013). Characteristic elements of authenticity of heritage structures and conservation for integrity in geotechnical engineering. In *ATC19 Workshop Part II, UNESCO*, Paris.

Iwasaki, Y., M. Fukuda, M. Iizuka, R. McCarty, T. Nakagawa, & V. Ly (2021). The authenticity and integrity of soil and foundation of bayon temple in angkor, cambodia. In *IOP Conf. Ser.: Earth Environ. Sci. 856. Second International Conference on Geotechnical Engineering*.

Iwasaki, Y., A. Zhussupbekov, & A. Issina (2013). Authenticity of foundations for heritage structures. In *Proceedings of the 18 International Conference on Soil Mechanics and Geotechnical Engineering*, Paris, pp. 3111–3114.

Klinke, A. & O. Rena (2002). A new approach to risk evaluation and management: Risk-based, precaution-based, and discourse-based strategies. *Risk Analysis (Society for Risk Analysis) 22*(6), 1071–1094.

Kérisel, J. (1975). Old structures in relation to soild conditions. *Géotechnique 25*(3), 433–483.

Kérisel, J. (1985). The history of geotechnical engineering up until 1700. In *11th International Conference on Soil Mechanics and Foundation Engineering (San Francisco)*, pp. 3–93. Balkema. Golden Jubilee Volume.

Lambe, T. W. (1973). Predictions in soil engineering. *Géotechnique 23*(2), 149–202.

Larsen, K. E. (Ed.) (1995). *Nara conference on authenticity in relation to the World Heritage Conservation*. UNESCO, ICCROM, ICOMOS.

Lassus, J.-B. & E.-E. Viollet-le Duc (1843). *Notre-Dame de Paris. Projet de restauration. Rapport.* Paris: Imprimerie de Mme de Lacombe.

MacIntyre, A. (1998). *A short history of ethics. A history of moral philosophy from the Homeric Age to the twentieth century*. London & New York: Routledge.

Montanari, T. (2015). *Privati de Patrimonio*. Torino: Giulio Einaudi editore.

Nervi, P. L. (1997). *Savoir construire (original: Costruire corretamente. Milano:Hoepli, 1995. Translated from the Italian by M. Gallot)*. Paris: Ëdition du Linteau.

Paquet, P. (1931). Le ciment armé dans la restauration. In *Conférence d'Athènes sur la conservation artistique et historique des monuments*.

Peck, R. (1969). Advantages and limitations of the observational method in applied mechanics. *Géotechnique 19*(2), 171–187.

Pevsner, N. (1943). *An Outline of Western Architecture*. Pelican.

Porter, K. A. (2003). An overview of peer's performance-based earthquake engineering methodology. In *Ninth International Conference on Applications of Statistics and Probability in Civil Engineering (ICASP9) July 6–9, 2003, San Francisco*.

Przewłócki, J., I. Dardzińska, & J. Świniański (2005). Review of historical buildings' foundations. *Géotechnique 55*(5), 363–372.

Riegl, A. (1903). *Der moderne Denkmalkultus, sein Wesen, seine Entstehung*. (Translation: K.W. Forster and D. Ghirardo, 1982. The modern cult of monuments: its character and origin. Oppositions 25, 20–51). Wien und Leipzig: W. Braumüller.

Rudofsky, B. (1964). *Architecture Without Architects: A Short Introduction to Non-Pedigreed Architecture*. Garden City, New-York: Doubleday & Company.

Ruskin, J. (1849). *The Seven Lamps of Architecture*. London: Smith, Elder, and co.

Saint Augustine (1991). *Confessions*. Oxford: Oxford University Press. (translated by Henry Chadwick).

Sandel, M. (2012). *What money can't buy. The moral limits of markets*. Allen Lane.

Scolobig, A., R. Mechler, N. Komendantova, W. Liu, S. Dagmar, & A. Patt (2014). The co-production of scientific advice and decision making under uncertainty: Lessons from the 2009 L'Aquila earthquake, Italy. *GRF Davos Planet@Risk 2*(2), 71–76.

Settis, S. (2004). *Futuro del "Classico"*. Torino: Giulio Einaudi editore.

Settis, S. (2017). *Architettura e Democrazia. Paesaggio, città, diritti civili*. Torino: Giulio Einaudi editore.

Skempton, A. A history of soil properties, 1717–1927. In *11th International Conference on Soil Mechanics and Foundation Engineerin (San Francisco)*, pp. 95–121. Balkema. Golden Jubilee Volume.

Smars, P. (2018). Adobe constructions in Yún-Lín county, Taiwan. In I. Wouters, S. Van de Voorde, I. Bertels, B. Espion, K. De Jonge, and D. Zastavni (Eds.), *6th International Congress on Construction History 2018, July 9–13 2018, Brussels (Belgium)*, pp. 1237–1244.

Smars, P. (2022). Documenting traditional architecture in Yún-Lín county. In K.-C. Yang and P. Smars (Eds.), *Challenges of traditional architecture protection. Belgium, Bulgaria, India, Taiwan, UK. (provisional title)*. (to be published).

Smars, P. & T. Patrício (2016). Ethical questions around structural interventions in archaeological sites. In K. Van Balen and E. Verstrynge (Eds.), *10th International Conference on Structural analysis of historical constructions (SAHC 2016), Sept. 13–15, Leuven, Belgium*, pp. 986–993.

Smars, P., T. Patrício, M. Santana, & A. Seif (2010). Archaeological site of Baalbek, Structural Risk Management Strategy. In *International Conference on Disaster Management and Cultural Heritage, 12–14 December, Thimphu (Bhutan)*.

Stanley-Price, N. (2009). The reconstruction of ruins: Principles and practice. In A. Richmond and A. Bracker (Eds.), *Conservation: Principles, Dilemmas and Uncomfortable Truths*, pp. 32–46. London: Elsevier/Butterworth Heinemann.

Torroja Miret, E. (2000). *Razon y Ser de los Typos Estructurales* (10° ed.). Textos Universitarios. Madrid: Consejo Superior de Investigaciones Cientificas.

UNESCO (2021). *Operational Guidelines for the Implementation of the World Heritage Convention*. Paris: UNESCO.

Verstrynge, E. & D. Van Gemert (2018). Creep failure of two historical masonry towers: analysis from material to structure. *Int. J. Masonry Research and Innovation 3*(1), 50–71.

Yang, K.-C. (2018). World heritage in the middle of political conflict: A discourse analysis of mixed reactions in response to the nomination of sites of Japan's Meiji industrial revolution. In N. Prothi Khanna and S. Burke (Eds.), *ICOMOS 19th General Assembly and Scientific Symposium "Heritage and Democracy"*.

Zevi, B. (1948). *Saper vedere l'architettura. Saggio sull'interpretazione spaziale dell'architettura*. Torino: Einaudi.

Zhu, Y. (2017). Use of the past: negotiating heritage in Xi'an. *International Journal of Heritage Studies*. DOI: 10.1080/13527258.2017.1347886.

Geotechnical Engineering for the Preservation of Monuments and Historic Sites III – Lancellotta, Viggiani, Flora, de Silva & Mele (Eds)
 ISBN 978-1-032-35998-4

Form and construction. The domes of the Baptistery and Santa Maria del Fiore in Florence

P. Matracchi
Department of Architecture, University of Florence

ABSTRACT: Each building requires specific surveys and appropriate instrumental investigations which are often indicated during the course of the analyses. The activities that form part of the architectural and structural diagnostics must aim to identify the actual construction site activities, in operational terms, that over time have generated and altered the architectural structure.

In this regard, the dome of the Baptistery of Florence is particularly significant. It has been the subject of countless studies that have provided different interpretations of the structure. The initial highly idealised approach provided schematic constructive structures. The gradual refinement of the inspections defined a completely different construction concept of the dome from that established up until recently.

This new information also had a significant impact on the interpretation of Brunelleschi's dome, which showed in an even more compelling way characteristics similar to that of the Baptistery.

1 INTRODUCTION

A feature common to some octagonal segmented domes is a curved intrados profile. A curvature whose radius is almost equal to 4/5 of the diameter of the base was identified in the corners of the dome of Florence Baptistery (Aminti 1996). A pointed profile was also used for the dome of Orcagna's Tabernacle (1359), constructed inside the church of Orsanmichele in Florence (Pisetta & Vitali 1996).

Giovanni di Gherardo da Prato, supervisor of the dome of Santa Maria del Fiore, in 1426 traced on a parchment a drawing with comments criticising the use of the 4/5 radius of the diameter of the base that Brunelleschi was creating in the dome. The precision surveys of the intrados revealed deviations from this curvature (Fondelli 2004). However, for long stretches of the intrados profile, there are strong similarities between the dome of the Baptistery and that of Santa Maria del Fiore (Giorgi 2004, pp. 161–163).

The fact that the curve drawn through 4/5 radius of the diameter was a well-known and widespread device was confirmed by Francesco di Giorgio Martini who mentions it in his Treatise (1479–1486) (Trattati di architettura 1967). The same curvature is also found in the dome of the church of Santa Maria delle Grazie al Calcinaio in Cortona, again designed by Francesco di Giorgio; but the dome was executed by the Florentine architect Pietro di Norbo between 1509 and 1514 (Matracchi 1992).

The two larger domes were built with the initial part in stone and the rest in brick; the other two are entirely made of brick, and stone slabs were added to extrados of that of the tabernacle.

But a mere geometric comparison, also taking into account the building materials, is clearly paradoxical (Figure 1). The different scale of magnitude makes it necessary to acquire more information on the specific construction solutions adopted, which always characterize each building. The construction choices were not only affected by the size of the buildings, but also by local traditions, solutions that could have been brought by builders who grew up elsewhere. The construction phase experimentation that went on above all in buildings of great importance is no less important, and

DOI 10.1201/9781003329756-8

there was no shortage of accidental factors that could influence the choices, especially in long-term construction sites.

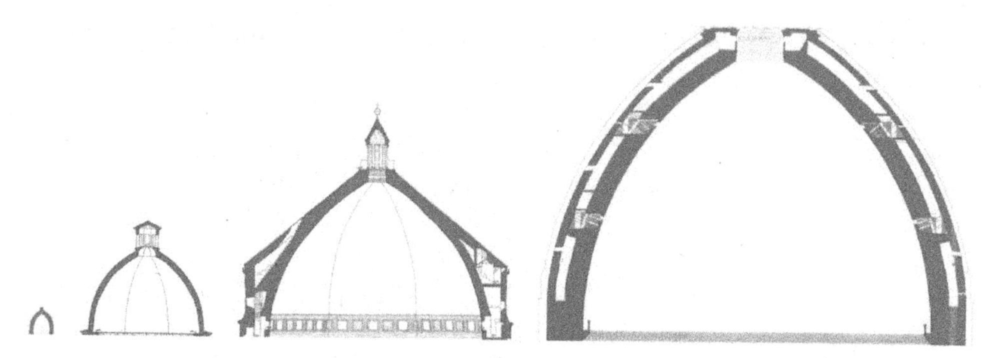

Figure 1. Comparison (from left) at the same scale between the domes of Orcagna's tabernacle, the church of Santa Maria delle Grazie al Calcinaio, the Baptistery and Brunelleschi's dome.

In this regard, a particularly interesting case is Florence Baptistery which, despite having been studied by scholars from the mid-19th century, continues to be analysed in order to obtain important information, to the extent that the constructive structure is evaluated with a profoundly renewed concept.

2 THE BAPTISTERY DOME

Knowledge and in-depth analysis of complex and large dome structures, like that of the Baptistery of Florence, is necessarily a long process linked to the available investigative tools, the conceptual approach guiding the study and the aims to be achieved. If the channel of interest pursued is that of conservation, some construction choices that link the parts to the whole in a holistic vision attentive to the actual act of construction that takes place at the construction site and results in the architectural structure must be explored in-depth. The actions that architecture undergoes over time, which can lead to alterations that are sometimes even substantial, but difficult to identify, are just as important.

The studies of the structure of the Baptistery can be divided into two different seasons. The first is marked by an intuitive approach, which represented the constructional device according to schematic and idealised drawings, which were at times even partial; these representations might have been accurate in graphic terms, but in general they were based on few measurements. They represented what they thought they had understood.

In the drawing by Heinrich Hübsch (1862) the ribs of the dome are envisaged extending until they join the diagonal ribs and the corners of the octagon are solid, but both hypothesis are wrong; the intuition of the box structure at the base of the dome is interesting, with dashed arches indicating the vaulted roofs of each compartment, but there are clear approximations in the dimensions of the masonry (Degl'Innocenti 2017).

The portion of the baptistery dome, depicted by Josef Durm (1887), shows some of the main construction elements, even focusing attention on less striking details. The wall offsets of the extrados at the base of the dome, the transversal walls that divide the various spaces of the garrets, the perimeter wooden tie of the dome, and the vaulted structures supporting the pyramidal roof of the dome are shown. The detail of the internal wall offset was also added to the attic wall.

Auguste Choisy (1899) depicted the dome in a more schematic way in an axonometric drawing, but it has the advantage of showing the entire building highlighting the difference between the section on the sides and on the corner of the octagon; the first shows a considerably hollow structure below the dome due to the presence of columns on the ground floor and the matronea on the level above; the second highlights the massive section of the corners of the octagonal plan, but does not show the hollowing of the wall mass in the area adjacent to the haunches of the dome.

Walter Horn (1943) effectively combines a half section with an axonometric section of one side of the octagon; he thereby highlights the partition walls that flank the dome, the partition walls of the matroneum and, further below, the columns of the internal perimeter (Degl'Innocenti 2017, p. 96).

The drawings prepared by Hübsch, Drum and Choisy are examples, others have also been prepared, of an approach that includes real elements, intuition and interpretations, according to a vision still limited to the arrangement of presumed key construction features, which depict a structural scheme of the building with a high degree of abstraction.

The 1970s marked the start of a new season of studies conducted with increasingly extensive and accurate survey campaigns, more attentive to the analysis of the architectural structure as a whole, the materials and phenomena resulting in the alteration of the latter and the structures. Over time increasing attention was paid to understanding the relationship between the parts and the whole of the architectural device and assessing the construction features, which can vary even in situations of apparent symmetry. The wealth of information collected over time was gradually refined until a cognitive and interpretative synthesis was found based on architectural diagnostics.

So the drawings indicate the portion of the dome, starting from the springer, built with stone in rows that were initially horizontal and that incline from a certain height (Pietramellara 1973). The brick section of the dome is indicated from the wooden tie. In one plan, executed just below the roofs, the partition walls that stand on the extrados of the dome are clearly represented, with the extension of the rampant vaults that support part of the conical roof. Observations on the continuity of the rows of masonry in the garrets, which place the dome, the transversal walls and the external perimeter wall in a close constructive relationship, are particularly important. This structure is considered a "double dome", or a "cellular structure", highly significant concepts, the implications of which have not been explored in-depth, and to which we shall return later on.

Further investigations highlighted the uniqueness of the vertical connections between the monolithic columns, two on each side of the octagon, except for the one where the scarsella is present, and the partition walls of the matronea above, which continue into the garret until becoming the ribs of the dome. In the corners the ribs form a Y-shaped fork that creates the hollowing of the angular buttress at the level of the dome; at the level below, the corner buttress is crossed by a narrow passage necessary to pass through the matroneum, on the ground floor it is solid (Rocchi Coopmans de Yoldy 1996a). The corner buttresses of the east side where the spiral staircases are located are a case of their own. Therefore a system of orthogonal ribs abuts the extrados of the dome at the sides of the octagon, which stand on the intermediate partition walls of the matronea and, further below, on the monolithic columns placed alongside the perimeter walls. The columns had independent foundations with respect to the wall perimeter, with which they slightly overlap. It should be added that the transition between the parallel bearing walls of the matronea and the columns is borrowed from transversal stone lintels, whose modest span between the column and external wall prevented them from being damaged. The Y-shaped ribbing corresponds to the corner pillars.

The construction organisation of the partition walls, in terms of their arrangement and dimensions, nevertheless shows considerable originality linked to the architectural structure of the Baptistery. The three entrance doors imposed a greater intercolumniation than the intervals between the columns themselves and the corner pillars. Indeed, in each of the remaining sides the spaces between the piers are essentially the same. On the sides of the matroneum above the entrances the central mullion windows are considerably wider than those beside them. The accentuation of the central mullion window was repeated on the sides without entrances, moving the parallel transversal walls to the outer edge or the wider lintels below resting on columns; this was achieved by the lintels being wider than the partition walls. Whereas, on the sides of the matroneum above the entrances the gap between the central space and the side spaces reduces, moving the partition walls towards the centre of the side (Figure 2). As a result, the pillars corresponding to the partition wall of the matroneum are not in axis with the columns below (Rocchi Coopmans de Yoldy 1996a, p. 45).

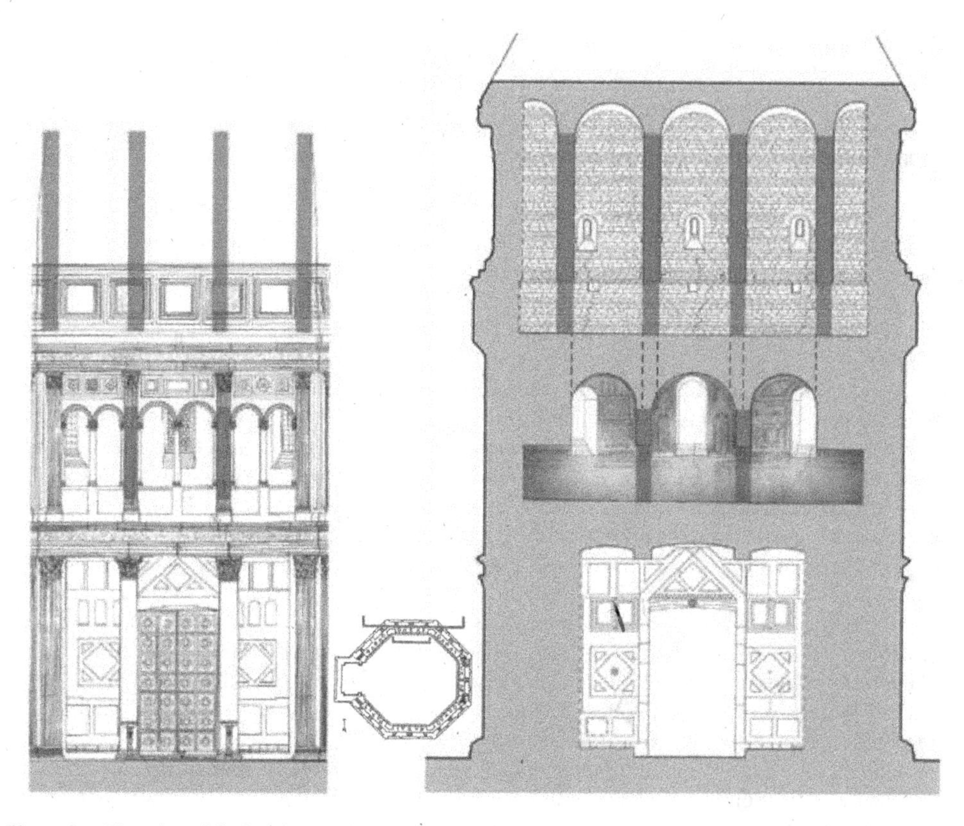

Figure 2. Elevation of the interior north side (left) indicating (in red) the position of the parallel bearing walls; longitudinal section of the north side (right) executed in the intermediate space highlighting the partition walls (in red); the section intercepts the ground floor, the matroneum and the garret.

But these changes in position between structures at different levels are more accentuated in the garret where three compartments of more or less the same width are created on each side (except for the north-west side where the central compartment, with respect to the side ones, is wider by around 30 cm).

Taking the north side into consideration, the two parallel bearing walls of the garrets have an accentuated inward shift, and are positioned outside of the partition walls of the matronea by almost 1/2. The misalignment of the partition walls is borrowed from a sort of wall base which is around 120 cm thick, placed between the transversal barrel vaults of the matronea and the bearing surface of the dome. The difference in the positioning of the structures in this case cannot be visually perceived as it is disguised by the dome, but it is clearly highlighted by the overlapping of the plans and the longitudinal section of the north side. The strong vertical misalignment between the columns and the partition walls is the result of the choice to create three compartments with a similar width close to each side of the dome.

The scarsella side also required specific adaptations, where the continuity of the perimeter wall structure is interrupted. At the base of the dome there are three open windows on each side of the octagon; but on the side above the scarsella it was decided during the works not to complete the two side openings and to infill them (Giorgi & Matracchi 2017). The intention was at least to reinforce the base of the dome, which however exceeded the thickness of the frontal arch of the scarsella. The transverse partition walls and the external perimeter wall completely extended beyond this arch (Giorgi 2004, p. 158). Such impressive structures must not have been placed directly on the vault of the scarsella, but more likely on the relieving arches which rest on the massive adjacent corner pillars of the octagonal perimeter (Figure 3).

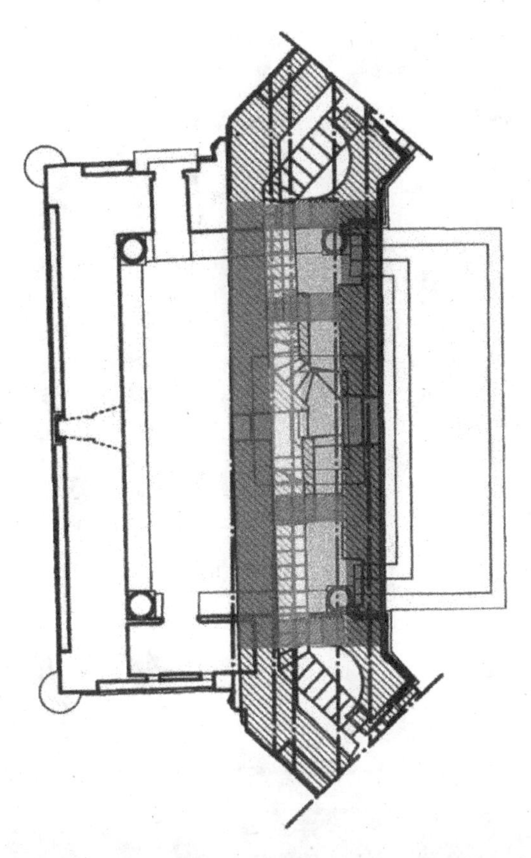

Figure 3. Detail of the plan of the Baptistery with the position of the walls weighing on the vault of the scarsella: the dome (blue), the parallel bearing walls (green), the external wall of the octagonal perimeter (red), (by Giorgi 2004).

Calculation simulations have shown that if the dome had been placed directly on the original vault it would not have been stable due to the small section of this vault (Miceli & Papi 2004). Despite the probable use of reliving arches, the scarsella must have been a vulnerable area from a construction perspective; so the groin vault on columns placed in the corners of the scarsella was added (Figure 4). The geophysical investigations corroborated the existence of two vaults, confirming the use of bricks for the added vault, while the earlier one is made of stone (Morelli 2004).

The further study campaign recently promoted by the Opera di Santa Maria del Fiore produced additional significant cognitive and diagnostic results on different aspects, including some regarding the structures.

The first section of the dome was made with stone ashlars placed in horizontal rows, which with progressive overhangs create the pointed curve of the intrados, while the extrados is vertical (Figure 5). This creates a characteristic reduction of the thickness at the base of the dome: at the springer it is 107 cm, at the summit of the vertical wall it is 130 cm. This level marked the start of a section with offsets on the extrados, until the dome achieves the constant thickness of approximately 103 cm. In the transition zone with wall offsets, the rows of stone start to incline; above, in the north web of the dome, there are small recesses on the extrados which evidence the use of wooden elements used to position the inclination of the rows. The curvature of the intrados established by the centring also generates the inclination of the rows where it is decided to align the rows with the centre, or centres, of curvature of the intrados. This is not the case of the Baptistery dome, where the line of the intrados curve can be distinguished from inclination of the rows.

Figure 4. Transversal partition wall above the scarsella; the crack (in yellow) sealed by vertical plates is visible; the extrados of the dome can be seen on the left.

Figure 5. Detail of the dome at the level of the springer. The photo shows the horizontal rows of the masonry positioned with a progressive overhang on the intrados, as shown by the vertical line of the laser.

These latter, in correspondence to the recesses, incline less than the radius of curvature of the intrados of the dome.

The inspection of the top part of the garrets showed that the transversal bearing walls are not adjacent to each other, but are connected with the dome in continuous rows. This results in unique feature: to obtain a better connection between the inclined rows of the dome and the partition walls, for a stretch the rows of the latter are also arranged on an inclined plane (Figure 6), and then become horizontal again, creating masonry in continuous rows with the external wall of the Baptistery (Giorgi & Matracchi 2017, p. 198).

Figure 6. Detail of a partition wall (right) and the extrados of the dome (left) of the Baptistery; the intersection between the vertical and horizontal red lines of the laser highlight the part of the partition wall with masonry in inclined rows perfectly connected with those of the dome.

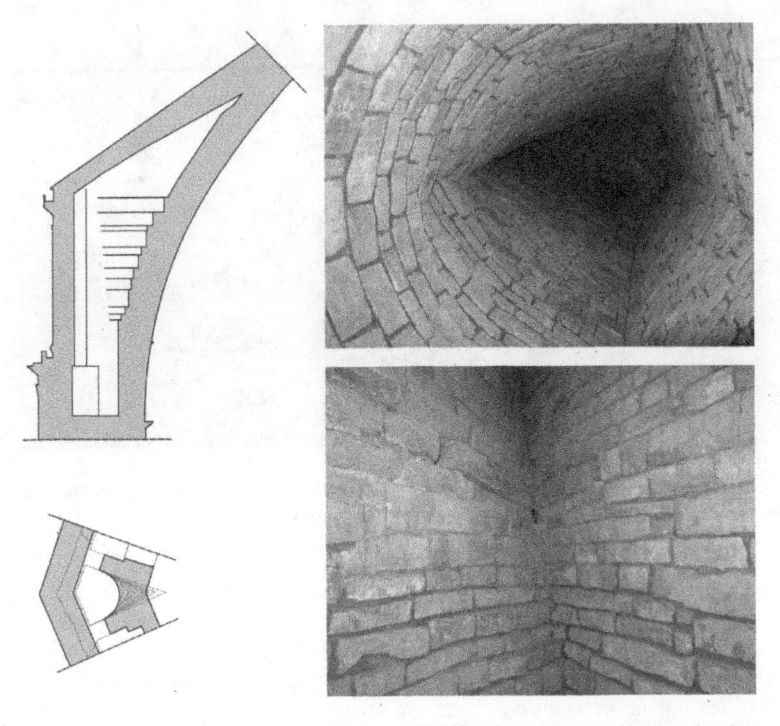

Figure 7. Detail of a corner space at garret level; plan and section with images showing the changes in shape and the continuity of the masonry rows.

The internal wall of the corners of the octagon is shaped as the arc of a circle, a form that follows that of the corners accommodating spiral staircases alongside the east door. In correspondence to the circle arcs of the corners, the thickness of the dome reduces with respect to the sides of the octagon; the dimensions, starting with the south-east side and proceeding in a clockwise direction up to the north-east one, are as follows: 91/90/75/100/89/88 cm. Continuing upwards, the semicircular part becomes triangular, assuming an even more acute angle until creating the Y-shaped fork of the corner ribs. These modifications of the corner spaces are created by maintaining the continuity of the rows for the entire perimeter, except clearly for the base area with the garret crossing points (Figure 7).

The dome, partition walls and external wall therefore create a box structure with continuous rows including the three spaces on each side and the corner compartments. The intention to create a box structure is further confirmed by the fact that this device was extended as far up as possible by means of a constructive expedient created in the corners between the rampant vaults, transversal partition walls and the external wall. The partition walls are covered by stone rampant barrel vaults, on which part of the conical roof of the Baptistery rests, which abut the perimeter wall of the attic. The aim was to prevent the box structure from being interrupted between the partition walls and the external walls at the springer of the barrel vault, in that the radial ashlars of the latter could not have been placed in continuity with the horizontal masonry rows of the lunette of the external wall. Thereby, also exploiting the fact that the vault is rampant, skewed horizontal rows with a progressive overhang were created, which connect to the transversal partition wall and a good part of the edge of the lunette, raising the box structure almost up to the top of the external wall (Figure 8).

Figure 8. Garret of the Baptistery; detail of the top of the transversal partition wall with the skewed masonry (red) connecting to the edges of the lunette of the outer wall of the attic.

Another distinctive solution linked to the box structure can be found in the external wall: it was reduced to a thickness similar to that of the transversal partition walls (it is just below) and was moved inward creating an overhanging wall offset ending at the bottom with a quarter-circle cornice. The layout with a box structure with symmetrical compartments on each side of the pavilion was pursued with clear intention by the builders, who in order to obtain it had to accept the consequent lack of continuity with the partition walls of the matronea below (Figures 9,10).

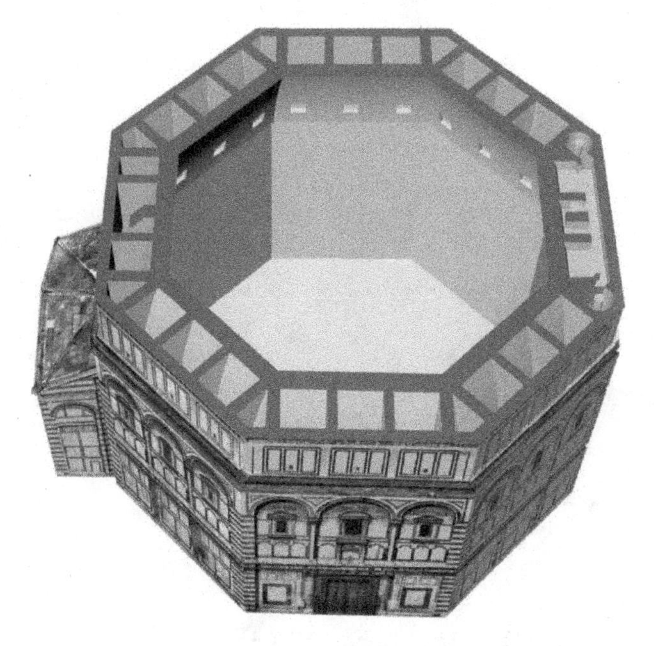

Figure 9. The box structure of the dome at attic level (by Giorgi & Matracchi 2017).

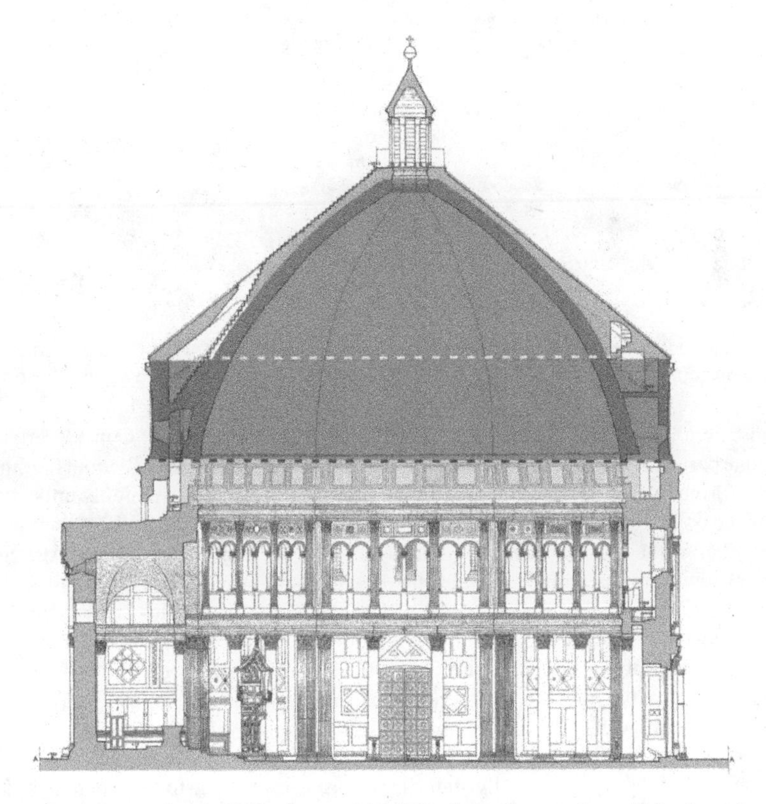

Figure 10. In the dome section the following are highlighted: the first section with horizontal rows (cyan), the box structure part, the final section in brick (red); the wooden tie (yellow dotted line) and the metal tie added in 1514 (brown dotted line) are also indicated (basic drawing Giovannini 1996).

Where the box structure ends, the dome continues with the use of bricks, for the most part salvaged material most likely obtained from the dismantling of the Roman walls of Florence (Giorgi & Matracchi 2017, p. 205). In the transition from the box structure to the single shell structure, a wooden tie is placed along the extrados of the pavilion crossing the transversal partition walls (Figure 11). The original connections visible between the wooden elements show tooth connections with the use of wooden nails. The tests of the state of tension of the wooden tie revealed a modest amount of tensile strength (Negri et al. 2017); the brick dome, with its springer on the inclined row of the stone one, transmits the thrust mainly to the box structure of the stone dome.

Figure 11. The wooden ties intersect at the corners of the octagonal plan of the Baptistery; the image shows the change in material of the dome at the level of the tie: below the masonry is stone, and above brick.

The construction device of the dome is therefore particularly complex. Considering the vertical rise, the stone part and the brick part are almost the same, but if we take into account the surface of the intrados the former would be clearly larger. In the stone part, there are two highly important aspects. The first is the box structure, which extends for 35% of the vertical rise of the dome; the second is the masonry below the box part, which includes an initial section of the dome constructed entirely of horizontal rows and interrupted both by the openings facing onto the inside of the baptistery, and the passages between the compartments of the garret. In view of this construction aspect, the position of the wooden tie at the base of the brick dome formed of a single shell appears justified.

The identification of the box structure profoundly changes how we consider the construction concept of the dome of the Baptistery, which until recently was considered to be a pavilion reinforced by transversal bearing walls, and allows us to understand more fully the position and extent of the system of cracks that has formed.

The dome has limited and essentially stable cracks; this is also due to the metal tie added in 1514 in the wall offset at the base of the attic level (Blasi et al. 2017, pp. 124–125). This tie is highly effective given that it is positioned at the base of the box structure. Moderate traces of cracks have been found in the corners of the intrados of the dome, up to 1-2 cm wide, which are barely visible in the mosaics where they are hidden by plaster, and more evident in the band of the medallions of the saints close to the springer of the dome, where deviations can be observed between the slabs of the stone cladding (Blasi et al. 2017, p. 120).

The masonry masses of the structure below the dome reduce a great deal on the inner sides of the Baptistery, where there are columns on the ground floor and mullion windows on the matroneum level. It should be noted that the partition walls of the garret are at times strongly misaligned with those of the matroneum, as on the north side. It is no surprise that these construction features together have resulted in the largest cracks in the matroneum area (Figure 12).

More extensive cracks have been observed in particular in the sides of the entrance door (up to 4–5 cm wide). On the inner side of north wall of the matroneum there is a highly visible vertical crack, which has been repaired with brick fragments and extends to the barrel vault above and the wall abutting the mullion window. The small column of the mullion window itself in turn generates a crack in the lintel below supported by columns (Giovannini 1996). On the outer sides clad in marble the cracks have been repaired by inserting marble plugs which restore the continuity of the horizontal cornices (Blasi et al. 2017, pp. 121–122).

The first pioneering modelling of the dome of the baptistery with the calculation of the elements dates back to the mid 1990s (Blasi & Papi, 1996). The topic was addressed there after several times with increasingly refined numerical mode salso thanks to the recentavailability of an accurate laser scanner survey (Bartoli et al. 2017). It would be very interesting and appropriate to further characterise the behaviour of the dome with modelling that takes into account the box structure of the dome. It would be just as useful to check the possible effects caused by the vertical misalignment of the partition walls, considering in particular the displacements between the transversal bearing walls of the box part of the dome and those below of the matroneum.

Figure 12. Detail of the outer wall of the north side of the attic; the obvious repaired crack is shown (in red) (by Giovannini 1996).

3 SHARED CHARACTERISTICS OF THE BAPTISTERY DOME AND BRUNELLESCHI'S DOME

For Brunelleschi, the Baptistery dome was an object to study and consider in order to develop construction solutions for the much larger dome of Santa Maria del Fiore. Several construction solutions shared by the two buildings have been observed, such as the use of stone in the lower part and then continuing with brick; the wooden tie revived with anchoring in the ribs; the use of two intermediate ribs in each web of the pavilion. Moreover, the roof partly resting on the rampant vaults of the baptistery could have suggested the creation of a double shell dome (Rocchi Coopmans de Yoldy 1996b).

Leaving aside the many and complex aspects linked to the brick masonry structure of Brunelleschi's dome, the strengthening of the connection between corner ribs and intermediate ribs by means of sub-horizontal arches is a significant innovation (Figures 13,14). The corresponding portion of the external ribbed dome is also strengthened by these sub-horizontal arches. To make the cohesion between such elements more effective, the sub-horizontal arches are made up of a segmental arch in vertical bricks to support the bricks arranged in a homogeneous position to that of the bricks of the ribs and of the external dome (Giorgi & Matracchi 2018). It still needs to be ascertained whether the solution of the sub-horizontal arches with variable sections (larger on the corner ribs side) also arises from the need to conceal the metal ties positioned in the external area of the ribs; indeed, a 24-sided polygonal tie could have had sections transversal to the corner ribs arranged partly inside the sub-horizontal arches (Figure 15). In any case, the corner areas of the pavilion of the dome of Santa Maria del Fiore are a clear construction innovation which creates a unitary system consisting of corner ribs, intermediate ribs and sub-horizontal arches (Rocchi Coopmans de Yoldy 2006).

However, perhaps the most significant constructive aspect that Brunelleschi could have taken from the dome of the Baptistery lies behind all of this. Reference is made to the box structure of the stone part, which Brunelleschi reworked with a double dome, intermediate ribs and stiffening arches close to the corners that reinforce the connection between adjacent ribs and at the same time innervate the external dome (0.80 m thick) which is much thinner than the internal one (about 2 meters thick) (Figure 16).

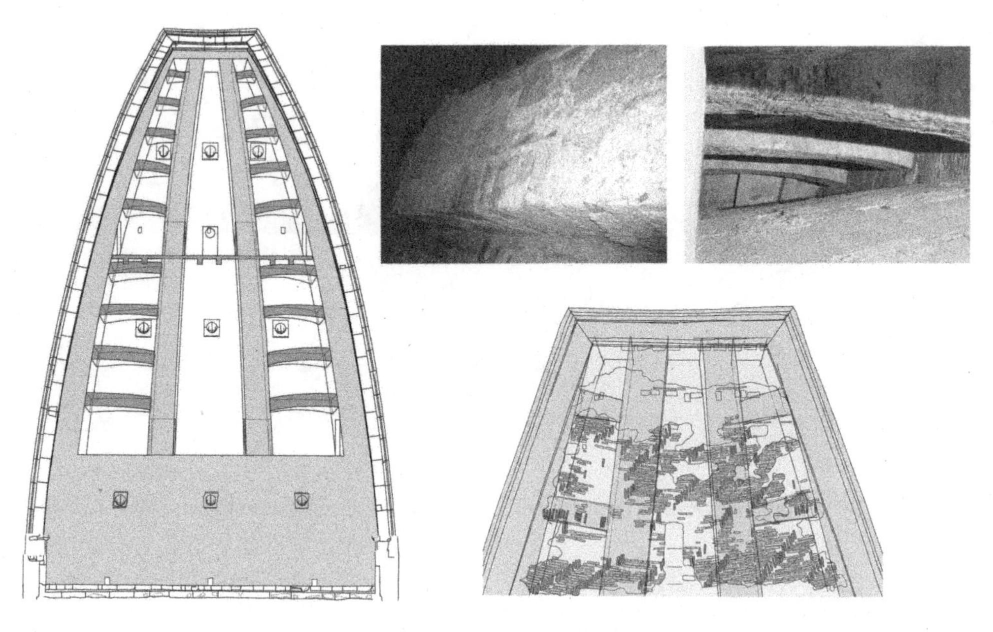

Figure 13. N-NE web of the dome of Santa Maria del Fiore highlighting the ribs and corbel-arches; (left), survey of the masonry structure of the NE outer dome highlighting the bricks (in red) of the corbel-arches placed vertically; (top right) details of corbel-arches with vertical bricks visible in the lower part (by Giorgi & Matracchi 2006).

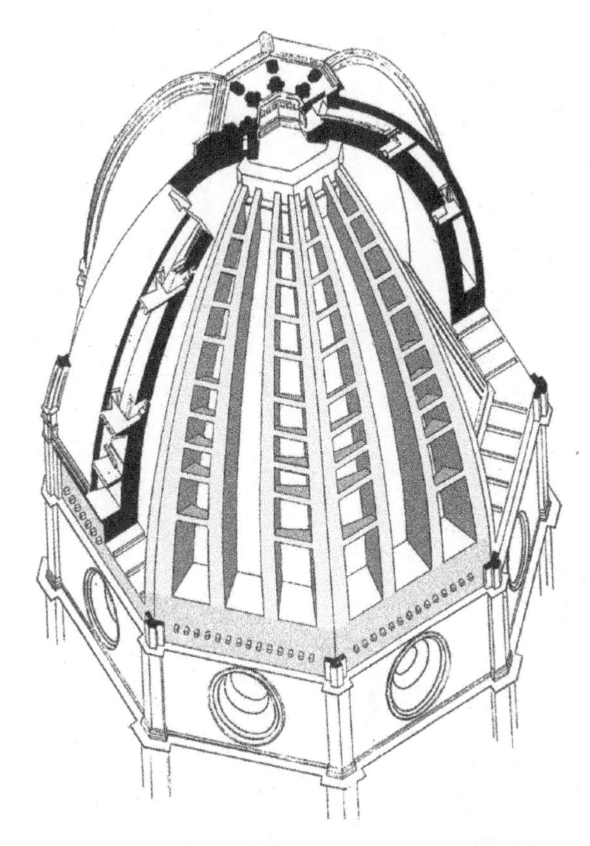

Figure 14. Axonometric diagram of the dome of Santa Maria del Fiore highlighting the connections between the corbel-arches and ribs (by Coompans de Yoldi 2006).

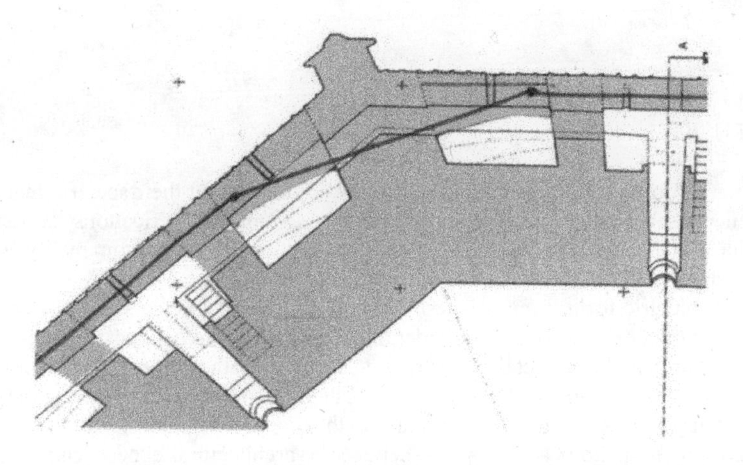

Figure 15. Excerpt of a plan of the dome with the corner ribs, intermediate ribs and corbel-arches (in yellow); hypothesis of a tie (in red) placed transversely to the corner ribs and corbel-arches.

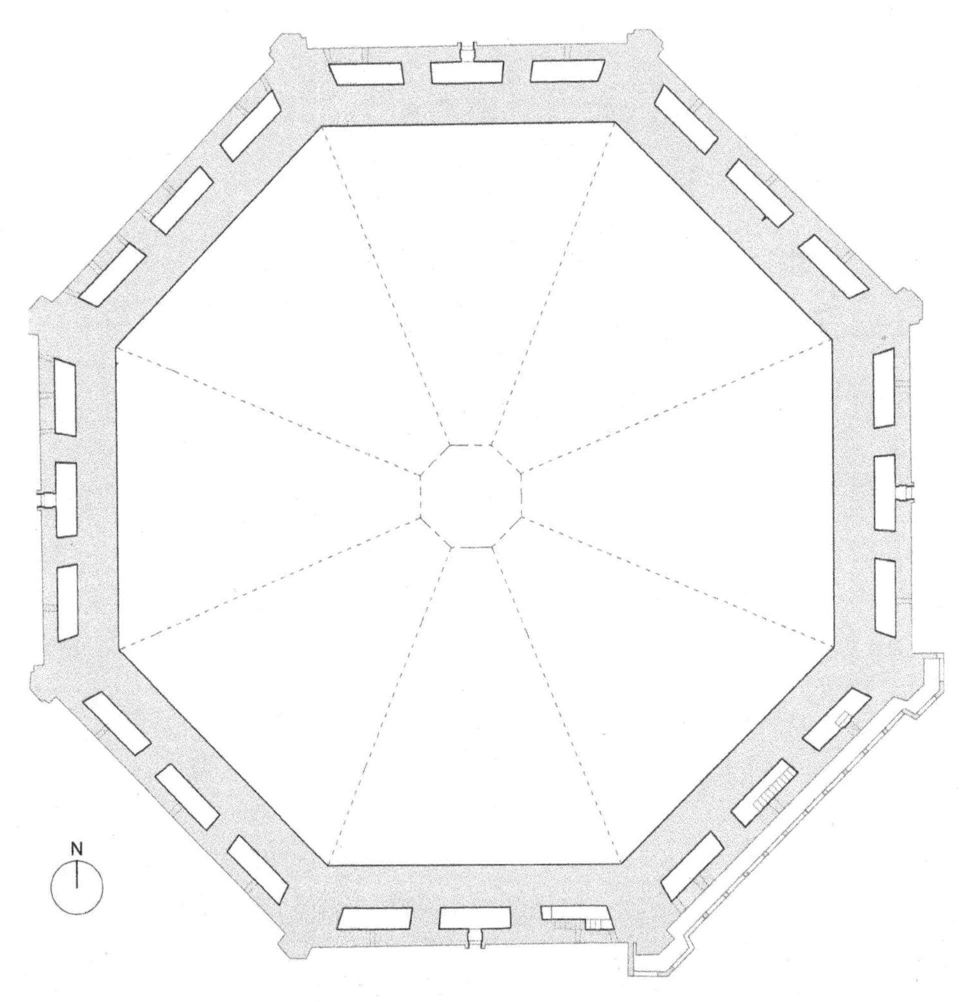

Figure 16. The plan of the Brunelleschi's dome shows the box structure formed by the inner dome, ribs and the outer dome.

4 CONCLUSIONS

The cognitive approach to the buildings should seek to understand their specific features, which are almost never self-evident, without making generalisations, simplifications, or assuming preconceived ideas that could place our vision of the construction device far from reality, in a (perhaps refined) metaphysical dimension.

The mere geometric form, even if based on highly detailed surveys, should be considered an indispensable but not exhaustive tool. In order to understand the actual construction elements it is necessary to identify and interpret the construction activities in the context of the construction site practices, or the construction sites that followed one another over time, considering the consequent specific operating methods and the choices made in the production and installation of the materials used. The construction site is a place par excellence of architectural experimentation, where the countless possible options reach executive synthesis. And this is not exempt from rethinking, adaptations and sometimes contradictory choices.

In the case of the Baptistery of Florence, there was no shortage of adaptations on the scarsella side, where the continuity of the octagonal perimeter wall is interrupted. A further vault was therefore added to the scarsella space and, on this same side, a decision was made during the works to close some openings which at the base of the dome would have alternated with mosaics depicting saints. The vertical misalignments between the transversal partition walls on different levels are no less significant, in particular between the mezzanine level and the garret, due to determining architectural choices and the decision to create compartments of a similar width on each side at the base of the dome.

It is customary for an ancient building to be made up of a complex system of construction layouts, which reflect the complexity of the construction history, with conditions of possible local vulnerability. This is made even more complex by the long timeframe of ancient architecture, which introduces further variables as a result of successive transformations, sudden external factors such as earthquakes, or the degradation of the materials themselves.

Knowledge of complex ancient buildings is almost always recognised as a process. This is well exemplified by the identification of the box structure in the stone part of the dome of the Baptistery of Florence, which today provides us with a profoundly different construction concept than the one considered until recently. This finding also has obvious repercussions in the interpretation of Brunelleschi's dome, which in turn is entirely recognised as a box structure with an outer dome innervated by sub-horizontal arches close to the corners of the octagonal pavilion. The recurring interpretation difficulties posed by the construction layouts of the architecture therefore require due caution in making diagnostic interpretations and in the consequent conservation projects, which could be based on reductive or even misleading information.

REFERENCES

Aminti, P. 1996. Rilievo e determinazione della curvature della cupola, analisi sui rapporti dimensionali tra principali elementi strutturali interni, in Giuseppe Rocchi Coopmans de Yoldy (ed.), *S. Maria del Fiore. Piazza, Battistero, Campanile*, Firenze, Il Torchio, p. 104.

Bartoli et al. 2017. Bartoli G., Betti M., Monchetti S., Modellazione numerica ed analisi strutturale del Battistero di San Giovanni a Firenze, in Francesco Gurrieri (ed.) *Il Battistero di San Giovanni. Conoscenza, Diagnostica, Conservazione*, Firenze, Mandragora, pp. 135–157.

Blasi C. & Papi R. 2004. Geometria e struttura, in Giuseppe Rocchi Coopmans de Yoldy (ed.) *S. Maria del Fiore. Piazza, Battistero, Campanile*, Firenze, Il Torchio, pp., 125–127.

Blasi et al. 2017. Blasi C., Ottoni F., Coïsson E., Tedeschi C., Battistero di San Giovanni in Firenze. Note su dissesti, lesioni e catene, in Francesco Gurrieri (ed.) *Il Battistero di San Giovanni. Conoscenza, Diagnostica, Conservazione*, Firenze, Mandragora.

Choisy A. 1899. Histoire de l'architecture, Paris, Gauthier-Villars, T. II, p. 602.

Degl'Innocenti P. 2017. Misurare, disegnare, conoscere: dai rilievi del San Giovanni alle ipotesi storico costruttive, in Francesco Gurrieri (ed.), *Il Battistero di San Giovanni. Conoscenza, Diagnostica, Conservazione*, Firenze, Mandragora, pp. 90, 96.

Durm J. 1887. *Die Domkuppel in Florenz und die Kuppel der Peterskirche in Rom: zwei Grossconstructionen der italienischen Renaissance*, Berlin, Ernst & Korn, plate II.

Fondelli, M. 2004. La Cupola di Santa Maria del Fiore. Antiche testimonianze e nuove ricerche sperimentali, in Giuseppe Rocchi Coopmans de Yoldy (ed.), *S. Maria del Fiore e le chiese fiorentine del Duecento e del Trecento nella città delle fabbriche arnolfiane*, Firenze, Alinea Editrice, pp. 343–358.

Giorgi, L. 2004. Il sistema voltato del Battistero, in Giuseppe Rocchi Coopmans de Yoldy (ed.), *S. Maria del Fiore* e *le chiesefiorentine del Duecento e del Trecento nella città delle fabbriche arnolfiane*, Firenze, Alinea Editrice.

Giorgi L. & Matracchi P. 2006. Santa Maria del Fiore, facciata, corpo basilicale, cupola, in Giuseppe Rocchi Coopmans de Yoldi, *S. Maria del Fiore. Teorie e storie dell'archeologia e del restauro nella città delle fabbriche arnolfiane*, Firenze, Alinea Editrice, pp. 316–317.

Giorgi & Matracchi 2017. Le murature a cassone alla base della cupola del battistero e altri aspetti costruttivi, in Francesco Gurrieri (ed.) *Il Battistero di San Giovanni. Conoscenza, Diagnostica, Conservazione*, Firenze, Mandragora.

Giorgi L, & Matracchi P. 2018. I mattoni del Brunelleschi. Osservazioni sulla Cupola di Santa Maria del Fiore, in *Costruire in laterizio*, 2018, 176, pp. 53–61.

Giovannini, P. 1996. L'apparecchio murario e il rivestimento interno del piano terra e del matroneo. Materiali, tecniche di lavorazione e caratteristiche di messa in opera, in Giuseppe Rocchi Coopmans de Yoldy (ed.) *S. Maria del Fiore. Piazza, Battistero, Campanile*, Firenze, Il Torchio, pp. 87–89.

Matracchi, P. 1992. *La chiesa di Santa Maria delleGrazie al Calcinaiopresso Cortona e l'opera di Francesco di Giorgio,*Cortona, Calosci, pp. 18–19.

Miceli E. & Riccardo Papi R. 2004. Sulla statica del battistero alla luce delle nuove indagini e delle interpretazioni dei dati sperimentali, in Giuseppe Rocchi Coopmans de Yoldi (ed.) *S. Maria del Fiore e le chiese fiorentine del Duecento e del Trecento nella città delle fabbriche arnolfiane*, Alinea Editrice, Firenze, pp. 177.

Morelli A. 2004. Indagini geofisiche non invasive sulla volta della scarsella del Battistero, in Giuseppe Rocchi Coopmans de Yoldi (ed.) *S. Maria del Fiore e le chiese fiorentine del Duecento e del Trecento nella città delle fabbriche arnolfiane*, Alinea Editrice, Firenze, pp. 167.

Negri et al. 2017. Negri M., Fellin M., Ceccotti A., I legni della cupola del Battistero, in Francesco Gurrieri (ed.) *Il Battistero di San Giovanni. Conoscenza, Diagnostica, Conservazione*, Firenze, Mandragora, pp. 170–173.

Pietramellara C. 1973.*Battistero di S. Giovanni a Firenze*, Firenze, Polistampa, pp. 20–21, 33.

Pisetta, C. & Vitali G. M. 1996. Nuove acquisizioni sul tabernacolo di Andrea Orcagnaattraverso il rilievo interpretativo, in Diane Finiello Zervas (ed.), *Orsanmichele a Firenze*, Modena, Franco Cosimo Panini, p. 382.

Rocchi Coopmans de Yoldy, G. 1996a. Il Battistero di San Giovanni. Lo svolgimento della fabbrica, in Giuseppe Rocchi Coopmans de Yoldy (ed.) *S. Maria del Fiore. Piazza, Battistero, Campanile*, Firenze, Il Torchio, pp. 43, 47–48.

Rocchi Coopmans de Yoldy G. 1996b. Brunelleschi e il Battistero, in Giuseppe Rocchi Coopmans de Yoldy (ed.) *S. Maria del Fiore. Piazza, Battistero, Campanile*, Firenze, Il Torchio, pp. 64–65.

Rocchi Coopmans de Yoldy G. 2006. Il cantiere del complesso di Santa Maria del Fiore dall'epoca arnolfiano-giottesca a quella brunelleschiana, in Giuseppe Rocchi Coopmans de Yoldi (ed.)*S. Maria del Fiore e le chiese fiorentine del Duecento e del Trecento nella città delle fabbriche arnolfiane*, Alinea Editrice, Firenze, p. 265.

Trattati di architettura 1967. *Trattati di architettura, ingegneria e arte militare*/Francesco di Giorgio Martini, Corrado Maltese (ed.), Livia Maltese Degrassi (tr.), Milano, Edizioni il Polifilo, T. I, p. 92.

Geotechnical Engineering for the Preservation of Monuments and Historic Sites III – Lancellotta, Viggiani, Flora, de Silva & Mele (Eds)

Understanding the mechanical history of the burial monument of the Kasta tumulus at Amphipolis, Greece: A tool for documentation and design of restoration strategy

D. Egglezos
Geoper S.A. Heraklion, Crete, Greece

ABSTRACT: This paper provides an interpretation of the historical pathology of the Kasta burial monument in Amphipolis, Greece. The main goal is the documentation of its mechanical history, i.e. the recognition and interpretation of mechanical events which have left an imprint (damage) on the structure of the monument, via appropriate geostatic analyses at historical stages linking the damage to a sequence of specific mechanical causes (correlation with cause and time sequence of mechanical events). Mechanical historical analyses require data on changes in the geometry of the monument (building phases), changes in the strength of building materials (e.g. ageing), and significant mechanical stresses from the external environment (earthquakes, landslides, etc.) deriving from a variety of scientific fields. This type of analysis highlights the vulnerabilities of a monument's design and allows a reliable estimate of its response to future charges, constituting a powerful tool for taking appropriate protection and restoration measures.

1 INTRODUCTION

From the engineer's point of view, standing monuments are a very interesting case. The cumulative effect of the physical and mechanical events that have affected them during their long history is reflected in their current form and state of preservation. Their image is a "mechanical" historical chronicle, in symbolic "mechanical" script, with "ideograms" in the form of structural damage (fractures, cracks, loss of mass, permanent displacements, etc). For the engineer who undertakes the design of the protective measures and the planning of the restoration of the monument, the above data set as a basic scientific challenge the reading (decryption) of the monument's mechanical history with a double aim: on the one hand the documentation of the history of the monument and on the other the substantial understanding of its mechanical function as a basic condition a) for highlighting of any vulnerabilities and b) designing appropriate measures for its protection and/or restoration.

More specifically, in the context of the present paper, the identification and the interpretation of the mechanical stress (historical pathology) of the burial monument of the Kasta tomb of Amphipolis, Serres, is attempted. The main goal is the documentation of its mechanical history, ie the recognition and interpretation of mechanical events with an imprint (damage) on the structure of the monument. Through this process, an attempt is made to highlight the vulnerabilities in the ancient design of the monument and the (new) design of its long-term protection through effective restoration proposals.

The burial monument of the Kasta tomb is a very interesting case of a complete standing monument. Until its discovery it was completely unknown both archaeologically and historically.

At the present stage, the monument is in the process of documentation, implementation of protection measures and planning of restoration interventions for its presentation to the public. In light of the above, reading its mechanical history is the cornerstone of the development of a rational restoration policy.

2 THE BURIAL MONUMENT

2.1 *The discovery*

The burial monument in the Kasta tumulus in Amphipolis is part of a monumental complex that consists of the tumulus itself, a significant geotechnical work of antiquity, discovered by archaeologist D. Lazarides (1964), its imposing retaining wall almost 500m in circumference (Lazarides 1965; Lefantzis 2014), the burial monument and the surviving part of the foundation of a structure on the top of the tumulus which is interpreted as a burial marker or *sema,* (Lazarides 1972; Lefantzis 2013), located approximately 30 meters above the base of the tumulus (Figure 1). The burial monument extends along a chord of the circular plan of the tumulus. The original terrain in which the monumental complex was built consisted of two hills divided by a col. The ancient construction work made use of these two hills to create the extensive circular tumulus with the minimum fill (Syrides et al. 2016, 2017).

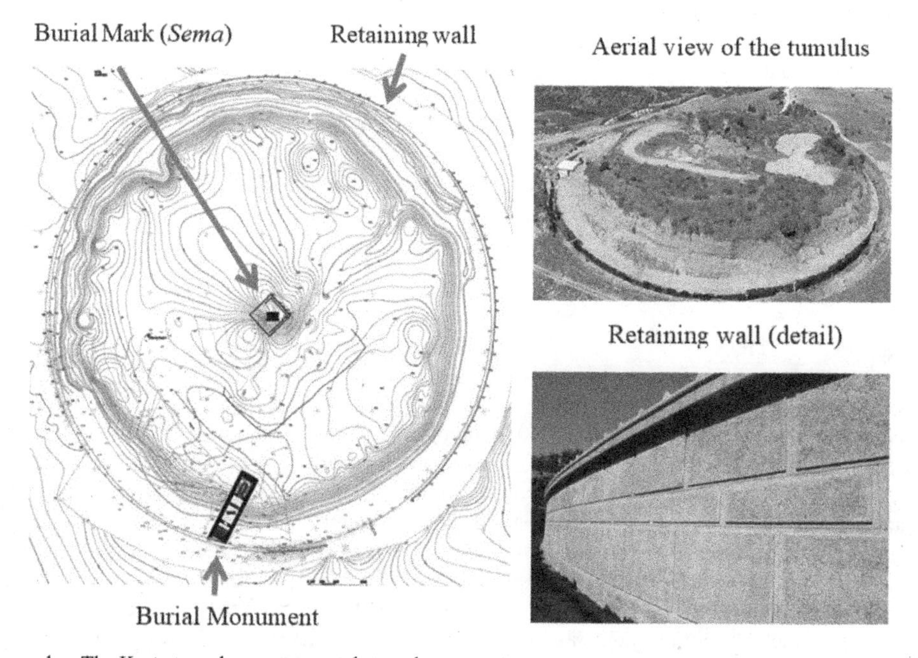

Figure 1. The Kasta tumulus monumental complex.

The monument (Figure 2) came to light during relatively recent archaeological excavations in the area of the Kasta tomb, as part of the excavation activities of the Ephorate of Antiquities of Serres in August 2014, under the supervision of archaeologist Katerina Peristeri. The process of uncovering all the areas of the burial monument was a complex technical work which was assigned to the author by the competent Ephorate of Antiquities and implemented during the period September – November 2014. The uncovering required the archaeological excavation of the exterior and particularly the interior of the monument, which lasted until mid-November 2014. Alongside the excavation, the temporary internal consolidation of the monument was carried out; this was completed in late December 2014.

During the course of the excavation, there were continuous on-site updates of the mechanical documentation of the monument (geometry, structure, materials and properties). An instrumental monitoring system set up for this purpose also produced useful data. It should be noted that a temporary shelter was erected to protect the burial monument during its excavation.

Figure 2. The entrance of the burial monument during the excavation (2014).

2.2 *Description of the monument*

The monument, of the barrel-vaulted Macedonian tomb type, consists of four distinct spaces (X1-X4) delimited by four partition walls (D1-D4) (Figures 3 and 4):

1. First space (X1): This is the antechamber of the monument, with vertical parallel walls and a staircase (which also functions as an internal support) descending to the level of the floor of the vaulted structure. The antechamber of the monument starts from the crown of the retaining wall.
2. Wall of the Sphinxes (D1): This separates the antechamber from the first chamber of the vaulted monument (X2). It consists of an opening flanked by pillars with an entablature. The pillars support a horizontal beam that forms the horizontal base of the statues of two sphinxes facing each other. Recent investigations into the structure of the pillars of Wall D1 show that they are fake walls hiding the supports inside (Lefantzis 2019).
3. Second space (X2): This is the first chamber of the monument, approximately 6.0m long. It extends from the Wall of the Sphinxes to the Wall of the Maidens.
4. Wall of the Maidens (D2): This separates the first chamber of the monument from the second (X3). It consists of an opening framed by two marble pedestals supporting two statues of maidens (korai), a pillar and an entablature. Above the entablature, a transverse wall (tympanum) of poros stones has been erected, in contact with the vaulted roof. On the west side of the transverse wall is an opening, resulting from the deliberate removal of stone blocks in antiquity.
5. Third space (X3): This is the second chamber of the monument, approximately 3.0m long, which has a smaller internal opening (approximately 4.30m). It extends from the Wall of the Maidens to the Wall of the Marble Door. It has a mosaic floor depicting the abduction of Persephone by Pluto, god of the underworld.
6. Wall of the Marble Door (D3): This separates the second chamber (X3) from the third (X4). It consists of an opening for the marble door between two pilasters, each formed of two adjoining

rectangular pieces of marble, with an entablature. Above the entablature is a tympanum of poros stone blocks, in contact with the vaulted roof. On the west side of the tympanum is a small rectangular hole, resulting from the deliberate removal of stone blocks in antiquity.

7. Fourth space (X4): This is the third chamber of the monument, approximately 6.0m long, with an internal opening measuring approximately 4.50m. The floor of the monument consists of poros stone blocks, while in its central part there is a deeper artificial trench, measuring approximately 4.0 X 2.10m and about 3.0m deep. The trench contains a cist grave. The geostatic investigation by the author has shown that the trench for the cist grave predated the construction of the monument and largely determined its longitudinal axis (Egglezos 2016).
8. End wall of the monument (D4): This is the north external transverse wall, which separates the fourth chamber from the external transitional embankment. Thorough architectural documentation of the structure and construction phases of the burial monument is presented in the corresponding architectural study.

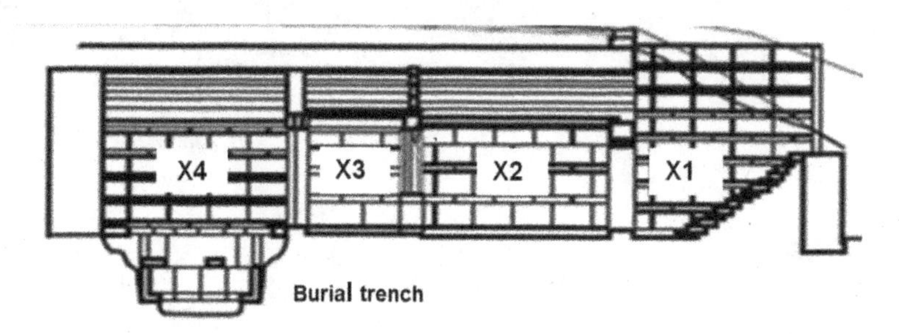

Figure 3. Longitudinal view of the burial monument and the burial trench (Drawing M. Lefantzis).

2.3 *Static function of the monument*

The construction of the monument is intended to support the external (vertical and lateral) geostatic loads from the soils of the transitional lateral embankment and the natural soil on the one hand, and from the overlying fill forming the tumulus on the other. To this end, the side walls act as gravity walls of suitable thickness to bear horizontal earth pressure, supported in places by partition walls and further stabilized by the oblique load transferred to their crown by the arch. The arch is a typical archway of voussoirs (which are in compression due to the pressure each exerts on the other), transferring the mainly vertical geostatic load and the self-weight to the crown of the side walls. As a structure, it is stabilized by the external load of the earth, provided of course that the size of the geostatic load does not exceed the limit of the compressive strength of the vault material, and that support on the vertical side walls is ensured.

The foundation of the monument can be considered as a continuous foundation, of insufficiently documented thickness, set on well-compacted fill which lies on the highly stiff natural soil (dense silty sand) and can safely bear high vertical pressures and reduce settlement to acceptable limits. The floor of the monument (Chambers X2 and X3) also functions, to some extent, in a stabilizing way as a continuous slab, since it consists of hard foundation mortar and gravel, which helps to prevent the internal convergence of the side walls. An exception is the floor of Chamber X4, which is differentiated due to the cist grave.

3 IMPORTANT CONSTRUCTION AND MECHANICAL DATA

For an in-depth understanding of the mechanics of the burial monument, its structural features should be taken into account: these relate to the bonding of its individual parts, the use of the monument in ancient times, the sealing and the internal anthropogenic fill in a later historical

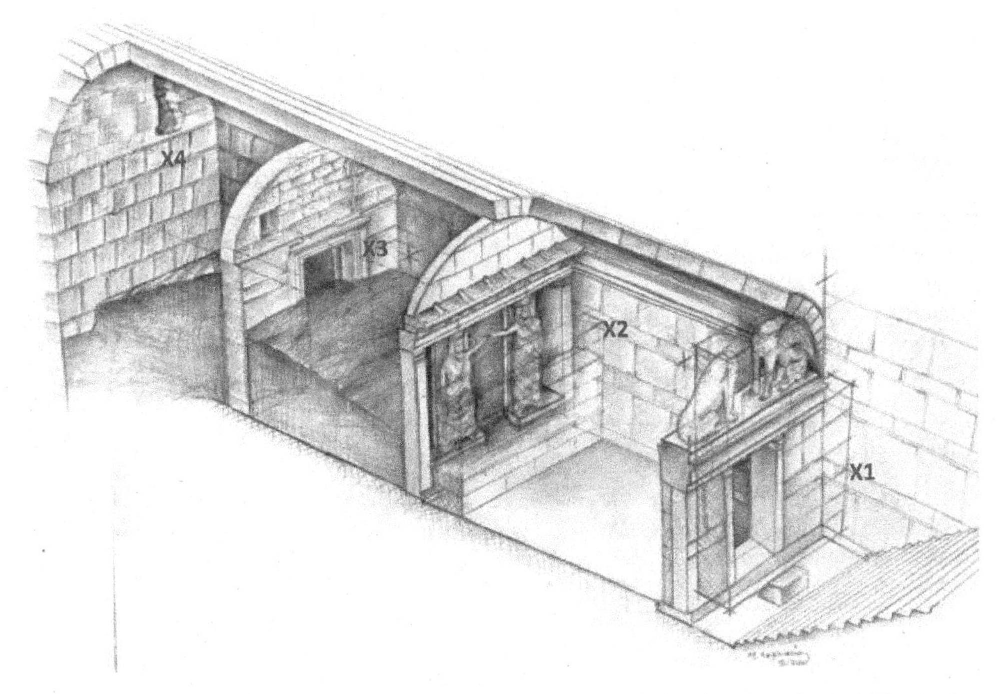

Figure 4. Perspective view of the burial monument. Drawing made during the excavation by M. Lefantzis.

period, and the discovery of metal joints in the marble blocks of the walls and in the vault of the arch. These issues are examined below:

3.1 *Bonding*

The following observations can be made regarding the bonding of the transverse walls with the main structural parts of the monument (walls and arch):

- The possible bonding of partition wall D1 with the longitudinal walls and the vault of Chamber X2 is unclear (Lefantzis 2019).
- Partition wall D2 is bonded with the longitudinal walls through the architrave, and with the arch through the tympanum. The bonding of the wall with the vaulted roof, according to M. Lefantzis, took place at a later construction phase.
- The front of partition wall D3 (marble door pillars) is bonded with the marble revetment of Chamber X3 and connected to the entablature of wall D2 through the surviving slab roof. The rear of the wall (initial phase) is engaged with the longitudinal walls of Chamber X4.

3.2 *Metal joining elements*

Metal joining elements are widely used both in the side walls and in the vault. This reinforcement may reflect the early period of construction of the burial monument (which is dated by the excavator to the last quarter of the 4th century BC, Peristeri 2016) and the concern of the ancient engineer for the stability of the structure.

A further fact worth noting is the great length of the vault (approx. 20m), with a relatively large internal opening (4.50m) between the side walls, and the strong geostatic shear load from the fill of the particularly extensive mound.

The voussoirs are longitudinally connected by lead-coated cast-iron dowels of rectangular cross-section (indicative dimensions, according to Lefantzis 2019: width x height x length = 2.5x7.5x14cm). The marble structural parts of the walls of the monument are also connected horizontally with metal joints. The use of metal joining elements is documented in other vaulted Macedonian monuments (Zambas 2016).

3.3 *Internal fill*

During its excavation, the monument was found to be full of anthropogenic fill in well-formed layers, consisting of relatively loose silty sand. The level of the fill varied along the length of the monument:

- Spaces X1 and X2 are/were completely filled
- Chamber X3 is/was largely filled
- X4 contained a moderate amount of fill (sloping surface from the crown of transverse wall D3 to the middle of Wall D4)

In addition to the fill, the monument was sealed by two sealing walls of poros stone, one at the entrance of Chamber X2 (in front of Wall D1 with the Sphinxes) and the other at the entrance of Chamber X3 (in front of Wall D2 with the Maidens). The sealing walls were dismantled during the work of uncovering the interior of the monument.

3.4 *Use of the monument:*

Systematic construction phases, structural damage and ancient repairs are observed, indicating systematic use before its final filling.

3.5 *Documented construction phases*

The burial monument in the Kasta tomb, according to architect M. Lefantzis (2019), includes characteristic historical building phases, associated with the mechanical history of the monument. A detailed analysis of the construction phases is presented in the architectural documentation study. The main building phases from its construction until its recent archaeological discovery can be summarized as follows:

3.5.1 *Pre-existing phase of the burial monument*

Construction of the cist grave in Chamber X4 (the cist grave predates the burial monument, into which it was then incorporated, Egglezos 2016). The time interval between the construction of the cist grave and the construction of the vaulted monument has not been clarified at present (whether it is a question, for example, of months or a century).

3.5.2 *Main construction phase*

3.5.2.1 Initial construction

i. Construction of the burial monument with the four spaces (X1-X4) using the cut and cover system: Space X1 is covered by a porch or portico, while Chambers X2-X4 are roofed with a continuous vault. The monument seems to have remained open (unfilled) to human presence, as indicated by ancient repairs in various places inside.
ii. Wall D1 consists of the entablature with the sculptures of the sphinxes which stand on a (currently undocumented) form of support.
iii. The tympanum of Wall D2 has probably not yet been constructed (Lefantzis 2019).
iv. In Chamber X3 the marble revetment of the side walls has not yet been constructed.
v. In Wall D3 the marble door has not been constructed for the sealing of Chamber X4.

3.5.2.2 Significant internal building modifications after the initial construction. It is not yet clear when these were implemented.

vi. The fake walls (pillars) are created to cover the columns in Wall D1.
vii. The tympanum of Wall D2 is constructed.
viii. Marble revetment is applied to the side walls of Chamber X3.
ix. Wall D3 is modified for the installation of the marble door.
x. An internal roof covering of Chamber X3 is constructed, using four parallel rectangular marble slabs supported on the entablature of Walls D3 and D2.

3.5.2.3 Significant internal anthropogenic damage. It is not yet clear when this occurred.

xi. the marble door and the floor around the perimeter of the burial case are destroyed
xii. the burial case is excavated/looted in antiquity before the inner filling.

3.5.2.4 Filling – sealing of the burial monument (Figure 6) The time at which it occurred has not yet been documented.

xiii. Holes are pierced in the west parts of the tympana in Walls D2 and D3
xiv. The roof slabs in Chamber X3 are destroyed (except the west one, which bridges the gap between the holes in the entablatures of D2 and D3)
xv. The spaces of the burial monument are filled with an increasing surface level of fill from the interior of the monument to its entrance.
xvi. Two internal sealing walls (partitions transverse to the longitudinal axis of the monument) are constructed in front of Walls D2 and D1 (Figure 5a).
xvii. The area outside the entrance is filled. The monument and (locally at least) the retaining enclosure are concealed

Figure 5. a. The Wall of the Maidens, as found, before the dismantling of the sealing wall, b. detail showing the vertical fracture in the west Maiden.

3.6 *Preservation status – structural damage to the burial monument*

The structural damage to the monument falls under two headings: a) changes to the original geometry of the monument, and b) damage due to the exceedance of the characteristic strength (compressive, tensile, shear and/or a combination of the above).

The damage is connected to the actions that caused it using a computational methodology including geostatic soil - structure interaction analyses in different stages of mechanical history. The main damage to the burial monument can be summarized as follows:

3.6.1 *Fractures – Cracks:*

- Fractures with loss of mass in voussoirs of the vault in Chambers X4 and X3, in characteristic longitudinal zones: western and eastern axis of the burial monument (Figure 6).
- Cracks, with virtually no mass loss, in voussoirs in Chamber X2.
- Fractures in the marble of the side walls of Chambers X3 and X4 and Wall D4, and cracks in the walls (especially the west wall) and the vault of Chamber X2. The phenomenon appears with increasing intensity and frequency from Chamber X2 to Chamber X4 (Figure 6).
- Vertical fractures in the central part of the horizontal beams of the entablature of Walls D1, D2 and D3 of the burial monument (architraves, lintels, etc.) (e.g. Figure 5a).
- Vertical fractures in the heads of the Maidens of Wall D2 (which support the overlying architrave). In particular, the front part of the head of the east Maiden has been completely cut off from the face, while the west head has extensive fractures in vertical planes and is preserved thanks to the binding applied during its uncovering (Figure 5b).
- Fractures in the lower part of the external pilasters and the walls of the front part of Wall D3.
- Systematic vertical cracks of the strong plaster mortar of entrance area X1, located symmetrically on the walls of this space. These fractures extend from the surviving crest to the point where the vault begins.
- Presumed fractures and detachment of the plaster mortar of the intrados of the arch, as a result of geometric deformations of the vault.

Figure 6. Structural damages to the voussoirs and walls of chamber a. X4, b. X2.

3.6.2 *Geometrical changes – permanent displacements:*

- Convergences and wall distortions: all the vertical side walls show a curvature in both the horizontal and in the vertical plane, towards the inside of the burial monument (plate function). The phenomenon occurs with increasing intensity from Chamber X2 to Chamber X4.
- Rotation of the partition walls: there is a strong rotation of D2 and the front (south) part of D3 (towards the entrance), which dates to a later construction phase (Lefantzis 2019). Wall D3 is in contact with D2 through the surviving roof slab (the roof is assumed to have initially consisted of four slabs, Lefantzis 2019). The rear (north) side of D3, which belongs to the initial construction phase, shows a small rotation. Wall D4 also shows a rotation towards the inside of the monument. In contrast to Walls D2, D3 (front) and D4, Wall D1 (with the Sphinxes) does not appear to show any rotation (at least noteworthy).
- The vault has longitudinal subsidence along its central part, gradually increasing from Chamber X3 (indicative reported subsidence 1.5cm in the middle of the chamber (Lefantzis 2019), to Chamber X4 (indicative reported subsidence 4.5cm in the middle of the chamber).
- The crown of the arch at the entrance of Chamber X2 has a displacement in the longitudinal axis greater than 1.0cm. with respect to Wall D1.

- Between the west and east axis of the longitudinal walls of the monument there is a differential settlement (historical settlement of the west axis with respect to the east) gradually increasing towards Wall D4.

The structural damage is presented in the plans of the documentation study of the burial monument (Lefantzis 2019). The side sections in Figure 7 provide a collective overview of the main structural damage to the monument.

3.7 *Undocumented areas of the burial monument*

Despite the systematic investigation into the structural documentation of the monument, there are still issues that need further clarification. These include, for example, the preservation status of the external surface of the monument, the geotechnical data of the transitional embankment, the bearing conditions of the entablature and sphinxes in Wall D1, and the systematic recording of the metal joints. To overcome these uncertainties, reasonable assumptions have been made for the monument simulations during the mechanical history analyses.

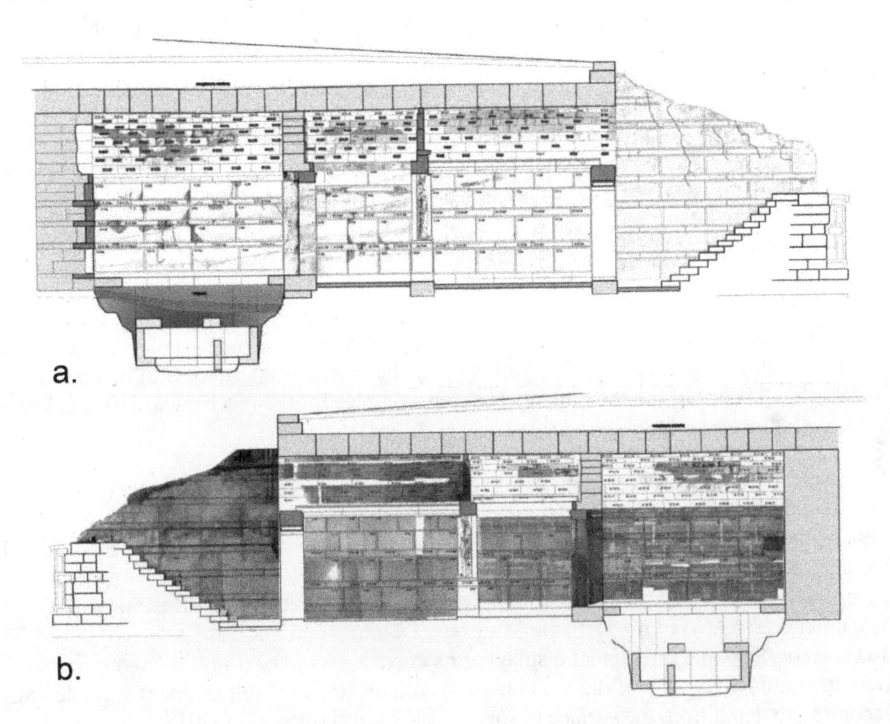

Figure 7. Longitudinal section views with the main structural damage to the burial monument a. west side b. east side (Lefantzis 2019).

3.8 *Actions after the unveiling and temporary fixing of the burial monument*

From the excavation of the monument to the present date (late 2021), a series of documentation and protection actions have been carried out. These include, briefly, studies for upgrading its protection, earthworks and rain protection works, as well increased instrumental monitoring. The monument has already been included in a financial program, the main goal being its systematic documentation, protection and (partial) restoration (Egglezos 2019c; Lefantzis 2019; Sotiropoulos et al. 2020b; Papadopoulos 2020).

4 ANALYSES OF THE BURIAL MONUMENT IN STAGES OF MECHANICAL HISTORY

During the elaboration of the study investigating the structural pathology of the burial monument,. analyses were carried out in stages of mechanical history in order to sufficiently correlate and document the observed structural damage with the actions that caused it. These analyses also made it possible to reach important conclusions on the response of the monument under complex loading conditions. The organization, implementation and utilization of these analyses are presented in detail as follows:

4.1 *Methodology*

The methodology for interpreting the mechanical stress of the monument includes appropriate geostatic analyses of the interaction of the soil environment and the monument at different historical stages, in order to adequately document the structural damage to the monument (correlation with cause and time sequence of mechanical events).

These analyses require data on changes in the geometry of the monument (construction phases), changes in the strength of building materials (ageing, reduction of strength due to mechanical stress, etc.) and documentation of significant mechanical actions from the external environment (earthquakes, earth pressures, etc.).

Data from various scientific fields were used for the elaboration of the historical mechanical stages: archaeology, history, architecture, topography, geology, geotechnics, geophysics, seismology, strength of building materials, instrumental monitoring, etc.

4.2 *Available data*

The discovery of the burial monument in the Kasta tumulus is a happy coincidence, with the available data allowing, with the proper utilization, the more or less successful reconstruction of the mechanical history of the monument. The main sources for the mechanical historical data are listed below:

- Architectural documentation study of the burial monument, recording its structural pathology and main construction phases (Lefantzis 2019)
- Geotechnical survey to determine the geotechnical conditions in the immediate vicinity of the burial monument (Sotiropoulos et al. 2020a)
- Topographical survey of the Kasta Tomb provided by the Directorate of Restoration of Ancient Monuments (DAAM) of the Hellenic Ministry of Culture and Sports
- 3-D scan of the burial monument displaying its current geometry (Prof. V. Pagounis)
- Geophysical investigation of the Kasta mound with detection of soil layers of variable stiffness according to depth from the surface (Tsokas et al. 2016, Tsokas et al. 2018)
- Structural investigation with laboratory tests to determine natural and mechanical properties of the building materials of the monument (Directorate of Research and Technical Support for Restoration Projects - DETYMEA 2019)
- Archaeological data for the assessment of the sequence of the mechanical phases of the monument, e.g. construction, internal filling, etc. (Peristeri 2016)
- Data from instrumental monitoring measurements, confirming assumptions about the mechanical properties of the individual structural parts of the monument (Egglezos 2015)
- Collection of historical seismicity data from the available literature, identifying events with an estimated impact on the monument (Egglezos 2019c)
- Study for the interpretation of the structural pathology of the burial monument with computational documentation of the damage to it, linking each type of damage with a cause (Egglezos 2019c). This essentially formed the basis of the present work.

The successful selection and synthesis of data from the above reports is the framework within which the analyses are developed in mechanical history stages: the data allow the most accurate simulation of the monument and its environment, the selection of appropriate (possibly time-varying) values for the natural and mechanical properties of the monument's materials, the assessment of relatively realistic historical actions, and the description of the evolution of the structure over time (building phases, etc.).

4.3 *Important historical mechanical stages of the burial monument*

The following historical mechanical phases emerge from the management of the available data for the burial monument:

- Digging of the burial trench in the natural hill
- Construction of the burial trench and the cist grave (in X4)
- Construction of the main vaulted monument with parallel construction of the transitional embankment
- External artificial filling of the monument
- Building changes inside the monument
- Internal filling and sealing of the monument
- Gradual degradation of building material properties (ageing)
- Stress from a strong earthquake (620 AD, with an estimated magnitude of 6.8 on the Richter scale and epicenter distance about 15 km) (Papazachos and Papazachou 2003, Pavlidis et al. 2016)
- Loading of the monument with additional external anthropogenic fill (loose soil accumulation) in the decades before the discovery of the monument
- Earthworks during the excavation phase (August 2014) (resulting to OC effects on the soil surrounding the monument)
- Use of construction machinery near the arch of the monument (2014)
- Removal of internal fill in phases (2014)
- Construction of temporary consolidation devices (struts-pillars) in phases (2014)
- Accidental events (loading of Chamber X4 by surface slip upstream) (2014)
- Earthworks for earth pressure relief (Chamber X4) (2014)
- Strong precipitation action (2015)
- Construction of a medium-term (conforming to normative requirements) redesigned fastening device (2019)

4.4 *Geostatic analyses for the documentation of the structural damages of the monuments of the Kasta tumulus*

The geostatic calculations include 2-D analyses with finite and distinct elements in four characteristic cross-sections of the monument. These cross-sections correspond to the central cross-section of the Chambers X2, X3 and X4 and a simplified indicative simulation of its longitudinal central cross-section. The use of 2-D analyses is acceptable (certainly for the cross-sections of the chambers), as conditions of plane strain prevail (taking into account the width-to-length ratio of the burial monument). Regarding the longitudinal section, a case in which conditions of plane strain clearly do not strictly apply (3-D problem), the following measures are adopted in order to make the simulation realistic:

1. Vertical and horizontal springs are inserted, corresponding to the contribution of the vault (as an arch) and the side walls as a bearing element of the vault (compression only vertical elastic springs and friction horizontal springs).
2. The stiffness of the partition walls is proportionally reduced in order to effectively represent the real 3-D geometry, which includes large openings.

The simulation of the geometry of the burial monument is based on the architectural data (Lefantzis 2019). The soil simulation is based on the geotechnical data produced by the geotechnical investigation in the environs of the monument (Sotiropoulos et al. 2020a). All geomaterials, building materials as well as interfaces (joints between structural elements and interfaces with the surrounding soil) are simulated elastoplastically based on the Mohr-Coulomb and Hoek-Brown failure criteria.

It should be noted that the quality geological strength index (GSI) on the micro-scale of the stone is used for the quantitative assessment of the effects of ageing on the strength of stone blocks and their interface. The GSI values applied to account for the ageing of materials and interfaces at different historical stages of the monument are shown in Table 1. The properties of the geomaterials, the structural materials and their interfaces are presented in Tables 2, 3 and 4 respectively.

The values for the initial (reflecting the construction stage of the monument) normal and shear stiffness of the interfaces are selected based on the relevant (staged) values of the elastic modulus of the materials in contact. The degradation of interface stiffness values, in the following historical stages, is based on Barton's formula for estimating joint stiffness in rockmass (Barton, 1972).

Table 1. Quantification of Ageing effect on structural materials (GSI application in microscale).

Main historical stage	GSI values porous stone		GSI values marble	
	Initial	Residual	Initial	Residual
4th century B.C.	95	85	95	90
Before the internal filling	90	80	90	95
2nd half of 20th c. (extra soil)	80	75	85	90
Before the excavation 2014	75	70	80	85

The corresponding GSI value from Table 1 is used for the application of Barton's formula at each stage of the analysis. The elements of the temporary support structure (struts and columns) were simulated using appropriate linear structural elements.

Table 2. Structural material properties in relation to historical stages of the monument.

	Dolomite POROUS Stones*				MARBLE Stones*			
	σ_{cm}**= 13 MPa				σ_{cm}**= 39 MPa			
GSI_{ini}	95	90	80	75	95	90	85	80
GSI_{res}	85	80	75	70	90	85	80	75
$\sigma_{cm,ini}$ (MPa)	10,66	8,38	5,58	4,71	30,45	24,10	19,55	16,25
$\sigma_{cm,res}$ (MPa)	6,75	5,58	4,71	4,05	24,0974	19,5521	16,2541	13,81
m_i	9	9	9	9	9	9	9	9
E_i (MPa)	7500	7500	7500	7500	57375	57375	57375	57375
$E_{RM,inI}$ (MPa)	7351	7190	6603	6123	56236	55001	53163	50510
$E_{RM,res}$ (MPa)	6949	6603	6123	5496	55001	53163	50510	46838
γ (kN/m^3)	23	23	23	23	26	26	26	26

* Intact Rock Stone properties: GSI=100, E_i, ** Compression Strength

Moreover, in order to examine the possible effect of the metal joints on the vault of the arch and the marble members of the side walls, analyses were performed both with and without the metal joints.

Based on the above, eight analyses were performed in total, attempting to represent the mechanical history of the monument as realistically as possible.

Table 3. Interface properties in relation to historical stages of the monument.

	Interfaces*					
	P-P	M-M	P-M	P-GC	M-SM	P-SM_
Normal stiffness, kn, ini (kPa/m)	$75x10^6$	$550x10^6$	$75x10^6$	200000	150000	150000
Shear stiffness, ks, ini (kPa/m)	$7.5x10^6$	$55x10^6$	$7.5x10^6$	40000	30000	30000
c (kPa)	0	0	0	0	0	0
ϕ,ini (o)	32	35	32	25	22	22
ϕ,res (o)	32	35	32	25	22	22

* P = POROUS FACE, M= MARBLE FACE, GC = Transitional Backfilling FACE, SM = SOIL FACE

Table 4. Kasta Tumulus: geo-material properties.

	I1	II2	III3	IV4	V5_
AUSCS	GM/GC	SM/SC	SM/SC	SM-SC	SM
γ (kN/m^3)	21 1	9-21^2	19	19	20-21
c (kN/m^2)	3	0	0	1	5-25*
ϕ (°)	35	30-32	30	30	32-35
ϕ_{res}(°)	32	30	30	30	30-32
c_u(kN/m^2)	-	-	-	-	-
v	0,30	0,30	0,30	0,30	0,25
E_i (MPa)	0,30	0,30	0,30	0,30	0,25
DR(%)	60 - 80	50-70	30-40	40-50	65-90*
OCR	44652	1-5	1	1	1-3

* Depending on the depth
(1) Transitional artificial Embankment - GC
(2) External ancient undisturbed soil filling for the tumulus formation: SM/SC, γ=21 for fully saturated soil
(3) External disturbed loose cover soil deposition on the tumulus: SM/SC, γ=21 for fully saturated soil
(4) Internal anthropogenic loose filling of the burial monument: SM/SC
(5) Natural soil deposit -Silty SAND: SM

4.5 *Results of analyses in mechanical history stages:*

The results of the 2-D analyses include (among others):

- the permanent deformation of the geometry of the burial monument (horizontal – vertical displacements)
- the settlement of the foundations of the monument
- the stress state of the structural elements of the monument (zones of plastic condition/cracking)
- the earth pressures from the surrounding soil on the monument

The results of the analyses show a good matching of the computational results with the observed structural condition of the monument. An indicative overview of results of the above analyses in mechanical loading history stages is presented in the following figures (Figures 8–12), in which the computational estimate is contrasted with the observed structural condition of the chambers of the monument.

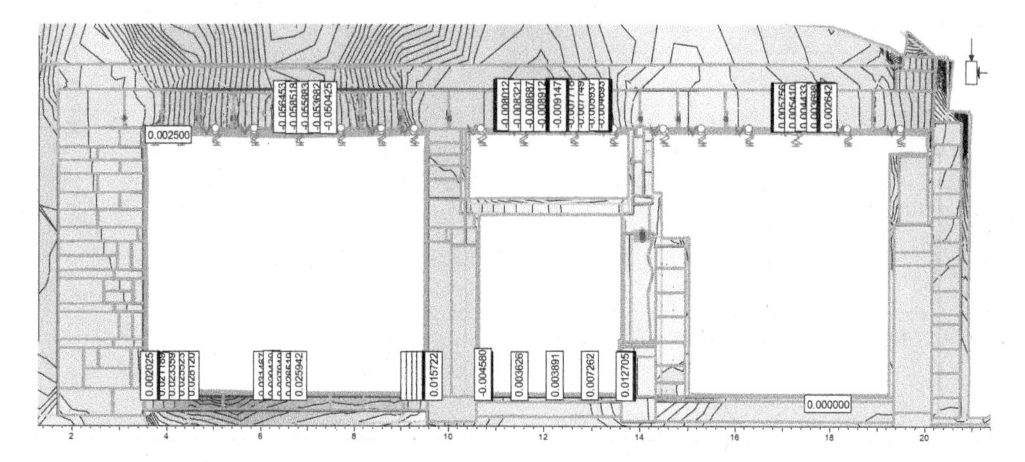

Figure 8. Calculated subsidence of the vault keystone. X4. 5.5cm calculated subsidence vs 6cm measured, X3: 0.9cm calculated subsidence vs 1.5cm measured, X2: 0.2-0.5cm calculated subsidence vs <0.5 cm measured.

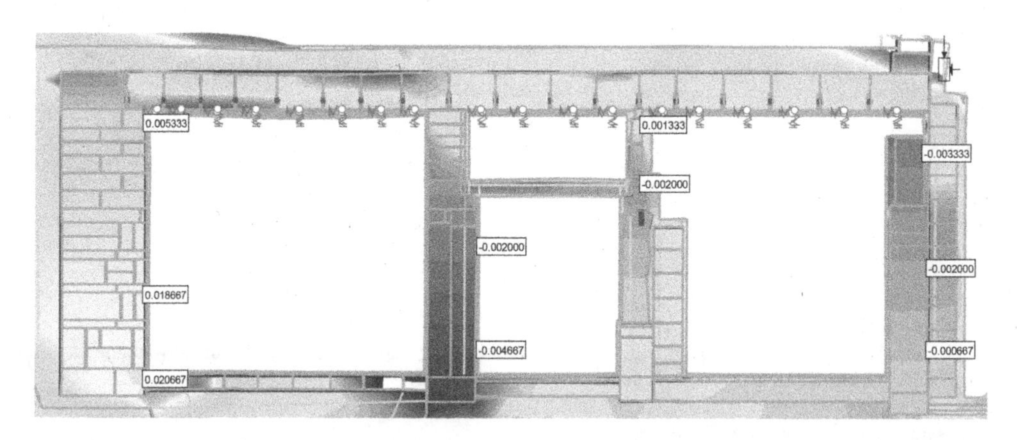

Figure 9. Permanent horizontal deformations of the partition walls (Di-D4). The measured rotations from 3D scanning are in good agreement with the calculations.

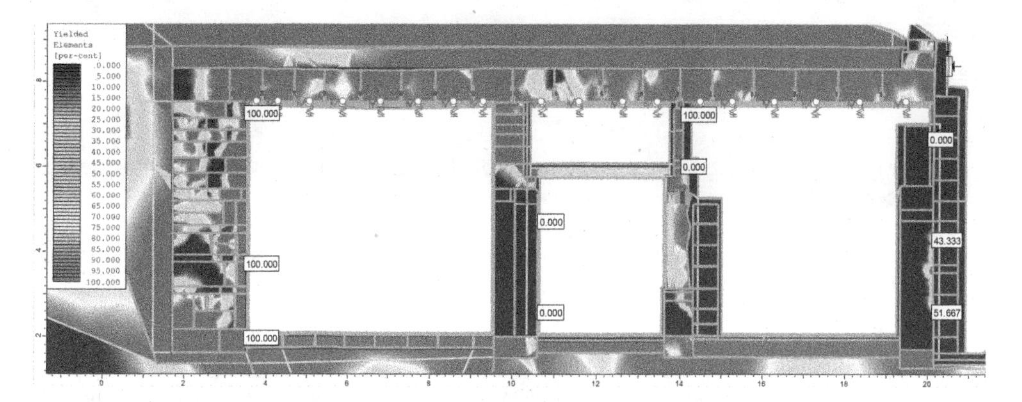

Figure 10. Longitudinal section views showing the main structural damage to the burial monument. For comparison see Figures 6 and 7.

5 EVALUATION OF RESULTS

The geostatic analyses in stages of mechanical history have produced conclusions regarding the ancient design of the monument, the interpretation of its structural pathology, the calibration of the simulation of the burial monument, and the guidelines for the design of the restoration program. These conclusions can be summarized as follows:

- The ancient design (based on the estimated geometry of the tumulus around the monument) is considered absolutely satisfactory. In particular, there is computational stability of the monument and low stress of the structural elements. The foundation of the monument is also considered sufficient, despite the (calculated and observed) differential settlement (approx. 1-1.5cm) on the west side of the monument in relation to the east. This settlement occurred on completion of the construction of the monument and the tomb. The settlement is considered acceptable even by modern regulatory standards.
- The presence of metal joining elements, as the analyses show, does not differentiate the resulting structural damage (permanent geometric deformations and main zones of cracking of building materials) on the macro scale. On a micro scale, it seems to affect the concentration of damage to the stone blocks and vault in the vicinity of the joints, due to the imposed constraint on the deformation. Similar observations may have led ancient engineers to abandon the use of joints in structures of this type (vaulted roofs).
- The strong historical earthquakes which struck the site of the monument (Figure 11) do not seem (computationally) to have substantially burdened its structural elements, as is to be expected with buried structures, (in this case with internal fill, which offers additional resistance). Of course, had the monument not been filled, it would have suffered greater shear strain due to the seismic action of the external fill.
- The main stressing factors on the monument are due to recent mechanical events: the accumulation of high soil deposition coming mainly from the top of the mound (during the 20th century) and the asymmetric earthworks during the 2014 archaeological excavation of the monument (Figures 10–11).

The analysis in stages of mechanical history is used for the evaluation of the attempted simulation of the burial monument. The result shows a good correlation of computational predictions with the observed structural image of the monument. This makes the simulation reliable for the realistic assessment of the response of the monument to adverse future stresses and the design of protection - restoration measures (adoption and location of appropriate measures based on real needs).

5.1 *Proposal for interpretation of the structural damage of a burial monument:*

The results of the geostatic calculations in mechanical history stages show the following connections between damage type and cause:

- The fractures of structural members and the permanent deformations of the monument walls in Chamber X4 arise computationally from the high loose fill, which accumulated anthropogenically over the burial monument in the last decades of the 20th century. The observed fractures of the voussoirs in Chambers X4 and X3 are due to the same cause. The damage to the walls in Chambers X2 and X3 is computationally attributed to the external asymmetric earthmoving configuration during the uncovering of the burial monument (Aug. 2014). It should be noted that the observed damage to the vault (from the vertical component of the external soil pressure) matches the general qualitative interpretation scheme in relevant works on Macedonian tombs (Athanasiou et al. 2012; Zambas 2016).
- The differential subsidence of the west walls of the monument with respect to the east walls is due to a historical differential settlement, which results computationally from the very beginning of the construction of the monument and its overlap with the earthworks of the tumulus. This subsidence is attributed to the fact that the west side is filled to a slightly greater height than the

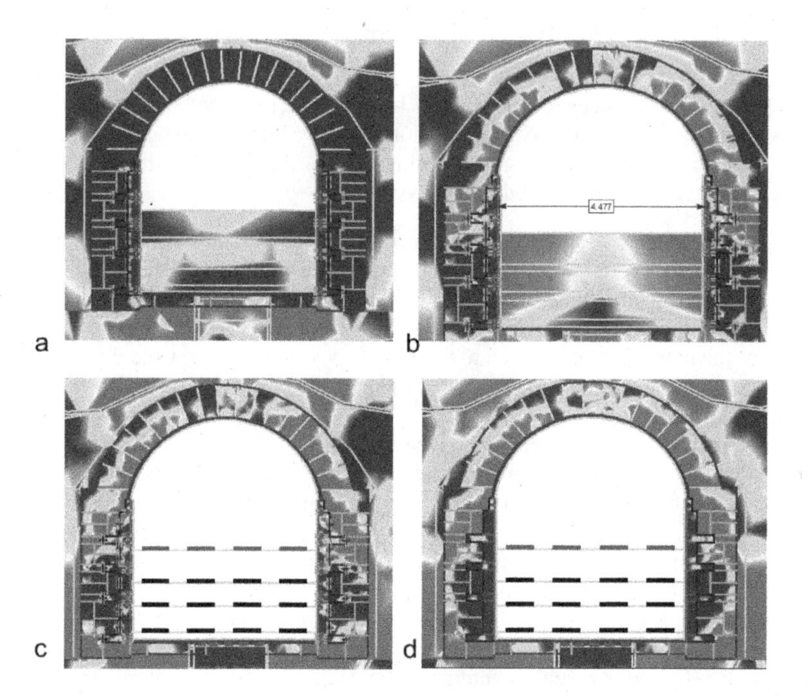

Figure 11. Computational evolution of the stress state of the voussoirs and the side walls of Chamber X4, from the analyses in historical stages (metal joining elements are included) a) post 7th century earthquake b) soil accumulation of the 2nd half of the 20th century on top of the monument c) current state, d) current state from a staged analysis without metal joining elements.

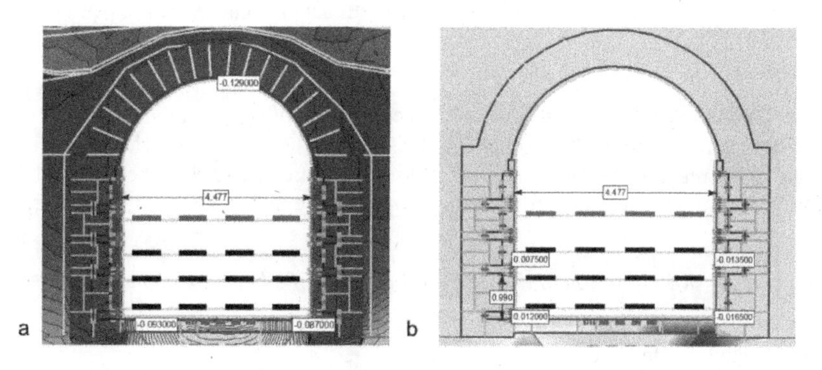

Figure 12. Permanent deformations in the mid-section of Chamber X4 a) subsidence of the vault b) convergence of the side walls. The measured rotations from 3D scanning are in good agreement with the calculations.

east side, as a result of the original topography at the site of the monument. The fracture of the (poros stone?) architrave in Wall D1 must also be attributed to this settlement. It should be noted that this subsidence has been exacerbated by the loading of high loose fill from the excavations on the top of the mound in recent decades.

- The rotation of Wall D2 towards the entrance of the monument occurs computationally as a result of the shear action of the slope on the arch and the bonding of the tympanum of D2 with the arch. Since, according to the architectural study, the tympanum is placed in a construction phase following the completion of the construction of the slopes of the tumulus, it appears computationally that the mechanical events which affected the rotation of Wall D2 are the strong earthquakes (late 6th - early 7th c. AD) focused with epicentres on Mt. Paggaion, and the high loose fill which has accumulated anthropogenically over the burial monument in recent decades.

- The rotation of the front of Wall D3 is logically and computationally related to the rotation of D2, since the two partition walls are connected to the roof slab through which the imposed stresses are transferred by friction from one to the other (mainly from D2 to D3). The absence of rotation from the rear of D3 (towards X4) should be attributed to the bonding of that part with the longitudinal walls.
- Wall D1 does not present rotation, because there is no bonding with the vault of the monument.
- The extensive vertical cracks in the coating in Space X1 must be attributed firstly to the out-of-plane bending of the walls of Space X1 (quasi-triangular plate function), and secondly to their compression between the main vaulted monument and the circular retaining perimetric wall of the tumulus (most probably as a result of the strong seismic events of the 6th and 7th centuries AD). The contribution of the differential settlement between the west and east wall during the initial construction phase should also be taken into account.
- The rupture of the architrave and the heads of the Maidens is computationally attributed to the settlement of the tympanum and the flexural stress of the architrave, combined with the asymmetric loading as a result of from the west hole in the tympanum. The fractures of the heads / supports of the Maidens are due to indirect tension, resulting from the compressive stress imposed by the entablature. The observed breakage of the marble head of the east Maiden, with the loss of its anterior half, is probably due to an event dating back a very long time. This is also indicated by the state of preservation of the preserved surface of the marble. Based on the above, it is possible that the initial rupture was due either to a strong historical (unknown) earthquake or to anthropogenic action, but in any case it occurred before the internal filling and sealing of the monument. This supposition is based on the fact that during the excavation of the monument in 2014, the broken part of the head of the east Maiden was not found.
- The rupture of the coating of the vault is attributed to its deformation (mainly transversal). The discovery of a large volume of coating fragments on the surface of the fill of Chambers X2 and X4 indicates that the event occurred predominantly after the internal filling of the monument. Since the basic deformation of the vault is due to the strong earthquakes of the 6th and 7th centuries AD and the high external fill of recent decades, the coating failure should be associated with these mechanical events.
- The observed longitudinal shear deformation and/or displacement of the arch towards the entrance of the monument is the cumulative effect of the shear action of the slopes of the mound on the arch under static and seismic conditions (initial displacement on the completion of the slopes of the mound, next established displacement with the strong earthquakes of the 6th and 7th centuries AD, and finally additional longitudinal displacement due to the high fill of recent decades).
- Based on the geostatic calculations, there is no structural damage to the monument from the use of earthmoving equipment on the mound during the uncovering of the monument.

The structural damage to the monument in correlation with its causes is presented in Table 5.

5.2 *Key vulnerabilities of the burial monument:*

The evaluation of the results of the analyses in stages of mechanical history led to the following conclusions on the vulnerabilities of the burial monument (and of corresponding structures in general):

- The height of the external artificial fill (according to the typical ancient design of the tumulus) determines the external lateral and vertical pressures on the vaulted structure, the two basic actions for the stability of the monument. It is also responsible for the longitudinal action on the monument (through frictional forces). Exceedance of the geostatic intensity of the ancient design can lead to significant permanent structural damage.

- The mechanical and natural properties of the external artificial fill (achieved density, compressibility, water permeability) are directly related to the magnitude of the pressures on the monument (geo- and hydrostatic) and to the action of physicochemical factors responsible for the degradation of structural material strength (water and/or moisture action).
- The asymmetry of the external fill: in the case of asymmetry along the transverse axis of the monument, strong stress is expected in case of seismic action.
- The bonding of the partition walls with the main structure (vault and side walls) creates conditions for their intense out of plane stress. The bonding of the partition walls D3 and D2 with a marble slab, thus forming a frame structure, is an effort to improve their out-of-level stability.
- The presence of metal joining elements, without being decisive for the imposed stress state on the structural members, locates the main fractures around the joining area. This conclusion mainly concerns the marble blocks of the side walls and to a lesser extent the voussoirs of the vault. Finally, the trench of the cist grave is crucial to the stability of Chamber X4 (it corresponds to a deep excavation at the edge of a structure with surface foundations).

Table 5. Correlation of structural damages with causative agent.

DAMAGE	Tumulus construction			Ancient sealing walls and internal fill			Earthquake 7th century A.C			Recent external anthopogenic soil deposition upon the monument - 20th c.			External earthmoving configuration - 2014			Use of Construction Machinery - 2014		
	LOW	MEDIUM	HIGH	LOW	MEDIUM	HIGH	LOW	MEDIUM	HIGH	LOW	MEDIUM	HIGH	LOW	MEDIUM	HIGH	LOW	MEDIUM	HIGH
Fractures - Cracks in the WALLS																		
X1							√			√								
X2														√				
X3							√								√			
X4							√					√						
Fractures - Cracks in the VOUSSOIRS																		
X2										√								
X3											√							
X4												√						
Rotation / Flexural stress PARTITION WALLS (IN AND OUT OF PLANE)																		
D1	No rotation due to lack of bonding with vault and side walls																	
D2	√							√			√		√					
D3	√							√			√		√					
D4	√							√				√						
Horizontal shear displacement of the VAULT																		
	√							√			√							
Horizontal transverse displacement of the VAULT																		
X2							√								√			
X3							√								√			
X4							√											
Subsidence of the VAULT																		
X2							√			√				√				
X3							√					√		√				
X4							√					√						

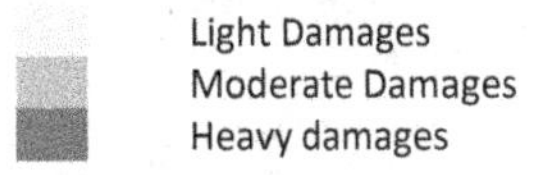

Light Damages
Moderate Damages
Heavy damages

5.3 *Utilization of the results from the analyses in mechanical historical stages for the design of protection-restoration measures of the monument:*

Understanding the mechanical behavior of the burial monument and highlighting its vulnerabilities determined the main directions of intervention for the internal and external protection of the monument. In particular, the studies for the external protection of the monument (Sotiropoulos et al. 2020b), propose the removal of the asymmetric earth fill on the west side of the burial monument. The favorable effect of this intervention is confirmed by the calibrated simulation of the monument (based on the analyses in historical stages). It is also planned to restore the external fill with lightweight geomaterials with reduced inertial characteristics, to deal with the regulatory seismic action, while provisions have been made to prevent water flow to the interior of the burial monument by creating a geotechnical sealing layer. The structural restoration study of the monument (Papadopoulos 2020) also proposes that the dense fixing support of Chamber X4 should be preserved until the arch has been fully restored, based on macroscopic observations and computational predictions of the existence of extensive zones (largely invisible) of cracked/fragmented voussoirs. Finally, it is worth mentioning the usefulness of the analyses in mechanical historical stages, in determining the optimal time sequence of the external protection and structural restoration work, in order to ensure minimal impact on the burial monument.

6 CONCLUSIONS

This paper presents the general methodology for the comprehensive qualitative and computational treatment of the mechanical history of monuments. The application of the above methodology in the case of the burial monument of the Kasta Tomb allows a convincing interpretation of its preservation status. In particular, the investigation of its mechanical stress provided basic conclusions on the historical development of its structural pathology: it proved possible to connect damage with a sequence of specific mechanical causes. The vulnerabilities of the design of the monument were also highlighted. This, in combination with the calibrated - through the observed damage - simulation of the burial monument, allows a reliable assessment of its response to future actions. Therefore the successful reading of the mechanical history of the burial monument (and indeed of every monument) is a powerful tool in the toolbox of the researcher / restorer, allowing the appropriate measures to be taken for its effective protection and promotion.

REFERENCES

Athanasiou F., Malama V., Miza M., Sarantidou M., Papasotiriou A., 2012 The restoration of the Macedonian tomb of Makridis Bey in Derveni, Thessaloniki, *Proceedings of the 3rd Panhellenic Congress of Restoration ETEPAM*, 2012 (in Greek).

Barton N. R., 1972, A model study of rock-joint deformation, *Int.J. Rock. Mech. Min. Sci.* 9. 579-602

GEOPER 2015, *Study for the redesign of the temporary fixing device for the support of the Kasta Tomb burial complex*, Unpubl. Study. Archive of Ephorate of Antiquities of Serres (in Greek).

DETYMEA, 2019, *Examination of stone and marble samples from the surrounding area of the Kastas mound in Amphipolis*, Serres, Unpubl. Study, Athens. Archive of the Directorate of Research and Technical Support for Restoration Projects of the Hellenic Ministry of Culture and Sports-DETYMEA Archive (in Greek).

Egglezos D., 2015a, *Construction and Computational support of KH EPKA for the uncovering of the monument of the Kasta Tomb and ensuring its stability by designing temporary measures: Contractual submission of a Technical Consultant for the period 1-9-2014 to 31-10- 2014*, Unpubl. Study. Archive of Ephorate of Antiquities of Serres (in Greek).

Egglezos D. 2015b, *Geostatic stability study of the Kasta burial complex*, Unpubl. Study. Archive of the Directorate for the Restoration of Ancient Monuments of the Hellenic Ministry of Culture and Sports-DAAM Archive (in Greek).

Egglezos D. 2016, The Burial Complex of the hill-tomb Casta from the side of a Civil Engineer, *Proceedings of the 29th Meeting of the Archaeological Project of Macedonia-Thrace (AEMTH)*, Thessaloniki (in Greek).

Egglezos D. 2019a, The effect of the excavation process on monumental structures interacting with the surrounding soil, *Proceedings of the 5th Panhellenic Restoration Conference (ETEPAM)*, Athens (in Greek).

Egglezos D. 2019b, Preliminary Geostatic analyses in stages of "mechanical history" for the interpretation of the structural condition of standing monumental structures: the example of the monuments at the Kasta mound in Amphipolis, *8th Panhellenic Congress of Geotechnical Engineering*, Athens (in Greek).

Egglezos D. 2019c, *Issues of structural pathology of the monumental complex of the Casta mound*, Unpubl.Study. Archive of the Directorate for the Restoration of Ancient Monuments of the Hellenic Ministry of Culture and Sports -DAAM Archive (in Greek).

Lazaridis D. 1964, Excavations and research in Amphipolis, *Proceedings of the Archaeological Society (PAE)*, 35-40 (in Greek)

Lazaridis D. 1965, Excavations and research in Amphipolis, *Proceedings of the Archaeological Society (PAE)*, 47-52 (in Greek)

Lazaridis D. 1972, Excavations and research in Amphipolis, *Proceedings of the Archaeological Society (PAE)*, 63-72 (in Greek)

Lefantzis M. 2013, New research on the base of the Lion of Amphipolis, *Proceedings of the 26th conference of the Archaeological Project of Macedonia-Thrace (AEMTH)*, Thessaloniki (in Greek).

Lefantzis M. 2014, The architecture of the Casta Tomb, *Proceedings of the 27th conference of the Archaeological Project of Macedonia-Thrace (AEMTH)*, Thessaloniki (in Greek).

Lefantzis M. 2019, *Architectural study of the tomb of Kasta of Amfipolis, Part B: The burial monument*, Unpubl. Study. Archive of the Directorate for the Restoration of Ancient Monuments of the Hellenic Ministry of Culture and Sports- DAAM Archive (in Greek).

Papadopoulos Ch. 2020, *Static Study of Structural Restoration - Restoration of the Tomb Monument of Kasta Tomb*, Unpubl. Study. Archive of the Directorate for the Restoration of Ancient Monuments of the Hellenic Ministry of Culture and Sports (DAAM Archive)

Papazachos B. and Papazachou K. 2003, *The earthquakes of Greece*, Thessaloniki, Ziti Publications (in Greek).

Pavlides S., Chatzipetros A., Syrides G., Lefantzis M. 2016, Tectonic structure and paleoseismology of Kastas Hill and the wider region of eastern Macedonia, *Proceedings of the 29th Meeting of the Archaeological Project of Macedonia-Thrace (AEMTH)*, Thessaloniki (in Greek).

Peristeri K. 2016, Excavation of the Kasta tomb of Amfipolis 2014, *Proceedings of the 29th Meeting of the Archaeological Project of Macedonia-Thrace (AEMTH)*, Thessaloniki (in Greek).

Peristeri K., Lefantzis M., Corso A. 2016, Study of scattered marble reliefs from the wider area of the tomb of Kasta Amfipolis, *Proceedings of the 29th Meeting of the Archaeological Project of Macedonia-Thrace (AEMTH)*, Thessaloniki (in Greek).

Sotiropoulos and Associates ATE, 2020a, *Presentation of geotechnical research and geotechnical evaluation*, Unpubl. Study. Archive of the Directorate for the Restoration of Ancient Monuments of the Hellenic Ministry of Culture and Sports -DAAM Archive (in Greek).

Sotiropoulos and Associates ATE, 2020b, *Soil Technical - Geostatic Study for the protection of the Burial Monument from the surrounding soils*, Unpubl. Study. Archive of Directorate for the Restoration of Ancient Monuments of the Hellenic Ministry of Culture and Sports (DAAM Archive)

Syrides G., Pavlides S., Chatzipetros A., Tsokas G., Lefantzis M. 2016, The geological structure of Kastas (Amphipolis), *Proceedings of the 29th Meeting of the Archaeological Project of Macedonia-Thrace (AEMTH)*, Thessaloniki (in Greek).

Syrides G., Pavlides S., Chatzipetros A. 2017, The geological structure of Kastas hill archaeological site, Amphipolis, Eastern Macedonia, Greece. *Bulletin of the Geological Society of Greece*, 51, 39–51.

Tsokas G., Tsourlos P., Vargemezis G., Fikos I., 2016, Course of geophysical surveys on the Kasta hill, *Proceedings of the 29th Meeting of the Archaeological Project of Macedonia-Thrace (AEMTH)*, Thessaloniki (in Greek).

Tsokas G. N., Tsourlos P. I., Kim J. H, et al., 2018, ERT imaging of the interior of the huge tumulus of Kastas in Amphipolis (northern Greece), *Archaeological Prospection*, 1–15.

Zambas K. 2016, Remarks on the Design and Construction of Macedonian Tombs, *Honorary Volume for Professor Manolis Korres*, Athens: Melissa Publishing House (in Greek).

Geotechnical Engineering for the Preservation of Monuments and Historic Sites III – Lancellotta, Viggiani, Flora, de Silva & Mele (Eds)

Structural health monitoring of historic masonry towers: The Case of the Ghirlandina Tower, Modena

Donato Sabia
Politecnico di Torino, Department of Structural, Geotechnical and Building Engineering, Torino, Italy

Giacomo Vincenzo Demarie
Structural Engineer, Torino, Italy

Antonino Quattrone
Politecnico di Torino, Department of Structural, Geotechnical and Building Engineering, Torino, Italy

ABSTRACT: Masonry towers are an important architectural heritage, whose conservation and maintenance requires a deep understanding of their structural behaviour. To this end, monitoring the dynamic response to ambient and service loads is a fundamental source of information. By repeating the data acquisition over the time, it is moreover possible to check for variations in the structure's response, whose entity may be correlated to the appearance or growth of a damage (e.g. following some exceptional event as an earthquake or as a consequence of materials and components ageing). The complexity of some existing structures and their interaction with the environment claim for a detailed monitoring plan, to support an evidence-based decision process. If the sensor network acquires data continuously over time, the evolution of the structural behaviour may be tracked continuously as well. This process needs the proper methods and algorithms to manage the large amount of available data and extract actionable information from it. This paper presents a methodology for the automatic structural long-term monitoring, which relies on existing methods from the Machine Learning and Data Mining fields. The results of its application to the real-world case of an ancient masonry bell tower, the Ghirlandina Tower (Modena, Italy) are also discussed.

1 INTRODUCTION

The main goal of the Structural Health Monitoring is to transform the experimental data into information for the assessment of the structural conditions (Farrar CR & Worden K 2007). To this extent, two general approaches can be applied: physics-based and data-driven.

The physics-based approach involves building a model of the for analysis and prediction, starting from the "first principles" of Physics. In this case, the model formulation must be appropriate enough to characterize the actual structure.

In the data-driven approach no assumptions are made about the system generating the data, the model from the data only and has not any specific physical meaning. Such models are very general in their nature and well suited when large amount of data is available or the structural behaviour is too complex to be described from basing on physical principles (Ying EJ et al. 2013).

The purpose of long-term monitoring systems is generally to track over time the status of a structure and to answer questions about the safety and the serviceability after specific events, or as a consequence of the materials and components ageing (Cross EJ et al. 2013). In areas with

DOI 10.1201/9781003329756-10

high levels of seismicity, for example, the assessment of the structural integrity can be paramount during the post-event activities.

The long-term monitoring produces a large amount of data and proper methods and algorithms are required to extract valuable and reliable information. In this context the methods and algorithms rooted in the Machine Learning and Data Mining fields have proven to be extremely effective (Farrar CR & Worden K 2013; Worden K & Manson G 2007).

The objective of this research is to define a method for the automatic identification of the vibration modes of a structure, so that they can be tracked continuously over a long period of time. The proposed approach unfolds into four steps: the model selection and validation, the system identification, the clustering and the classification steps. The natural frequencies, damping factors and mode shapes are automatically identified from the measured data and monitored over time to detect possible changes in the health state of a structure. Moreover, the observation of the environment parameters, such as the temperature, helps in classifying the stream of data and recognizing structural novelties.

The long-term monitoring of an ancient masonry bell tower, the Ghirlandina Tower (Modena, Italy), has been selected as a real-world case application for the proposed method. The structural characterization is performed by identifying the first modes of vibration, whose evolution over time has been tracked.

2 MACHINE LEARNING APPROACH

The methodology presented in this work aims to characterize the structural response, assuming the starting point in time as the normal condition with no damage and following its evolution over time. The time tracking is performed by detecting deviations from the normal condition (Demarie G & Sabia D 2019).

The proposed method is implemented as a four-step process: Model selection and validation, System Identification, Clustering and Automatic Monitoring.

2.1 *Model Selection and Validation*

A linear model is used to describe the relationship among the signals at different time instants. For the case of a single channel available, this assumption translates in the following equation:

$$y_t = w_0 + w_1 y_{t-1} + \cdots + w_p y_{t-p} \tag{1}$$

For N different instants leads to the following system of linear equations:

$$\begin{cases} y_{t_1} = w_0 + w_1 y_{t_1-1} + \cdots + w_p y_{t_1-p} \\ y_{t_2} = w_0 + w_1 y_{t_2-1} + \cdots + w_p y_{t_2-p} \\ \ldots \\ y_{t_N} = w_0 + w_1 y_{t_N-1} + \cdots + w_p y_{t_N-p} \end{cases} \tag{2}$$

The system of equations 2 can be written in matrix form:

$$\{y\} = [Y]\{w\} \tag{3}$$

For the case where n_{ch} signals are acquired an expression similar to equation (1) holds. Specifically, the $i-$th signal at a certain time instant can be obtained as a linear function of itself and all the remaining signals at p previous instants.

$$\begin{aligned} y_{i,t} = {} & w_{i,0} + w^{(1)}_{i,1} y_{1,t-1} + \ldots + w^{(1)}_{i,n_{ch}} y_{n_{ch},t-1} + w^{(2)}_{i,1} y_{1,t-2} + \ldots + w^{(2)}_{i,n_{ch}} y_{n_{ch},t-2} + \ldots \\ & + w^{(p)}_{i,1} y_{1,t-p} + \ldots + w^{(p)}_{i,n_{ch}} y_{n_{ch},t-p} \end{aligned} \tag{4}$$

For N different instants of time we can write the following linear system:

$$[Y_{current}] = [Y_{past}][W] \tag{5}$$

The proper model order "p" and the signal length in samples "N" are estimated from a subset of the measured data.

2.2 *System Identification*

The linear model coefficients are repeatedly estimated from the measured signals by solving a sequence of linear regression problems.

The data produced by the monitoring system can be thought of as a continuous succession of blocks of data, each consisting of n_{ch} signals made of $N + p$ samples.

For each block of data the linear system of equations 5 can be set up and the matrix $[W]$ estimated from it. The process can be performed by applying it each time a new block of data is acquired. It results in a succession of matrices $[W]_1$,$[W]_2$, ... $[W]_i$, ... which constitutes the basis for the long-term monitoring of the structure. The coefficients of the matrix $[W]$ can be re-organized in a way that allows for the eigenmodes of the structure to be evaluated. To this aim, Equation 4 can be re-stated in the following way:

$$\{y_t\}^T = \{w_0\} + [W^{(1)}]\{y_{t-1}\}^T + [W^{(2)}]\{y_{t-2}\}^T + \ldots + [W^{(p)}]\{y_{t-p}\}^T \tag{6}$$

The system of equations 6 can be written in matrix form:

$$\{z_t\} = \{w_0\} + [A]\{z_{t-1}\} \tag{7}$$

The system is characterized by $n_{ch}.p$ modes of vibration.

2.3 *Clustering*

The modes of vibration identified over a limited period are clustered based on the natural frequencies, damping values and the mode shapes. A decision is made on each cluster if it must be considered important and worth to be monitored. Such a decision is taken on top of a domain-specific knowledge and engineering judgement and it is intended to limit the monitoring only to the modes that are deemed as relevant for the practical case at hand.

2.4 *Automatic Monitoring*

A classifier is built on top of the clusters found as the outcome of the previous phase. Each new identified mode of vibration is automatically classified and, by repeating the process over time, the monitoring of the relevant modes is accomplished.

The first and third phases are performed at least once, but in general very few times, to properly define the model order and the clusters (that are the class types or relevant modes). The second and fourth steps, instead, are executed on every acquisition as soon as it is recorded or, alternatively, off-line.

The proposed method has been applied to the data provided by the monitoring system installed on the Ghirlandina Tower (Modena, Italy). The results obtained are discussed in detail in the Section 3.

3 LONG-TERM MONITORING OF THE GHIRLANDINA TOWER

The Ghirlandina is the bell tower of the Cathedral of Modena, Italy (Figure 1). The construction was started around the year 1160 and completed on 1184, following the initial five floors project.

An additional sixth floor was built on 1261 and the gothic octagonal cusp closed the construction phase on 1319, reaching a final height of 89.3 m (Cadignani R & Lugli S 2010; Cadignani R et al. 2017).

During the first half of 2012 an experimental modal analysis was carried out so to characterize the dynamics of the tower by measuring its response to the ambient excitation. The results obtained showed the interaction with the cathedral and the important contribution of the soil-structure interaction on the tower's modes of vibration (Lancellotta R & Sabia D 2013; Lancellotta R & Sabia D 2014; Sabia D et al. 2015; Cosentini R et al. 2015).

Figure 1. The Cathedral of Modena and the Ghirlandina Tower.

During the same year a sensor network made of 12 capacitive accelerometers and 3 thermocouples was permanently installed on the tower, in order to implement a monitoring system acquiring data continuously over a long period of time (Figure 2). Since the end 2012 the network has been measuring the accelerations at the rate of 100 Hz.

The whole database of acquisitions from August 2012 to August 2013 has been considered and each phase of the 4-step process is addressed and detailed.

3.1 *Model Selection and Validation*

The correlation structure between the signals at the current instant and their past values is expressed through the linear system in Equation 4, whose dimensions depend from the system order p and the length in samples N.

A set of 50 couples of 1 hour long acquisitions have been randomly chosen from the entire August 2012 – August 2013 database. The selected signals have been used for the estimating the linear regression coefficients "w". The parameters p and N has been determined by averaging across the values obtained from each couple of signals in the set considered for the model selection and validation.

The optimal value of the system order "p", as defined above, turned out to be equal to 9 and the minimum signal length to be considered for the identification step has been found equal to 24 minutes.

Figure 2 compares the experimental signal acquired by the channels 10 and 12 respectively with the corresponding linear regression model fit. The good level of approximation given by the model fit prove that the dynamics of the Ghirlandina Tower has been well captured.

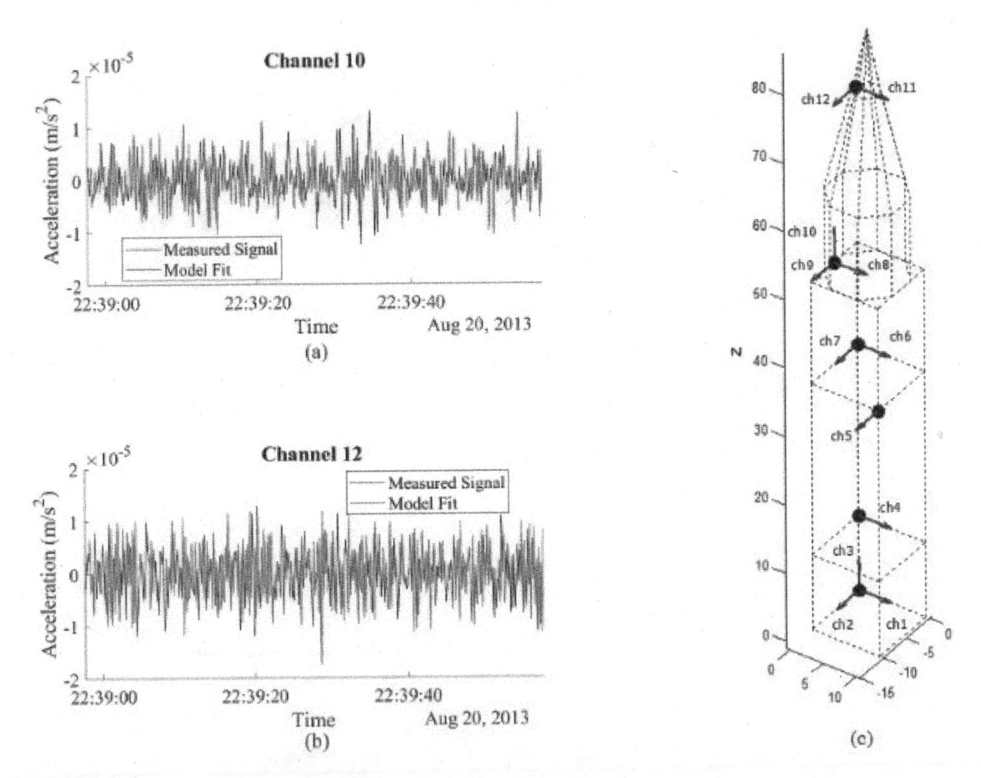

Figure 2. (a) Linear regression accuracy channel 10, (b) Linear regression accuracy channel 12, (c) Sensor positions and channel names (Modified, source: Demarie G & Sabia D 2019).

3.2 *System Identification and Clustering*

The database of signals acquired by the sensor network from August 1st 2012 to August 28th 2013 has been processed and the matrix $[W]$ of the linear regression coefficients has been repeatedly estimated according to the values of p and N determined in the previous section, for a total of 16944 times. Each time the matrix $[A]$ of the first order difference representation has been formed from $[W]$ and the eigenmodes of the systems evaluated. Provided that $p=9$ and the number of channels is 12, a database of 1829952 identified modes has been built.

Before clustering the data, a significant portion of the modes have been excluded based on the engineering criteria. A correlation coefficient higher than 0.975 has been assumed as the threshold for considering a modal shape as “almost” real. Furthermore, a threshold in the damping factor equal to 25% has been adopted, this leading to the exclusion of all the modes characterized by higher damping. At the end of the preliminary selection based on the engineering criteria 56537 modes of vibration have been retained.

The clustering process is essential to detect the modes that are worth to be tracked over time. To this end, the complete set of modes identified in the interval August – September 2012 has been clustered. Figure 3 summarizes some of the outcomes obtained from the identification of the signals in the August – September 2012 interval. The charts (a) and (b) also suggest the existence of six clusters. The figures 4 show some of the principal mode shapes identified.

The type of the mode along with the frequency and damping values are summarized in Table 1.

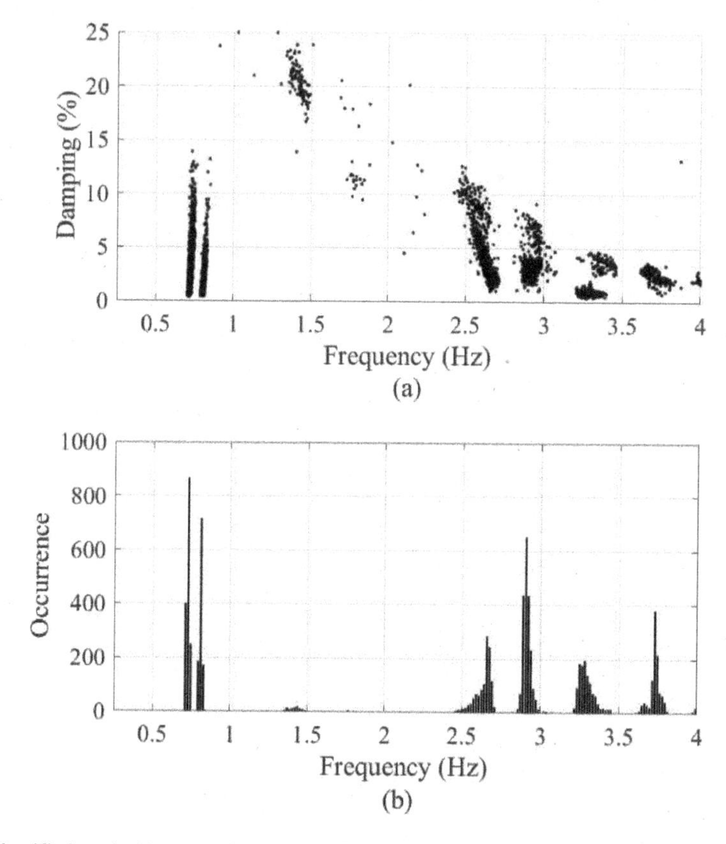

Figure 3. Identified mode (August – September 2012) (Modified, source: Demarie G & Sabia D 2019).

Table 1. Correspondence between clusters and mode types (Modified, source: Demarie G & Sabia D 2019).

Cluster #	Frequency (Hz)	Damping (%)	Type
1	0.73	3.6	1st bending (y-direction)
2	0.81	2.4	1st bending (x-direction)
3	2.65	2.7	2nd bending (y-direction)
4	2.91	2.9	2nd bending (x-direction)
5	3.28	0.8	1st torsional

3.3 *Monitoring*

The last step of the process implements the continuous time monitoring by making automatic the classification of the identified modes. To this end, a rule is needed so that each time a mode is identified it is possible to decide which mode it is, that is, which is the class it belongs to. This is implemented through a classification algorithm drawn from the Machine Learning literature, which for the case at hand is the k-nearest neighbor (kNN) classification (Gareth J et al. 2013).

The time evolution of the modal frequencies for four structural modes over the entire monitoring period is represented in Figures 5.

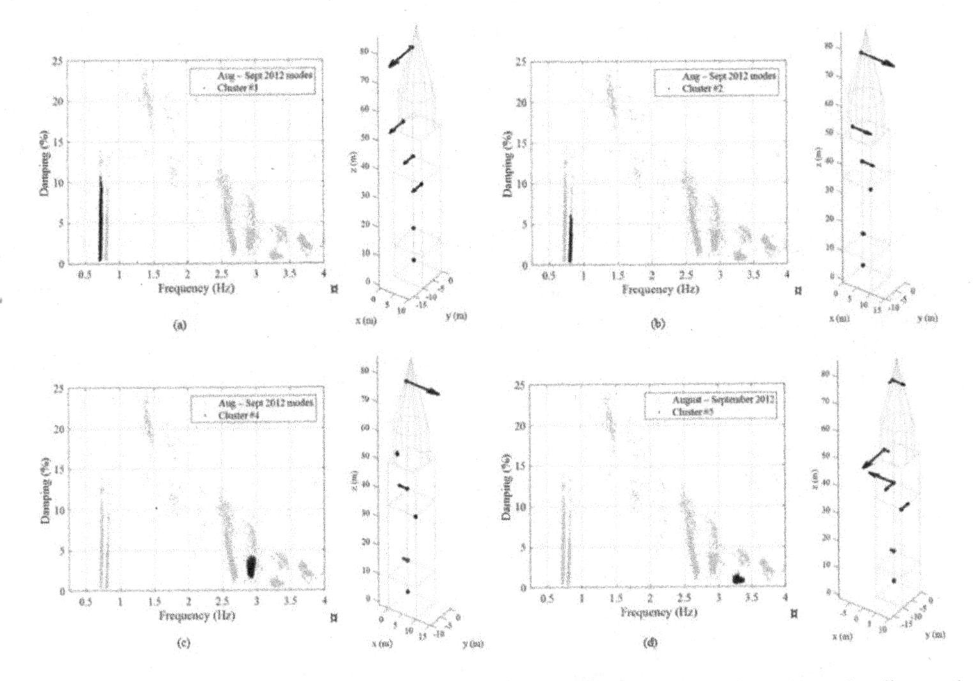

Figure 4. Clusters and mode shapes identified: (a) 1st bending mode (y-direction), (b) 1st bending mode (x-direction), (c) 2st bending mode (x-direction), (d) 1st torsional mode (Modified, source: Demarie G and Sabia D, 2019).

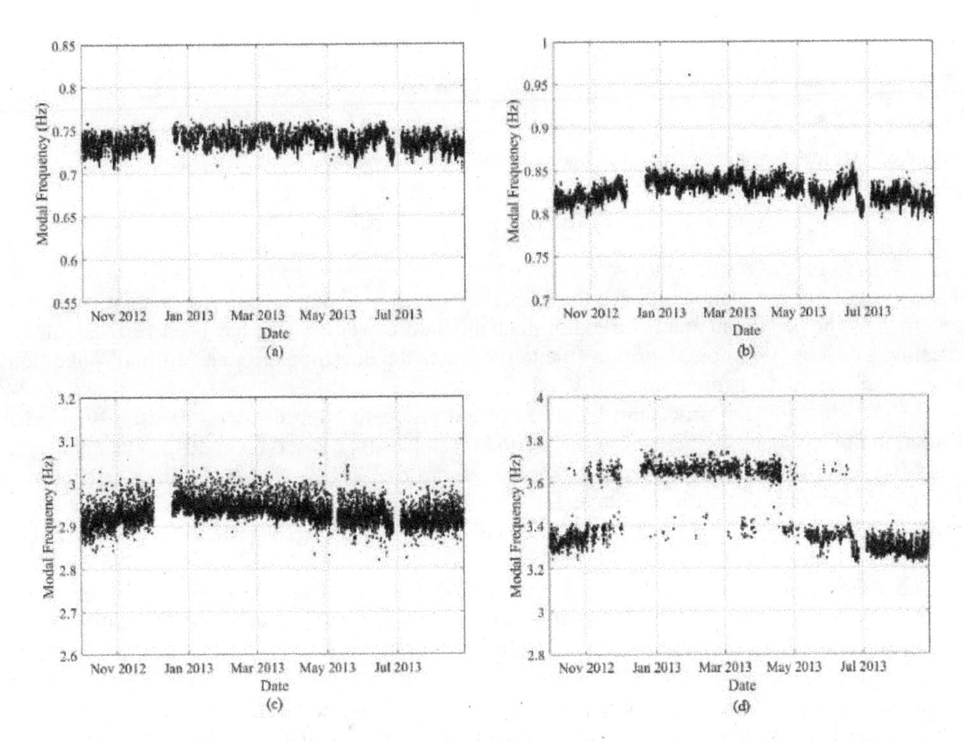

Figure 5. Modal frequency evolution: (a) 1st bending mode (y-direction), (b) 1st bending mode (x-direction), (c) 2st bending mode (x-direction), (d) 1st torsional mode (Modified, source: Demarie G & Sabia D 2019).

The modal frequencies, particularly those related to the bending modes along the x-direction and the torsional mode, clearly show a trend. The frequency values tend to increase from October to December 2012, remain stable on the average until April 2013 and then reduce to the initial values from July 2013 on. Such a variation is smooth for the bending modes, while it shows up abrupt for the torsional mode. The Figure 6 proposes once again the torsional modal frequency along with the time evolution of the temperature overlapped, showing the correlation between the quantities: the lower the temperature is, the higher the modal frequency are. Figure 7 shows the seasonal movements of the tower detected by a pendulum, and the trend of the temperature over the years (Lancellotta R & Sabia D 2013). The cause of such trends is likely to be found in the movement of tower towards the Cathedral when the temperature decreases, which is likely to affect the interaction between the two structures.

During the period considered for the monitoring a few relevant earthquakes happened, but no evidence of causing a change in the structural behaviour has been found in the identified modes.

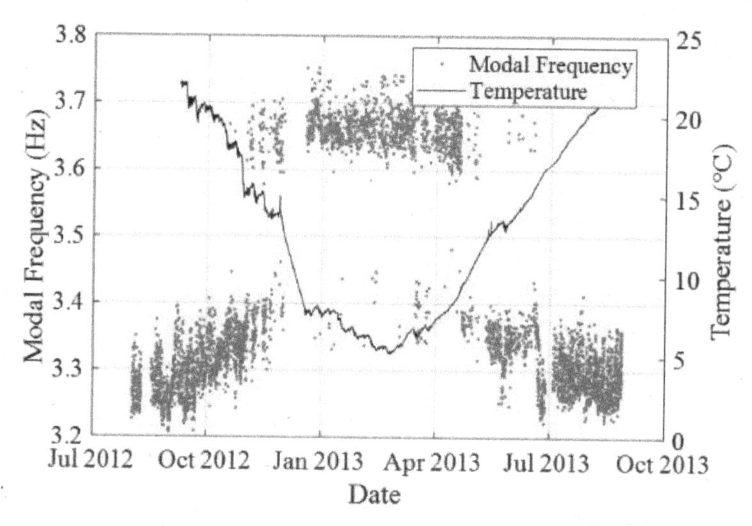

Figure 6. 1st torsional modal frequency and temperature time evolution (Modified, source: Demarie G & Sabia D 2019).

3.3.1 *The impact of temperature*

The hypothesis of correlation between the frequency variation of the tower, its seasonal movements recorded by the pendulum and the trend of the temperature over the year has been verified implementing a novelty detection algorithm able to recognize the deviations from a "normal" condition of data, i.e. the modal parameters.

The robustness of the procedure has been tested processing the data recorded in 2016. The evolution of the modal frequencies in this period resembles that observed in 2013. The algorithm classifies as an anomaly the variation of the 5th mode from 3.7 Hz to 3.35 Hz with respect of a training period of two month of data streaming. The Figure 8a shows the trend of the torsional mode in 2016, highlighting in red the detected novelties. Two distinct distributions of the variable frequency are clearly observable in Figure 8b.

Under the hypothesis of a correlation between the frequency variations and the temperature, the novelty detection algorithm has been finally trained using a one-year series of frequencies and temperature data, adopting a bivariate normal distribution and a 95% confidence interval as a rejection criterion. Introducing the temperature variable, the novelties are only detected when a mode shows a high frequency at warmer temperature and vice versa (Figure 9, red dots). The results confirm the interaction between the Ghirlandina tower and the Cathedral is highly influenced by the seasonal variation of temperature.

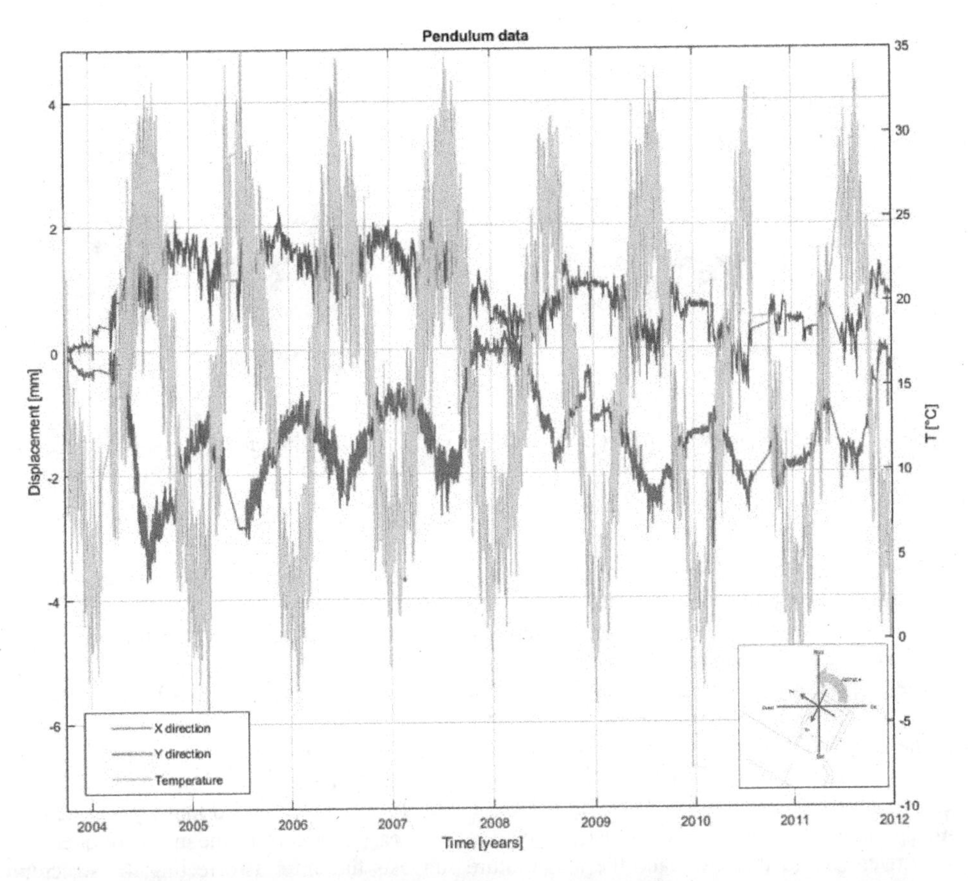

Figure 7. Seasonal movements of Ghirlandina tower and the trend of temperature.

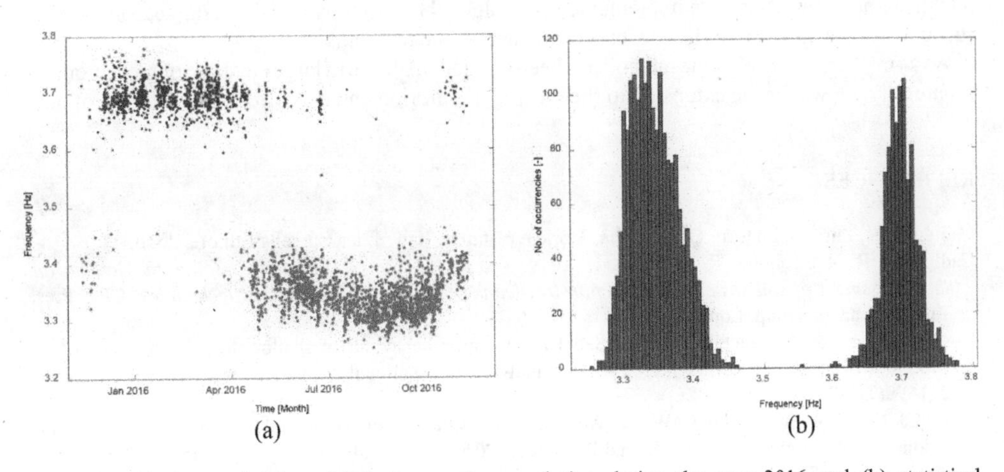

Figure 8. (a) 1st torsional modal frequency time evolution during the year 2016 and (b) statistical distribution.

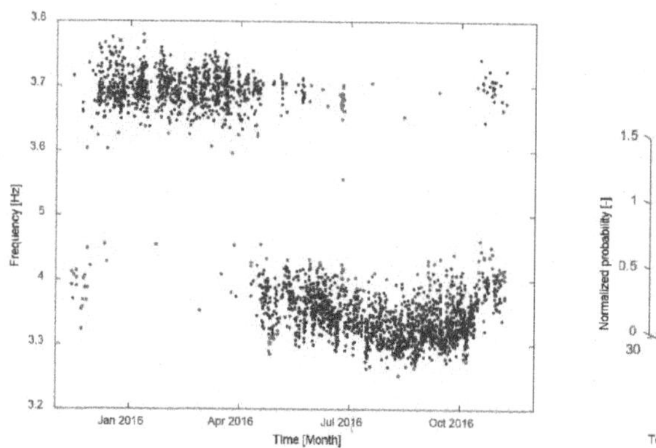

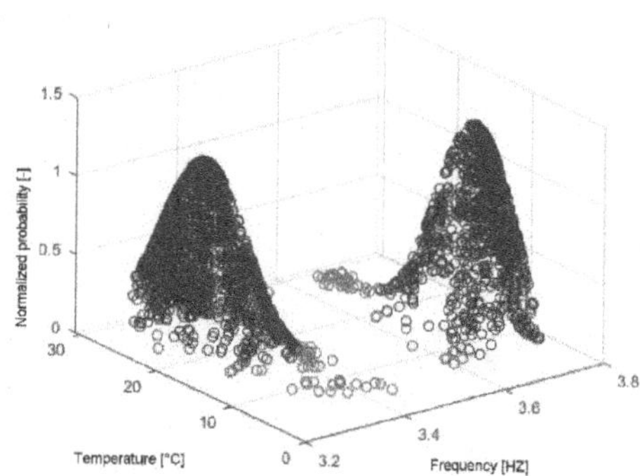

Figure 9. 1st torsional modal frequency evolution during the year 2016 with temperature correlation.

4 CONCLUSIONS

This paper introduces a novel method for the automatic identification of the vibration modes of a structure and implements a concrete approach for the long-term continuous structural health monitoring. The method belongs to the data-driven framework, it relies on some existing algorithms in the Machine Learning and Data Mining fields.

The long-term monitoring of an ancient masonry bell tower, the Ghirlandina Tower (Modena, Italy), has been selected as a real-world case application.

The obtained results prove the capability of the method not only to automatically identify the relevant structural modes with a very limited classification error, but also to highlight some long-term trends which have shown up during the August 2012 – August 2013 and January – December 2016 periods. In particular, the consistency between the seasonal trends of the modal frequencies, the movements of the tower and the temperature suggests the latter as affecting the structural interaction between the Tower and the Cathedral.

A novelty detection algorithm, based on a bivariate distribution which integrates the modal data and the temperature, has been implemented. The algorithm correctly classified the seasonal trends, recognizing the variation in frequencies correlated to the temperature.

As a final remark, even if the method has been applied off-line to a large database of measurement, it naturally allows for the extension to the on-line monitoring and classification of stream of data.

REFERENCES

Cadignani R, Lugli S. La torre Ghirlandina. Storia e restauro. Italy: Luca Sossella Editore, 2010.

Cadignani, R., Lancellotta, R. & Sabia, D. 2019. *The restoration of Ghirlandina Tower in Modena and the assessment of soil-structure interaction by means of dynamic identification techniques*. CRC Press, Taylor&Francis Group, London.

Cosentini RM, Foti S, Lancellotta R and Sabia D. Dynamic behaviour of shallow founded historic towers: validation of simplified approaches for seismic analyses. International Journal of Geotechnical Engineering, 2015; 9(1): 13–29.

Cross EJ, Koo KY, Brownjohn JMW and Worden K. Long-term monitoring and data analysis of the Tamar Bridge. Mechanical Systems and Signal Processing 2012; 35: 16–34

Demarie G, Sabia D. A machine learning approach for the automatic long-term structural health monitoring. Structural Health Monitoring, 2019(3): 819–837.

Farrar CR and Worden K. An introduction to structural health monitoring. Philos. Trans. Soc. A: Math. Phys. Eng. Sci. 2007; 365: 303–315.

Farrar CR and Worden K. Structural Health Monitoring: A Machine Learning Perspective. John Wiley & Sons Inc., 2013.

Gareth J, Witten D, Hastie T and Tibshirani R. An Introduction to Statistical Learning with Applications. Springer Text in Statistics, 2013.

Lancellotta R and Sabia D. The role of monitoring and identification techniques on the preservation of historic towers. Keynote Lecture in: 2nd Int. Symposium on Geotechnical engineering for the preservation of monuments and historic sites. London: CRC Press/Taylor and Francis Group, 2013.

Lancellotta R and Sabia D. Identification technique for soil structure analysis of the Ghirlandina tower. International Journal of Architectural Heritage, 2014; 9: 391–407.

Sabia D, Aoki T, Cosentini RM and Lancellotta R. Model Updating to Forecast the Dynamic Behavior of the Ghirlandina Tower in Modena, Italy. Journal of Earthquake Engineering, 2015; 19: 1–21.

Worden K and Manson G. The application of machine learning to structural health monitoring. Philos. Trans. Soc. A: Math. Phys. Eng. Sci. 2007; 365: 515–537.

Ying Y, Garrett JH Jr, et al. Toward Data-Driven Structural Health Monitoring: Application of Machine Learning and Signal Processing to Damage Detection. Journal of Computing in Civil Engineering. 2013; 27(6): 667–680.

Geotechnical Engineering for the Preservation of Monuments and Historic Sites III – Lancellotta, Viggiani, Flora, de Silva & Mele (Eds)
 ISBN 978-1-032-35998-4

Shake table testing of pillared historical stone constructions (mandapam) of South India

Arun Menon, Tamali Bhowmik, Shibu Samson & Jofin George
National Centre for Safety of Heritage Structures (NCSHS), Department of Civil Engineering
Indian Institute of Technology Madras (IITM), Chennai, India

ABSTRACT: The earthquake vulnerability of historical pillared dry-stack constructions in granite, of a typology which is widespread in south India, is studied through dynamic uniaxial shake table tests on a full-scale model prepared as per ancient traditional rules for proportioning, dimensioning and construction. The non-linear kinematic method based on the upper bound rigid body limit analysis provide collapse factor and maximum deformation for the governing overturning mechanism, which agree with the results of the dynamic shake table tests. Though these constructions do not have joints articulated between posts, plinths, and lintels, and rely on frictional resistance against seismic action, the capacity of these constructions appear to be adequate for the level of acceleration and displacement demands due to seismic ground shaking in south India.

1 INTRODUCTION

A *mandapam* is a multi-purpose trabeated hypostyle hall, a common archetypal feature of the ancient Dravidian temple architecture of south India (see Figure 1). *Mandapas* typically originated as a single storied entrance hall with multifunctional use near the sanctum sanctorum of the presiding deity in a temple (Manohar et al., 2020, 2021; Ronald et al., 2018). In larger temple complexes, the *mandapam* serves as a spatial linkage between various hierarchical spaces of a Hindu temple, with a series of intricately carved stone pillars (e.g., thousand-pillared hall of Madurai Meenakshi Temple in south India). The primary structural members of a *mandapam* are the monolithic stone pillars and corbels, stone beams, and slabs, all typically in granite, which is found in abundance in peninsular India.

Figure 1. Typical Dravidian style four-pillared and sixteen-pillared symmetrical granite mandapa.

DOI 10.1201/9781003329756-11

Extreme damage and loss of historical constructions of such architectural typology have been encountered in the past significant seismic events in the active regions in India such as Bhuj earthquake in 2001 (Mathews et al. 2003) and Barpak and Kodari earthquakes in Nepal in 2015 (Menon et al. 2017). A total collapse of pillared halls in sandstone triggered by the mass of the heavy domical roof in Rao Lakha Chattri in Bhuj (see Figure 2a-b) and complete overturning of the stone pillared construction in Nepal, despite the interlocking provided by the tenon and mortise joint (see Figure 2c-d) are common. Historical constructions in these seismically active regions do show evidence of seismic-resistant features such as positive interconnections between pillars and beams using iron dowels or articulated joints in stone. However, they prove to be inadequate in the presence of significant acceleration and displacement demands.

Although peninsular and south India is seismically less active than the northern and western regions, the complete absence of any seismic-resistant features such as interconnections at the post-lintel joints, and the reliance only on joint frictional resistance imply that these heritage structures may be highly vulnerable to even moderate earthquakes.

Figure 2. (a)-(b) Observed collapse in historical mandapa Rao Lakha Chattri in Gujarat (Bhuj. 2001) and (c)-(d) Collapse in a stone mandapa in Nepal (2015) with significant tenon-mortise joints.

Past investigation by the research team on historical constructions of the region aimed at the implementing a multidisciplinary methodology integrating seismic hazard, local site response, vulnerability, and exposure of historical building stock for seismic risk reduction in architectural heritage in India and Italy (Lai et al. 2009; Magenes et al. 2006; Palmieri et al. 2012). Limit analysis based seismic assessment of the mandapam revealed that their vulnerability is mainly due to the lack of proper interconnection between structural members (Ronald et al. 2018). The research work presented in the current paper is a culmination of the past research in dynamic shake table testing of a full-scale model of a mandapa conducted at the biaxial shake table facility at the structural engineering laboratory of IIT Madras. The earthquake ground shaking was simulated on a physical model designed and constructed by a traditional architect as per ancient Indian treatises on construction of temple structures.

2 SELECTION AND CONSTRUCTION OF EXPERIMENTAL MODEL

2.1 *Description of structural geometry*

In its simplest form, a mandapam is a free-standing unit constructed with in the post and lintel (or trabeated) technique in stone. The structural members are the monolithic pillars with pedestal, shaft and stone capital, corbels, and beams, typically in granite. In Dravidian architecture, these members are merely juxtaposed over each other with dry joints, and lime mortar at the interface is used only for purposes of levelling, implying a total absence of interlocking at the joints and a total reliance on frictional resistance and precompression for lateral resistance. The typical spacing of pillars vary between 2-3 m. Wide stone sunshades rest on the beams, which are held in place, by the weight of overlying stone slabs of the roof spanning between beams, lime concrete overlay as a weathering course and short parapet walls.

2.2 *Expected behaviour under seismic action*

Past research based on limit analysis revealed that the collapse of such structures under lateral action is due to instability governed by the overturning mechanism, and not sliding despite the absence of interlocking at joints. Significant resistance to sliding is generally present and readers can see detailed research on characterization of joint parameters in such constructions elsewhere (Naik et al. 2021). Overturning resistance on the other hand depends on the geometry of the structure. Shorter mandapam are more vulnerable than taller ones due to their high fundamental frequency (3.6 Hz and 2.5 Hz were the fundamental frequencies of the 4-pillared and 16-pillared short mandapas, respectively). This is evident from the graph showing factors of safety against the governing overturning mechanism, for four different geometries of *mandapas* all symmetrical in plan, to peak ground accelerations corresponding to return periods: 95. 475, 975 and 2475 years for the south Indian town of Kancheepuram in Figure 3 (Ronald et al. 2018).

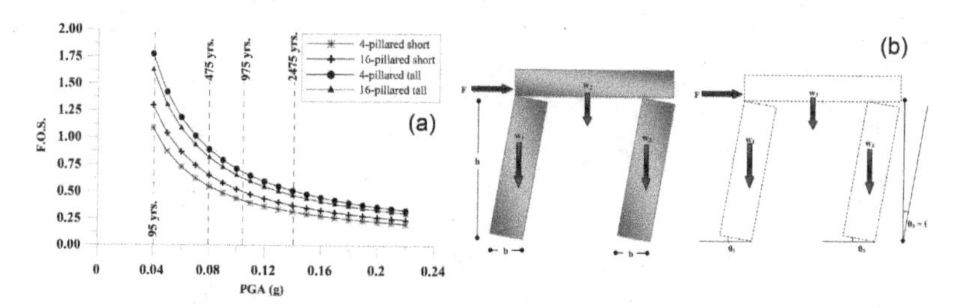

Figure 3. (a) Comparison of factors of safety against the governing overturning mechanism for the 4 mandapams to PGA corresponding to return periods: 95. 475, 975 and 2475 years for the south India town of Kancheepuram; (b) Idealisation of loads and geometry and kinematic action for overturning mechanism (Ronald et al. 2018).

2.3 *Description of Prototype and Model*

To experimentally study the structural typology using dynamic testing in the current research, a four-pillared granite *mandapa* was designed and constructed by a traditional architect (referred to as *Stapathy*) as per the traditional rules encoded in the ancient Indian treatises on temple architecture. To overcome the limitations imposed by the dimension of the shake table (i.e., 3.0 × 3.0 m: see section 3.1 for further details) and its payload of ten tons, necessitating a scaled model and consequent scaling of mass and time as per similitude requirements, a different approach was adopted based on the traditional rules for proportioning and dimensioning of structures. A real mandapa that could be accommodated within the table area, and within the maximum payload of the table was designed by the traditional architect (as the prototype) and its full-scale model was built on the shake table.

The plan dimension of the four-pillared mandapa worked out to 1.509×1.509 m and the total height of the structure worked out to 2.282m. The sectional elevation and plan of the mandapa are provided in Figure 4 showing the different components (plinth, stone pillars, corbels, beams, stone slabs, sunshade and lime concrete roof: respective traditional names are also provided in the figure) and the isometric view of the stone structure can be seen in Figure 5. The monolithic granite

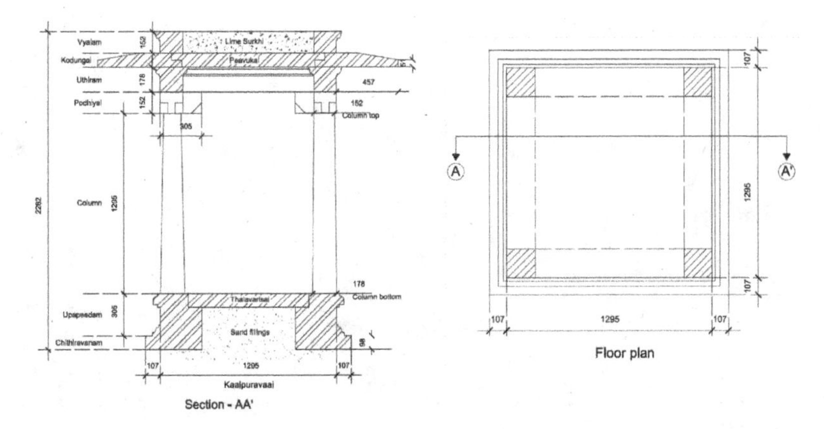

Figure 4. Cross-sectional elevation and plan of the full-scale model of the four-pillared mandapa.

Figure 5. Isometric view of the mandapa.

columns are of height 1.295 m and have a base square cross section with side dimension of 178 mm, tapering to 152 mm at their top. The central volume of the plinth is filled with graded sand.

The different structural components in granite (i.e., monolithic pillars, beams, corbels and stone blocks for the plinth and sunshades) were sculpted at the traditional architect's sculpting yard, transported to the laboratory, and assembled on the table, predominantly as a dry stack construction with minimal use of joint mortar (traditional lime) only to serve for purposes of levelling and not as a bonding agent. The different stages of construction of the mandapa can be seen in the sequence of photographs in Figure 6.

The stone blocks of the plinth are assembled over a saturated graded sand bed 150 mm deep confined using steel channels around its edge. The purpose of the sand bed is to allow for levelling of the different blocks of stone so that verticality of the construction is maintained as the construction progresses. The depth of the sand bed is not significant enough to alter the table motion characteristics before being transmitted to the specimen. The monolithic pillars, corbels are assembled on the completed plinth, and the monolithic beams and multiple stone blocks composing the sunshades are assembled thereafter. The stone slabs spanning between the beams lock the cantilevered sunshades in place (see cross-sectional elevation in Figure 4). A weathering course in brick jelly lime concrete is provided over the stone roof slabs, and together with the parapet wall constructed finally, act as the counterweight for the cantilevered sunshades.

Figure 6. Construction sequence of the mandapa on the shake table: (a) assembly of stone blocks of the plinth over a graded sand bed; (b) placement of pillars and corbels on the completed plinth; (c) assembly of sunshades over the beams; (d) provision of lime concrete weathering course over the stone roof slabs; (e) completed model with parapet wall as counterweight for the cantilevered sunshades.

3 EXPERIMENTAL TEST SET UP AND SELECTION OF GROUND MOTION

3.1 *Description of shake table facility*

The biaxial seismic test simulator (see Figure 7) is a shake table of dimension 3.0 m x 3.0 m in plan, operated by two 25-tonne servo-hydraulic actuators in two horizontal directions. The table has a maximum payload of 10 tons and can generate a maximum velocity of 1.0 m/s, a maximum stroke (displacement) of +/- 250 mm, and a maximum acceleration of 1.1 g (110% of g), all with maximum payload. The bare table can generate a maximum velocity of 1.4 m/s, and a maximum acceleration of 2.17 g (217% of g). Scaled or full-scale specimens of structural components and systems can be mounted on the table and subjected to real earthquake acceleration records or sine sweeps within a frequency range: 0-50 Hz. The table is supported on a reaction mass of 750 T created by in-situ concreting, going down to a depth of 4.5 m from the finished floor. The actuators are protected by a raised steel structural floor from falling debris of the test specimen. In the current testing program, only one actuator was active, while the second was kept locked.

Figure 7. Biaxial 3.0×3.0m shake table test facility at structural dynamics laboratory at IIT Madras.

3.2 *Instrumentation Plan*

Ambient vibration test (AVT) was carried out on the mandapa model prior to the seismic test to establish global dynamic modes and frequencies through Dynamic Identification (DI). The modal information from AVT has been estimated by using the Operational Modal Analysis (OMA) which is also referred to as output only modal analysis. AVT was performed using 8 numbers of PCB®piezo-electric accelerometers with sensitivity equal to 1.0 V/g and with a dynamic range of ± 5.0g and a Quantum X 840B 24 bit 8-channel analogue-to-digital converter system. The graphical user interface (GUI) for data visualization and real time control was achieved by CATMAN®– a commercial software. The identified natural frequencies of the model were 7.86 Hz (first translational mode) and 10.01 Hz (second translational mode) and 13.43 Hz (torsional mode).

A total of uniaxial 16 PCB®piezoelectric accelerometers with sensitivity equal to 1.0 V/g and with a dynamic range of ± 5.0g were mounted at different levels of the model, namely at the plinth, on the columns, on the beams and at the parapet level, to measure the response of the model during the seismic testing. In-built sensors of the shake table provide the table acceleration and displacement response. The positions of the uniaxial accelerometers along the loading direction are shown in Figure 8 below.

In addition, linear variable differential transducers (LVDTs) were placed to measure the model displacements, and five digital video cameras were positioned around the model to capture the response of the model.

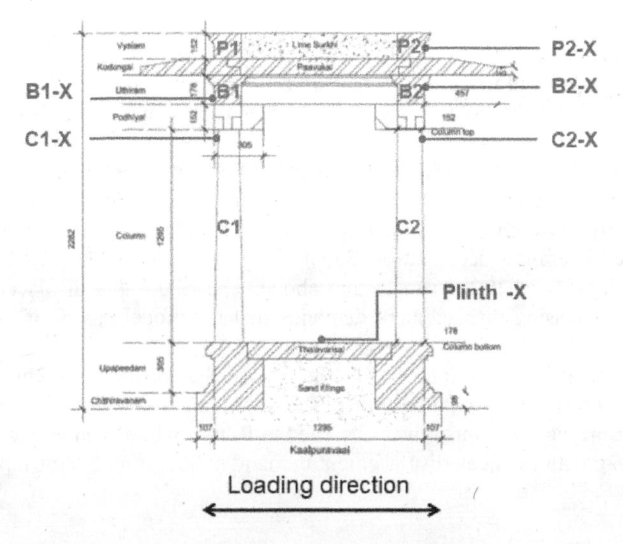

Figure 8. Positions of uniaxial accelerometers mounted on the specimen.

3.3 *Selected ground motion record*

Based on an estimate of the fundamental frequency of the model, the ground motion record was so chosen to ensure significant seismic energy around the structural frequencies of interest. In addition, a ground motion record with multiple significant oscillations is preferred to take the model through enough inelastic cycles.

In the absence of instrumented significant earthquakes from south India, the acceleration record pertaining to the Montenegro (ex-Yugoslavia) earthquake of 15th April 1979, registered at Bar-Skupstina Opstine station for a total length of 47.84 seconds, was adopted for the seismic test on the mandapa. The acceleration, velocity and displacement time histories of the record are reported in Figure 9. The PGA of the record was 0.35g.

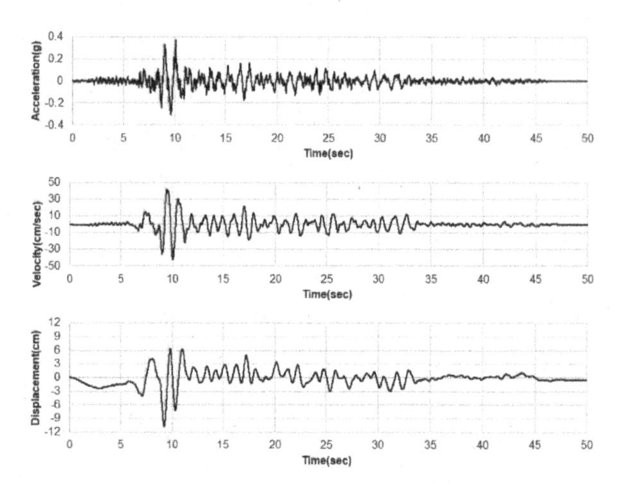

Figure 9. Input acceleration, velocity, and displacement time history (Montenegro, 1979 at Bar-Skupstina Opstine station).

4 DYNAMIC TESTING AND OUTCOME

A sufficient number of tests within the elastic range of the structural response was selected before acceleration levels capable of inducing mechanism formation and potential collapse. The necessary scale factors were selected based on analytical estimates of model response discussed in section 5 of the paper.

Table 1 reports the seven runs that were performed before the collapse of the model in the final run. The target peak table acceleration and the recorded peak table acceleration are reported showing significant disparity between the targeted and achieved peak accelerations, which is typically expected at low levels of input acceleration. Satisfactory convergence between target and achieved table accelerations are seen in the runs 5 and above. The peak table displacements provide an indication of the increasing displacement demands that the model has been subjected to in the seven runs.

The test was prematurely terminated after the loss of the specimen in Run-7 at a table peak acceleration of 0.146 g with a maximum displacement demand of 46 mm. The first instance of formation of overturning mechanism was observed in Run-7, while even in Run-6 when the table PGA reached 0.12g with the peak displacement demand at 30.5 mm, overturning was apparently not triggered.

Table 1. Shake table runs.

Test	Target peak table acceleration		Peak table acceleration (g)	Peak table displacement (mm)
	% Montenegro EQ	PGA (g)		
Run 1	2%	0.008	0.14	16.8
Run 2	5%	0.018	0.10	17.4
Run 3	10%	0.0326	0.15	16.5
Run 4	15%	0.055	0.13	12.0
Run 5	20%	0.06	0.086	20.23
Run 6	30%	0.11	0.12	30.5
Run 7	50%	0.175	0.146	46

The recorded response accelerations in the direction of seismic excitation are reported in Figure 10 at three of the instrumented locations, namely the top of the plinth, at the level of the beams

(B4-X) and at the level of the parapet (P3-X). Graphs in Figure 10 a, b & c report the response in Run-6, while Figure 10 d, e & f report the response in Run-7. The model showed overturning mechanism formation with hinges at the base of the pillars and on top of the corbels at the first significant oscillation of Run-7.

The peak response of the model at none of the levels exceeds 0.2g in Run-6 (table peak acceleration was at 0.12g), whereas in Run-7 with a table peak acceleration of 0.146g, the accelerations recorded at the plinth, beam and parapet levels are 0.6g, 0.6g and 1.0g, respectively, just before the collapse of the specimen.

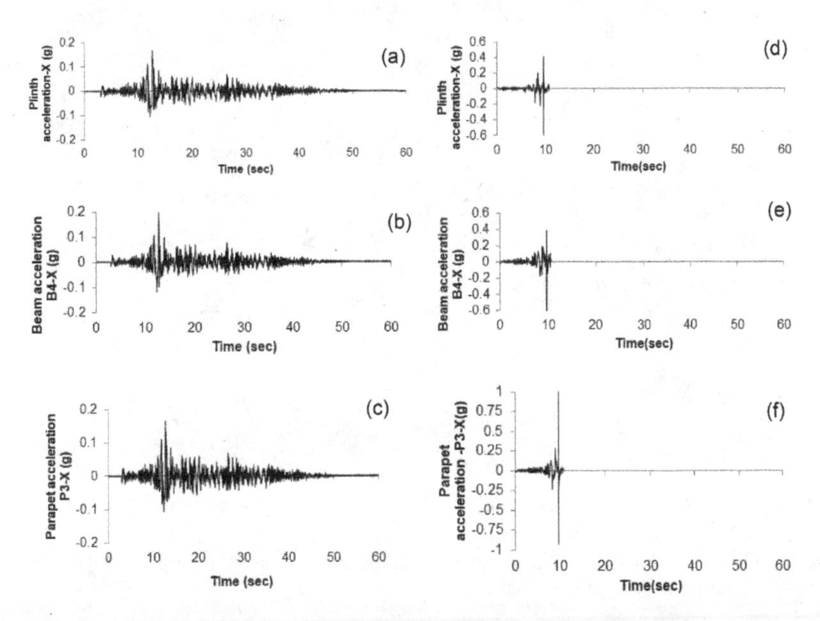

Figure 10. Run-6 (a, b, c) and Run-7 (d, e, f) response accelerations at top of the plinth, at the level of the beams (B4-X) and at the level of the parapet (P3-X) in the direction of the applied seismic excitation.

The series of images in Figure 11a-f are video captures of the model response in Run-7 from the different cameras positioned around the model. It is interesting to note how the monolithic pillar and the superimposed corbel acted as one integral unit with hinges at their base and top in the overturning mechanism at the rear end of the model, whereas the hinges have formed at the top and base of the monolithic pillar at the lead end. The reason for this lack of symmetry needs further investigation but might not be due to bonding from the mortar in the joint used for levelling purposes. The analytical model discussed in section 5 does not consider the difference in location of hinges.

Images in Figure 11e-f show the top view of the model confirming almost total response in the fundamental lateral mode. No slippage of the cantilevered sunshades or stone blocks of the parapet were observed during the tests, implying monolithic response of the construction. Disintegration was observed only at collapse of the model and upon impact of parts of the model with the protective steel scaffolding positioned around the model.

After the collapse of the model, examination of the structural members, particularly the stone pillars, revealed no crushing of the edges of the pillars implying rigid rocking response with no softening whatsoever. This is to be expected given the compressive strength of the granite, which is above 150 MPa. Softening can be expected in weaker varieties of stone such as sandstone and limestones, as observed in the Chattris of Bhuj in 2001.

No disintegration of the plinth was observed, nor was any sliding of the model. It must be mentioned here that the model was not restrained at the base of the plinth. The frictional resistance of the granite-graded sand interface at the base of the model was greater than the lateral force demand at that level in all the seven runs.

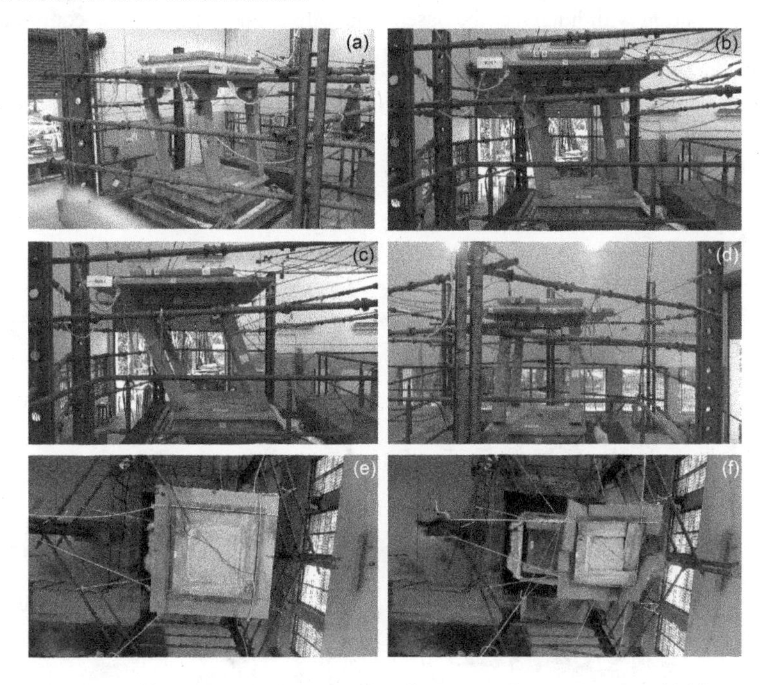

Figure 11. Images from video capture during Run-7 (a)-(b) Overturning mechanism with hinges at the bottom and top of the corbels in the lead pillars versus the rear pillars, respectively; (c)-(d) side views of the model at the verge of collapse; (e) top view of the model at initiation of mechanism; (f) top view at collapse showing disintegration of the stone slabs of the sunshades.

5 ANALYTICAL MODEL PREDICTIONS

As demonstrated from the experimental tests, the portion of the Mandapa above the plinth (i.e., *upapeedam*) is involved in the collapse mechanism. The system behaves like a chain of rigid bodies connected at the hinges at the onset of the collapse mechanism. Further, as mentioned earlier, there was no evidence of crushing of the stone observed at the hinges. Since the system deformations are restricted to the plane of loading, kinematic limit analysis can be used to validate the experimental behaviour.

The theory of limit analysis assumes that the developed kinematic chain formed by rigid bodies linked by hinges is on the verge of collapse if the equilibrium of external and internal forces is maintained and sufficient number of hinges form to turn the structure into a mechanism. This essentially means that the failure is by the formation of a collapse mechanism rather than exceedance of material strength (Clemente et al. 2010; Fanning et al. 2001; da Porto et al. 2007). The structure in this condition is now statically determinate. The theory of limit state analysis is based on the following fundamental assumptions.

- Equilibrium condition: As per the equilibrium condition, the external and internal loads should be balanced.
- Mechanism condition: Required number of hinges need to be generated to change the system to a mechanism.
- Yield condition: Material stresses should not be greater than the compressive strength (Block et al. 2006; Ochsendorf 2006).

The virtual work method is adopted, which proceeds by the identification of virtual displacements of each rigid block calculated at their center of gravity. The side view in Figure 12 shows the collapse configuration of mandapa geometry. The collapse configuration consists of a total of eight hinges, of which four symmetric hinges (HA, HB, HC, and HD) visible in the side view are shown in Figure 12. Hinges HA and HD allows rotation alone (θ) while the horizontal members of the kinematic chain such as beams, and the slab undergo translation without rotation.

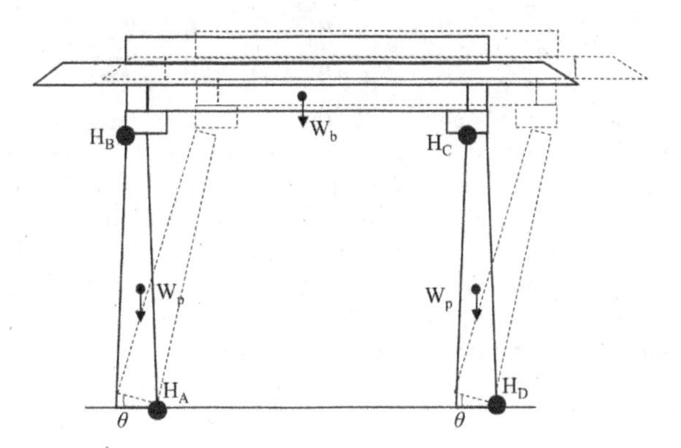

Figure 12. Collapse configuration of the mandapa.

The virtual displacements in each rigid block are calculated as follows. For the four pillars, the virtual displacements can be calculated as,

$$\begin{aligned} \delta x_{ij} &= (y_{ij} - y_i)\delta\theta \\ \delta y_{ij} &= (x_{ij} - x_i)\delta\theta \end{aligned} \tag{1}$$

Where x_i and y_i are the position coordinates of the rotating hinge, δx_{ij} and δy_{ij} are the virtual displacements of the pillar in the horizontal and vertical directions, respectively, and y_{ij} and x_{ij} are the centroidal coordinates of the pillars. For the four beams (i.e., *uthiram*), the virtual displacements remain identical as the member is subjected to translation alone, with no rotation. The virtual displacements can be written as:

$$\begin{aligned} \delta x_{jk} &= (y_j - y_i)\delta\theta \\ \delta y_{jk} &= (x_j - x_i)\delta\theta \end{aligned} \tag{2}$$

δx_{jk} and δy_{jk} are the virtual displacements of the beam in the horizontal and vertical directions, respectively. The self-weight of the slab contributed by stone slabs (i.e., *paavukkal*) and the sun-shades (i.e., *kodungai*) is equally distributed on to the four beams considering their tributary areas. The self-weight of the members in the kinematic chain is assumed to act at their center of gravity. The equilibrium equation of the Mandapam can now be written as:

$$\alpha = -\frac{4W_p\delta y_{ij} + 4W_b * \delta y_{ij} + W_{\frac{sl}{4}} * \delta y_{ij}}{4W_p\delta x_{ij} + 4W_b * \delta x_{ij} + W_{\frac{sl}{4}} * \delta x_{ij}} \tag{3}$$

Where W_p and W_b are the self-weights of the pillars, and beam (i.e., *pothiyal*) respectively, $W_{sl/4}$ is the self-weight contribution of the slab acting on a beam, δx_{ij} and δy_{ij} are the horizontal and vertical virtual displacement caused by the member of the kinematic chain in the horizontal and vertical direction, respectively. Unit weight of granite and the lime concrete are assumed as 2700kg/m^3 and 1300kg/m^3, respectively. The members above the plinth are only involved in the kinematic chain, as observed in the experimental campaign. Solving Eq.3 using linear kinematic analysis gives the value of horizontal collapse multiplier, α, which triggers the collapse mechanism as 0.246.

Formulation of the capacity curve for the collapse geometry essentially requires the elaboration of a nonlinear kinematic analysis procedure. The support rotation at HA is incremented by $\delta\theta$, and the corresponding rotations and displacements in the other members of the kinematic chains are identified using equations 1 and 2. The equilibrium equation can now be written as:

$$\alpha(\theta) = -\frac{4W_p\delta y_{ij} + 4W_b * \delta y_{ij} + W_{\frac{sl}{4}} * \delta y_{ij}}{4W_p\delta x_{ij} + 4W_b * \delta x_{ij} + W_{\frac{sl}{4}} * \delta x_{ij}} \tag{4}$$

Eq.4 is applied to the deformed geometry of the *mandapa* corresponding to the increment in rotation, $\delta\theta$. The developed capacity curve for the *mandapa* is shown in Figure 13 using linear and nonlinear kinematics. The ultimate displacement capacity (d_o), where the horizontal seismic resistance drops to zero, is obtained as 71mm. A performance limit is imposed on the capacity curve which is termed as a near collapse limit state (DL), which is the displacement threshold a little before overturning (Lagomarsino, 2015), adopted as $0.4d_o$. The seismic collapse multiplier corresponding to d_{DL} is obtained as 0.15, which agrees with the experimental results. The results from rigid body limit analysis are normally an upper bound.

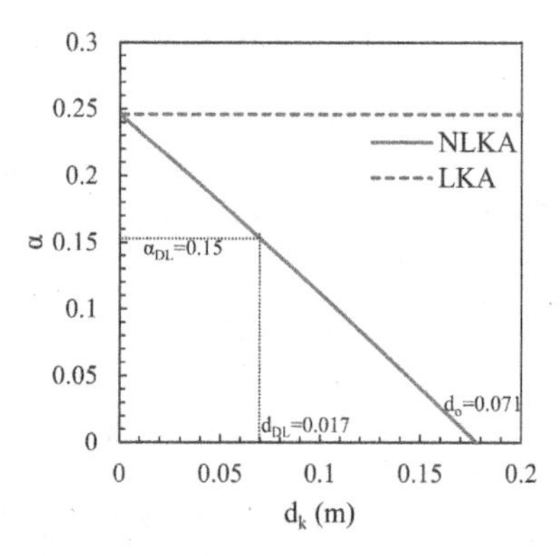

Figure 13. Capacity curve of the *mandapa* developed from non-linear kinematic analysis.

6 CONCLUSIONS

The seismic vulnerability of historical stone pillared constructions abundantly present in religious structures in India referred to as *mandapas* has been studied experimentally. Previous field-based investigations and analytical studies showed the vulnerability of the typology of dry stack stone pillared post and lintel constructions devoid of any positive interconnections between the monolithic posts and the plinth, and posts and lintels (or beams).

Dynamic shake table tests on a full-scale model in granite with a fundamental natural frequency of 7.86 Hz (natural period: 0.127 s) established from AVT, were conducted using scaled earthquake excitation from the Montenegro earthquake of 1979 with significant energy content between 0.1 and 1.0 Hz. The collapse of the structure is governed by the overturning mechanism with hinge formation at the base and top of the pillar or base of the pillar and top of the corbel, with the monolithic pillar and the corbel acting as one unit in the latter.

Mechanism formation was not observed even at a peak table acceleration of 0.12g and peak displacement demand of about 30.5 mm, which was the penultimate run of the dynamic test. The mechanism was triggered at a peak table acceleration of 0.146g and a peak displacement demand of 46 mm. These results are in agreement with those based the upper bound non-linear kinematic analysis from limit theory considering rigid rocking, which indicates that failure should be expected by overturning at a collapse factor of 0.15g and a displacement demand exceeding 28.4 mm.

These results show that the historical structural typology (mandapas) can be expected to have adequate safety against collapse for the level of seismic demand in south India but can be vulnerable in the active seismic zones of the country, where the resistance will depend on the strength and deformation capacity of the interlocking at the locations of hinge formation.

ACKNOWLEDGEMENTS

Research work presented in the paper was carried out with financial support from Dept. of Science & Technology (DST Grant No. SR/WOS-A/ET-103/2017), Govt. of India. The authors acknowledge infrastructural support from National Centre for Safety of Heritage Structures (NCSHS), Civil Engineering Dept., IIT Madras (Ministry of Human Resource Development, GoI D.O. No. 5-62013-TS-1). This work would not have been feasible without the assistance for design and construction of the *mandapa* from traditional architects (*Stapathy*) Mr. S. Kumaragurubaran and Mr. K. Gowri Sankar from Panrutti, Tamil Nadu.

REFERENCES

Block, P., Dejong, M., and Ochsendorf, J. 2006. *As hangs the flexible line: Equilibrium of masonry arches.* Nexus Network Journal, 8(2), 13–24.

Clemente, P., Buffarini, G., and Rinaldis, D. 2010. Application of limit analysis to stone arch bridges. *ARCH'10 – Proc. of 6th International Conference on Arch Bridges. Fuzhou, China, October 11–13, 2010.*

Da Porto, F., Franchetti, P., Grendene, M., Ranzato, L., Valluzzi, M., and Modena, C. 2007. Structural capacity of masonry arch bridges to horizontal loads. *Proc. of 5th International Conference on Arch Bridges (ARCH'07). Madeira, Portugal, 12–14 September 2007.*

Fanning, P. J., Boothby, T. E., and Roberts, B. J. (2001). *Longitudinal and transverse effects in masonry arch assessment.* Construction and Building Materials, 15(1), 51–60.

Jetson Ronald A., Menon, A., Prasad, A.M., Menon, D., & Magenes, G. 2018. *Modelling and analysis of a South Indian temple structures under earthquake loading*, Sādhanā, 43(74), 1–20.

Lagomarsino, S. (2015). *Seismic assessment of rocking masonry structures.* Bulletin of Earthquake Engineering, 13(1), 97–128.

Lai, C.G., Menon, A., Corigliano, M., Ornthamarrath, T., Sanchez, H.L. Dodagoudar, G.R. 2009. *Probabilistic seismic hazard assessment and stochastic site response analysis at the archaeological site of Kancheepuram in Southern India*, Research Report EUCENTRE 2009/01, IUSS Press, Pavia, pp. 250.

Magenes, G., Prasad, A.M., Dodagoudar, G.R., Lai, C.G., Macchi, G., Mathews, M.S., Menon, D., Menon, A., Pavese, A. and Penna, A. 2006. Indo-Italian joint research programme on seismic vulnerability of historical centres in south India, In P.B. Lourenco, P. Roca, C. Modena, S. Agrawal (ed.), *Proc. of the V International Conference on Structural Analysis of Historical Constructions, New Delhi, India, 2006,* pp. 1667-1674.

Manohar, S., Balamurugan, K., Shukla, S., Haneefa, M., Santhanam, M., Menon, A. 2021. *Multiscale fire damage assessment stone trabeated hypostyle halls*, International Journal of Architectural Heritage, Published online: 27 October 2021. DOI: 10.1080/15583058.2021.1992535.

Mathews, M.S., Menon, A., Chandran, S. 2003. Rehabilitation and retrofit of earthquake damaged monuments in Gujarat, in V. Jeyaraj (ed.) *Special volume on conservation of stone objects, The Commissioner of Museums, Chennai.*

Menon, A., Shukla, S., Samson, S., Aravaind, N., X. Romão, E. Paupério, A. 2017. Field Observations on the Performance of Heritage Structures in the Nepal 2015 Earthquake: Invited Lecture, *Proc. 16th World Conference on Earthquake Engineering, Santiago, Chile, January 9-13, 2017.*

Naik, P.M., Bhowmik, T. and Menon, A. [2021] *Estimating joint stiffness and friction parameters for dry stone masonry constructions*, Int. J. Masonry Research and Innovation, Vol. 6, No. 2, pp.232–254.

Ochsendorf, J. A. 2006. *The masonry arch on spreading supports.* Structural Engineer, 84(2), 29–35.

Palmieri M., Magenes G., Lai C.G., Penna A., Bozzoni F., Rota M., Macchi G., Auricchio F., Mangriotis M.D., Menon A., Meher Prasad A. & Murty C.V.R. 2012. Reduction of seismic risk of Roman and Hindu temples. In P.B. Lourenco, P. Roca, C. Modena, (ed.), *Proc. of VIII International Conference on Structural Analysis of Historical Constructions, 15–17 October, Wroclaw, Poland, 2012.*

Geotechnical Engineering for the Preservation of Monuments and Historic Sites III – Lancellotta, Viggiani, Flora, de Silva & Mele (Eds)

Site effects and intervention criteria for seismic risk mitigation in the ancient city of Pompeii: The case of the *Insula dei Casti Amanti*

L. de Sanctis, M. Iovino, R.M.S Maiorano & S. Aversa
University of Naples Parthenope, Naples, Italy

ABSTRACT: This paper deals with site amplification effects and possible strategies for seismic risk mitigations in the ancient Roman city of Pompeii, the well–known archaeological site near Naples, Italy, buried under 4–6 m of volcanic ashes and pumices during the eruption of Mount Vesuvius in 79 AD. The attention is focused on the restoration works of the *Insula dei Casti Amanti*, a block of masonry buildings in the city centre partly excavated in recent years, which included the stabilization of the excavation fronts, the preservation of the archaeological ruins and the replacement of the actual roofing system realised some decades ago. Two dimensional site amplification analyses carried out to quantify the seismic demand in the excavation fronts allowed to recognize the occurrence of remarkable aggravation phenomena at the crest of the slopes owing to the strong interference between morphology and topography. It is shown that a Newmark–type approach is adequate in this case to define the profile of the artificial slopes in a sustainable way. The focus is then set on the foundations of the new covering system of the *Insula*, a single span steel truss. A proper consideration of the load path followed until failure and the concept of interaction diagrams are fundamental ingredients to reduce as much as possible the impact of the new foundation system. As a final remark, the attention is placed on the likelihood occurrence of a double resonant mechanism between the subsoil and the archaeological ruins and on actions that have to be taken for the protection of the ancient city from future earthquakes.

1 INTRODUCTION

The ancient Roman city of Pompeii, near Naples, is among the most famous UNESCO World heritage sites. It was buried under 4 to 6 m of ashes and pumices after the eruption of Mount Vesuvius in 79 AD. This catastrophic event destroyed the city, killing about two thousand inhabitants but, at the same time, it paradoxically preserved its buildings, with their contents, for many centuries, until the breakthrough of its ruins, during the Reign of Bourbon King Carlo III, in 1748.

After the major earthquake of 62 AD, which caused widespread destruction in the bay of Naples, especially in Pompeii and surroundings, Mount Vesuvius erupted violently on 79 AD spewing forth a cloud of hot ashes and super–heated gases 14–32 km high as proved by Carey & Sigurdsson (1987). The eruption lasted 19 h from 1 pm of August, 24th to 8 am of August, 25th. On the first day, pumices and ashes began to fall down at an accumulation rate of 15 cm/h blanketing the city up to 2.8 m (Sigurdsson et al. 1985) and forming what is known as *Fall Units* (FU) or *Pyroclastic Fall.* Under the weight of the accumulated material, the roofs of many *Domus* collapsed, causing approximately 38% of the total casualties in Pompeii (Luongo et al. 2003). Yet rescues and escapes were still possible at that time. But the eruptive column collapsed and a rapid–moving, dense and very hot flow of molten rock, ashes and gas, referred to as *Pyroclastic density current* (PDC) or *Surge* of the Plinian event, began to slide downhill, knocking down every structure along its path

DOI 10.1201/9781003329756-12

and, what is more, burning and/or asphyxiating any living being. Unfortunately, many inhabitants had just come back in the city after fleeing from the pyroclastic flow to rescue their stuff, thus going to meet a dreadful death. On the morning of August 25th, the large part of the Pompeii buildings resulted almost completely destroyed and only some buildings were still standing. It goes noting saying that this was a tragedy. From a different perspective, the circulation of air in the deposit of *Fall Units* allowed to keep the buried remnants almost dry, while the overlying *Surge* acted as a waterproof layer against rain infiltration. This was a lucky condition that made it possible to protect the ancient city for many centuries. Nowadays, two third of the urban agglomerate within the city walls have already been excavated. The ruins brought to light are like 'story–tellers', since they tell us – with an extraordinary narrative intensity – the last hours of life in Pompeii and, above all, the dynamic of the catastrophic event. Such preliminary discussion has great relevance to plan and implement actions for the preservation of the archaeological site.

In this paper, the focus is set on the case study of the *Insula dei Casti Amanti*, a block of masonry buildings in the city centre brought to light by recent archaeological excavations. This choice comes from the fact that the area surrounding this block has the utmost density of in situ investigations within the city walls. This is a very lucky circumstance, as the chance to carry out boreholes and other in–situ investigations is strongly limited because of the very severe restrictions of the *Archaeological Park of Pompeii*. However, the geotechnical issues related to the restoration works in the *Insula* are widespread in the ancient city of Pompeii, so that this case study can be hold as a paradigm for future actions in the archaeological site. An attempt is finally made to fathom the main findings of this local study into general conclusions so as to stimulate the debate about the most appropriate actions to be taken in the future for the seismic protection of the site.

2 SUBSOIL PROFILE IN THE ANCIENT CITY

Figure 1 shows the plan view of the ancient city of Pompeii, with the position of the boreholes carried out for a major project granted by the *Italian Government,* aimed at the preservation, conservation and public utilization of the archaeological resources of the ancient city and referred to as *Grande Progetto Pompei*. Most of the available investigations is from the ground surface of *Regiones* I–III–IV–V–IX – that have been only partly excavated – and just outside the city walls, while ample areas are devoid of boreholes, where the archaeological excavations have already been completed (*Regiones* VI–VII–VIII).

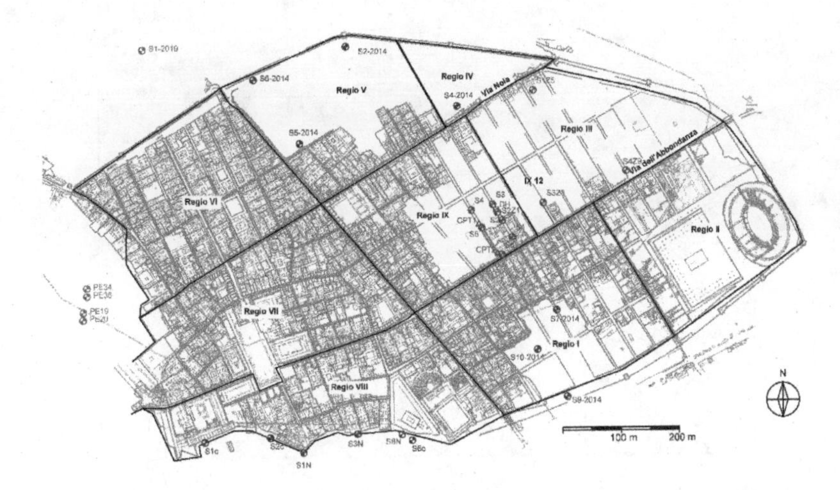

Figure 1. Plan view of the ancient city of Pompeii with the boreholes within the city walls.

The soil profile has been reconstructed in some of the *Regiones*. The typical subsoil conditions are shown in Figures 2–3. From ground surface, after a tiny layer of made ground, we first encounter the column of the Plinian eruption, consisting of the volcanic ashes of the so–called *Surge* underlain by the pumices deposit of the first phase of this eruption, until a depth of 4.7–6.8 m, which is the thickness of the materials blanketing the city. The level below the pumices consists of a paleosoil with varying thickness, interbedded with pyroclastic gravelly sand originated from the explosion of previous eruptive cycles, so that the separation between the paleosoil and the pyroclastic gravelly sand cannot be clearly distinguished. This layer was found until a depth varying between 8.0 and 9.7 m from ground level and, below it, a transition zone consisting of lava slags and dense sandy gravel can be considered the alteration of the underlying lava layer. The transition zone has often an erratic profile, as shown in Figures 2 and 3. Particularly, the morphological depression detected along the east alley of *Insula* XI–12 is worthy of note, as it will be explained in the ensuing.

In some boreholes (for example S10 in Figure 2), a chaotic level of building materials, including fragments of masonry, brick strips and travertine, mixed with pyroclastic sand and pumices, was found below the ancient paved road. In all likelihood, such levels are heaps of building ruins coming from the collapses caused by the major earthquake in 62 AD.

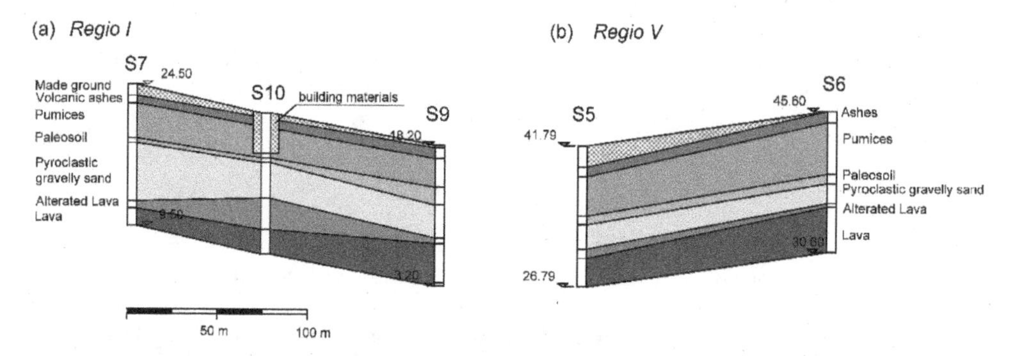

Figure 2. Soil stratigraphy in the area of *Regio V* (a) and *Regio V* (b). Ratio of elevation over distance is 1/5.

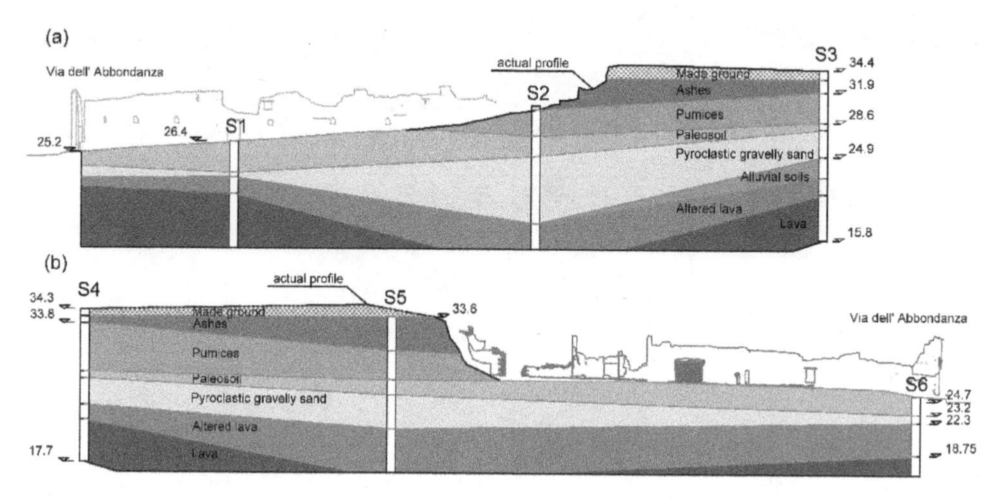

Figure 3. Soil stratigraphy in the area of *Insula* IX–12. Ratio of elevation over distance is 1/1.

3 THE CASE STUDY OF INSULA DEI CASTI AMANTI

3.1 *The Insula dei Casti Amanti: a magnificent story–teller*

The *Insula dei Casti Amanti* takes its name from a fresco on a wall in one of the houses located in this block, in which two lovers are kissing in a chaste way, in comparison with the subjects of other frescos in Pompeii (Figure 4). It is also referred to as IX–12, where the Latin number indicates the *Regio* in which the block is located. As shown in Figure 1, *Insula* IX–12 is delimited by the east–west paths of *Via Nola* and *Via dell'Abbondanza*, two of the main roads called *decumani*, and it is crossed by two alleys, which have been partly brought to light because of some recent archaeological excavations.

Figure 4. The fresco of the Chaste Lovers.

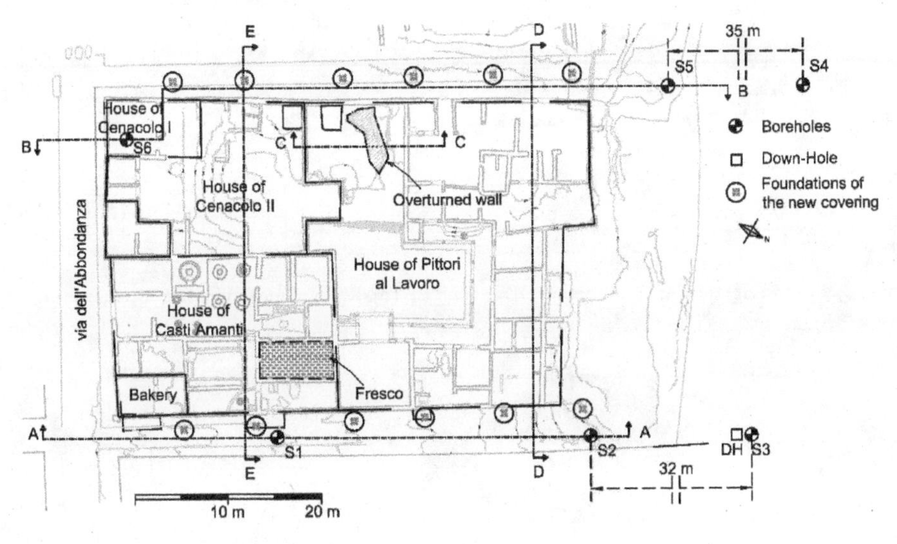

Figure 5. Plan view of the *Insula dei Casti Amanti*.

The archaeologist Vittorio Spinazzola began the excavations, which brought to light the *Insula* in 1911, starting from *Via dell'Abbondanza*. The basic idea of Spinazzola's project was to proceed along the main streets, in order to avoid problems generated by previous excavations strategies, such as the collapse of the *domus* upper floors, whose presence was proved by wall paintings recovered up until then. The idea of Spinazzola revealed as correct and, during the excavations in

Via dell'Abbondanza, the fronts of two– and three–storey buildings were recovered and, for the first time, the productive and commercial fabric of Pompeii, until then completely unknown, came to light. In 1924, the archaeologist Amedeo Maiuri followed the work of Spinazzola, continuing with the excavations in *Via dell'Abbondanza* and completely recovering all the *Insulae* of *Regiones I and II*. The *Insula dei Casti Amanti* was partly recovered in the period 1928–1934. It includes the following buildings: the *Casa del Cenacolo I and II*, the *Casa dei Casti Amanti*, partly devoted to bakery, the *Casa dei Pittori al Lavoro* (Figure 5). In all likelihood, at the time of the eruption in 79 AD, the *Casa dei Casti Amanti* was undergoing repair works after the damage produced by the 62 AD earthquake, while the *Casa dei Pittori a Lavoro* was distinguished by some preparatory works for wall paintings. Unfortunately, the *Casa del Cenacolo II* was seriously and irremediably damaged by a bombing raid in 1943, during World War II.

The soil profile near the faces of the artificial slopes was directly obtained from the visual examination of the outcropping soil. Along the western alley, the local stratigraphy clearly shows the transition between the pumices and the overlying volcanic ashes. There is also a perfect correspondence between the soil layering and the profile of the slope in that the front is steeper on the ashes layer and flatter on the pumices deposit, as shown in Figure 6a. As a noteworthy point, the ancient masonry walls on the boundary of adjacent *Insula* IX–11 improperly act as retaining walls, because of the excavation of west alley. Not surprisingly, these walls suffered severe damages, such as the partial collapse which brought to light the colonnade of *Insula* IX–11 and the pronounced tilting at the beginning of this alley, as shown in Figure 6b. A startling tale comes from a portion of wall still lying on a bed of pumices in a partially submerged room (Figure 7). This block collapsed by overturning under the impact of the pyroclastic flow and is a clear witness of the dynamic of the eruptive event. As highlighted by Luongo et al. (2003), the pyroclastic flow moved downslope along the direction perpendicular to *via dell'Abbondanza*.

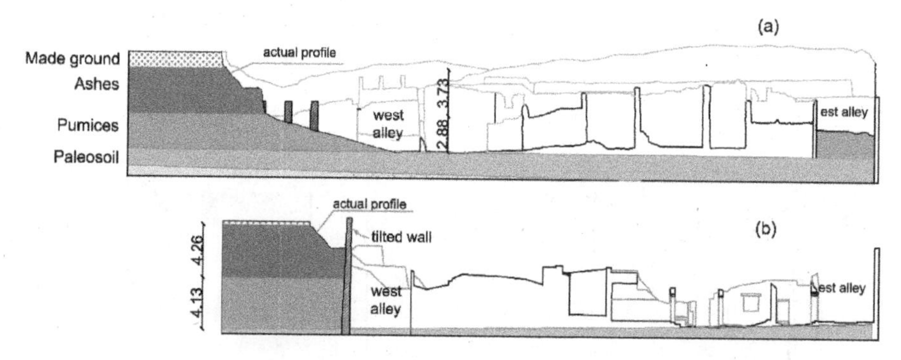

Figure 6. Soil stratigraphy in cross section DD (a) and EE (b). Ratio of elevation over distance is 1/1.

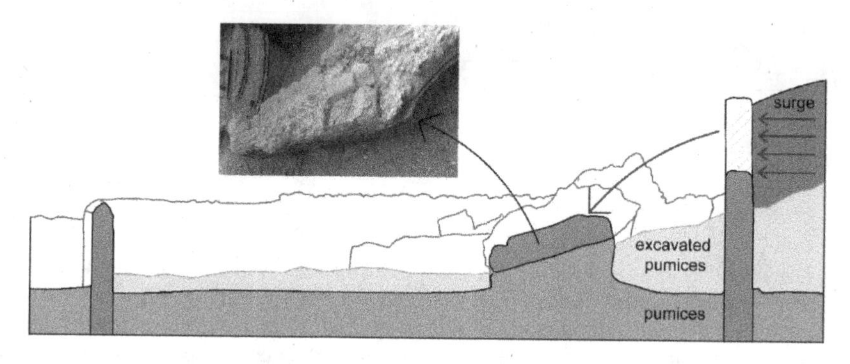

Figure 7. Cross section CC: overturned wall laying on the bed of pumices.

3.2 *Subsoil conditions in the area of the Insula*

The subsoil conditions in the area of the *Insula* have been already shown in Figure 3. They are in line with the general description of the subsoil layering supplied for the whole city in Section 2. The stratigraphy in this area may be summarized as follows: made ground from the soil surface until a depth of 1.3 m; volcanic ashes of the *Surge* up to a depth between 1.6 and 3.7 m; pumices deposit until a depth of 4.7–6.8 m; paleosoil with varying thickness up to a depth varying from 8 and 9.7 m, interbedded with pyroclastic gravelly sand originated from the explosion of the previous eruptive cycle; a transition zone consisting of lava slags and sand gravel; the bedrock of lava.

Figure 8 shows a schematic subsoil profile together with the results of the Down–Hole test, the Standard Penetration Test blow counts (N_{SPT}) and the geotechnical parameters obtained from all the available in situ and laboratory investigations carried out within the city walls (de Sanctis et al. 2019). It is worth mentioning that the cohesion of the volcanic ashes in Figure 8 is partly an 'apparent' term, due to the partial saturation of the soil. This quantity was back–figured by limit equilibrium analysis of a very steep front located in *Regio I* in which the safety factory was taken equal to 1.

From the roof of the altered lava downward, the average value of the shear velocity exceeds 800 m/s, so that this lava layer can be idealized as the reference bedrock. Figure 9 shows the decay of normalized shear modulus, G/G_0, and the variation of damping ratio, D, with the level of shear strain, γ, as obtained from soil samples retrieved from the volcanic ashes (S7C2) and the paleosoil (S7C5). For the purpose of the ground response analyses, the curves for the pumices and the pyroclastic gravelly sand are taken coincident with that of the paleosoil.

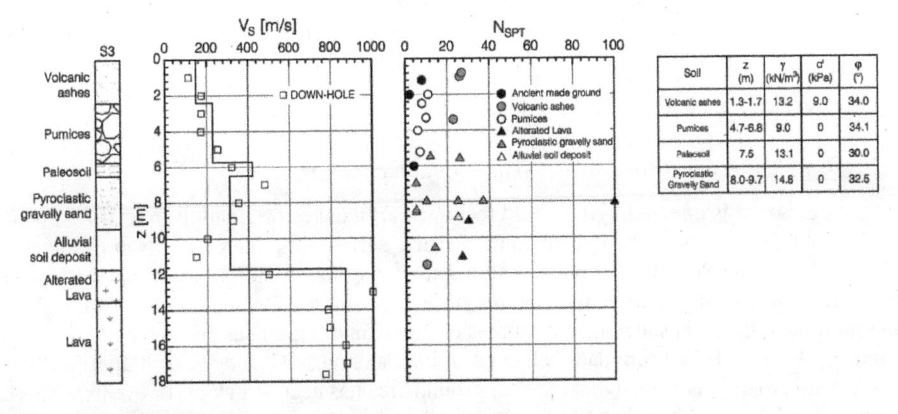

Soil	z (m)	γ (kN/m³)	c' (kPa)	φ (°)
Volcanic ashes	1.3-1.7	13.2	9.0	34.0
Pumices	4.7-6.8	9.0	0	34.1
Paleosoil	7.5	13.1	0	30.0
Pyroclastic Gravelly Sand	8.0-9.7	14.8	0	32.5

Figure 8. Subsoil properties.

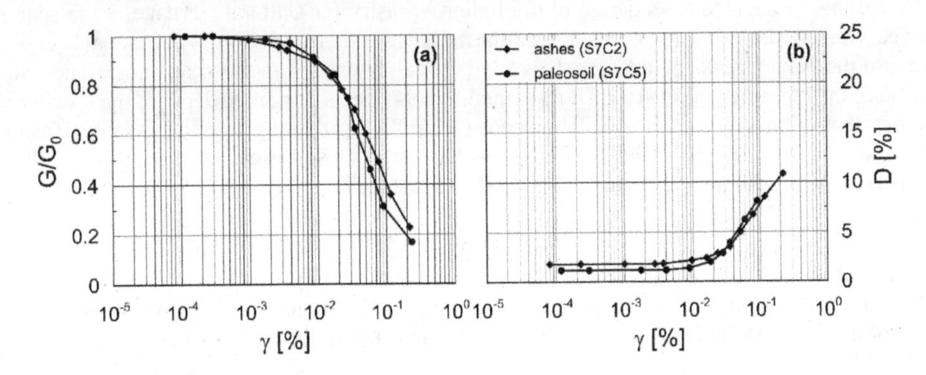

Figure 9. Results of the Resonant Column tests.

3.3 *Restoration works*

The restoration interventions of *Insula* IX–12, planned in the *Grande Progetto Pompei*, include the conservative restoration of masonry and paintings, safety measures for the excavation fronts and the design of the new *Insula* roof. The existing slopes surrounding the *Insula dei Casti Amanti* originated from previous archaeological excavations. Many small sliding events occurred in the past, indicating that these artificial slopes need to be stabilized. The excavation of the inner part of the *Insula*, not yet completed, was carried out in different phases. For this reason, the *Insula* was protected against rain and other meteorological events by a system of provisional roofs, each of them designed and built during the related excavation. Now this cover system is no longer adequate and it must be replaced with a single–span roof, which will permit the ground level of the whole *insula* to be left free of pillars. The third important purpose is the protection of the archaeological ruins from degradation owing to ageing effects and future earthquakes.

The geotechnical aspects considered in this projects are: (a) site amplification phenomena; (b) the stability of excavation fronts and elevator walls against the action of gravity and earthquakes; (c) seismic demand in the new covering and the limit state analysis of the pertinent foundations; (d) the seismic protection of the archaeological ruins. The geotechnical design was carried out taking into account the very relevant archaeological and historical value of the site, and taking advantage of what was defined by the ISSMGE – TC301 – *Preservation of Historic sites* (Aversa 2007, Burland & Standing 1997, Calabresi & D'Agostino 1997, Croce 1985, Flora 2013, Jappelli 1997, Viggiani 2013). Specifically, the following principles were adopted for the selection of the most appropriate solutions: (a) preservation of still–buried archaeological ruins; (b) minimum invasiveness of the excavations needed to re–profile the artificial slopes; (c) reduction of the impact of the new roof foundations; (d) reversibility and removability of the interventions; (e) reduction of the earth pressure on the walls of adjacent *Insula* IX–11; (f) full access to the archaeological site.

4 GROUND MOTION AMPLIFICATION

4.1 *Background seismicity and selection of natural recordings*

The seismic hazard is affected by far–field tectonic earthquakes originating from the Appennine chain, and by the activity of the two volcanic districts surrounding the city of Naples, the *Phlaegreas Fields* and the Somma–Vesuvius. The 6.9 moment magnitude (M_w) Irpinia earthquake in 1980 originated from the tectonic faults along of the Appennines (Meletti et al. 2008). A detailed representation of the main seismogenic sources in Campania Region has been recently reported by Ebrahimian et al. (2019). From data collected in the Parametric Catalogue of Italian Earthquake (Rovida et al. 2016), it is argued that the background area has been struck by 10 events with M_w in the range 6.1–7.2 over the period 1349–1980. Further details can be found in Licata et al. (2019) and Ebrahimian et al. (2019).

According to the recommendation of the Italian Ministry for Cultural Heritage, a life span of 50 years and a class of use IV were adopted for the excavation fronts. This is a reasonable choice, because the boundary of the excavated area will certainly expand in the near future. The return period of the life–safety limit state (ULS) is therefore 949 years. Since the depth of the Lava's roof is lower than 30 m, according to the Italian code (NTC 2018), adopting a similar soil classification criterion as Eurocode 8 EN–1998–1 (CEN 250 2003), the subsoil is classifiable as type E. Thus the maximum acceleration expected at the crest of the slope, a_{max1}, is:

$$a_{max1} = S_r \cdot a_r = 1.539 \times 0.168 = 0.259g \quad (1)$$

where S_r is the soil amplification factor and a_r the peak acceleration on outcropping rock. On the other hand, the life span of the new covering systems is taken equal to 100 years. With the same class of use as per the excavation fronts, the maximum acceleration at ground surface, a_{max2}, is

$$a_{max2} = S_r \cdot a_r = 1.415 \times 0.205 = 0.290g \quad (2)$$

On the other hand, the life span of the new covering system is taken equal to 100 years with the same class of use. The return periods of the life–safety limit states are therefore 949 and 1898 years, respectively. The same set of outcropping recordings defined by de Sanctis et al. (2020) is adopted for seismic site response analyses for both the excavation fronts and the new steel truss covering. In particular, the design earthquakes ($M_w = 5.5$–7 and epicentral distance, $R = 25$–70 km) were defined by a preliminary de–aggregation (partitioning) of the seismic hazard into selected magnitude and distance, so as to identify the modal contributions to the overall site hazard. Table 1 summarizes the main characteristics of the input signals (EQ) selected to define the response spectrum of the horizontal acceleration and the scaling factors (scal) corresponding to Equations (1) and (2).

Table 1. Input signals and scaling factors.

EQ	Code	Event	Date	M_w	R [km]	a_{max} [g]	scal (1)	scal (2)
1	290xa	Campano–Lucano	23/11/1980	6.9	32	0.216	0.777	0.950
2	290ya	Campano–Lucano	23/11/1980	6.9	32	0.323	0.521	0.636
3	292xa	Campano–Lucano	23/11/1980	6.9	25	0.060	2.804	3.428
4	292ya	Campano–Lucano	23/11/1980	6.9	25	0.060	2.805	3.429
5	949xa	Sicilia–Orientale	13/12/1990	5.6	29	0.061	2.754	3.367
6	4678ya	South Iceland	17/06/2000	6.5	32	0.053	3.146	3.846
7	5271ya	Mt. Vatnafjoll	25/05/1987	6.0	42	0.053	12.595	15.397

Figures 10a,b show the response spectra at a damping ratio ξ equal to 5% of the selected accelerograms scaled to 0.168g and 0.205g, respectively. Also shown for comparison is the mean spectral function and the spectrum specified by the Italian code for the life–safety limit state (SLV). In both situations, the mean spectral function compares well with the reference spectrum prescribed by the Italian code.

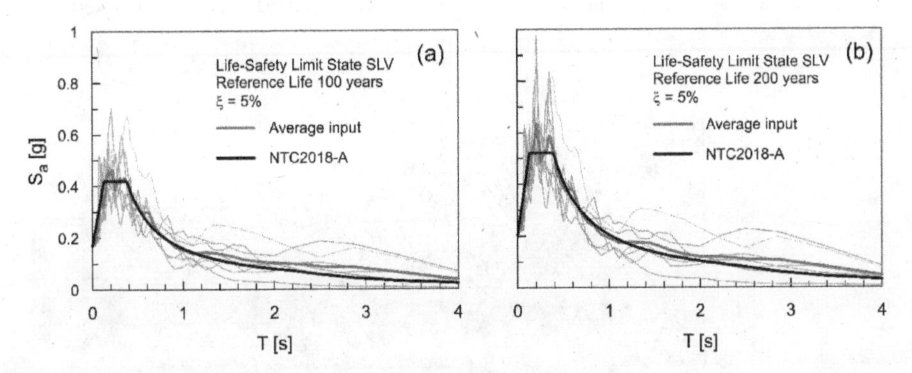

Figure 10. Response spectra of selected earthquake for excavation fronts (a) and the new covering (b).

4.2 *Excavation fronts: one– and two–dimensional wave propagation analyses*

Site amplification effects are evaluated by two dimensional (2D) analyses of wave propagation using the equivalent linear (EL) approach. The code LSR–2D (Local Seismic Response 2D rel. 4.3.1, Stacec) based on the finite element (FE) approach developed by Hudson et al. (1993) is adopted. Figures 11–12 illustrate the FE element domain for cross sections AA and BB, respectively. The natural recordings scaled to 0.168 g for seismic analysis of the excavation fronts are considered. They are applied to all nodes belonging to the lower boundary of the mesh, taking into account the change due to wave propagation from the outcropping rock to the lower bound of the model.

The energy dissipation due to radiation through the base of the model is accomplished by means of horizontal (c_x) and vertical (c_y) viscous dampers with properties:

$$c_x = \rho_b \cdot V_{sb}$$
$$c_y = \rho_b \cdot V_{pb} \tag{3}$$

where ρ_b, V_{sb} and V_{pb} are density, shear (s) and compression–dilation (p) wave velocity of the underlying lava layer. Theoretically, the displacement of the lateral boundary of the FE mesh should be equal to those in free–field. However, in case of low soil damping, the lateral boundaries should be far distant from the slopes to match the above conditions so that they are coupled with an isolated soil column with the same soil stratigraphy through a series of viscous dashpots, thus avoiding the re–introduction into the model of reflected waves.

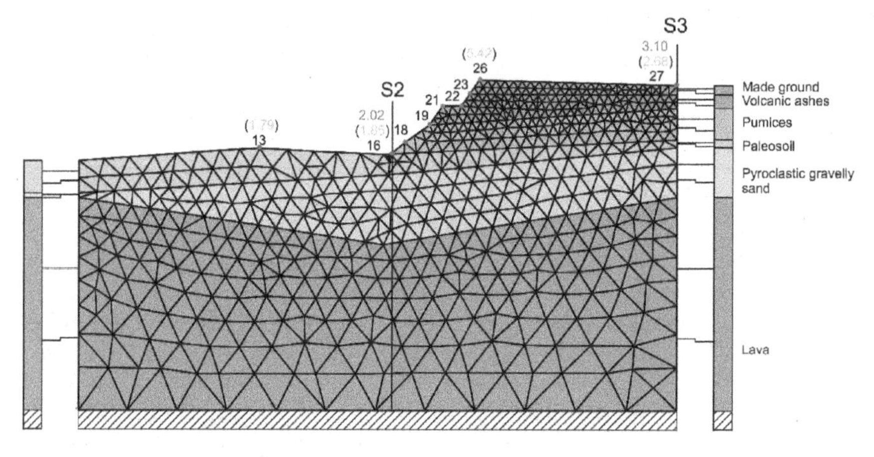

Figure 11. Section A, amplification function in 1D (out of brackets) and 2D analyses (in brackets).

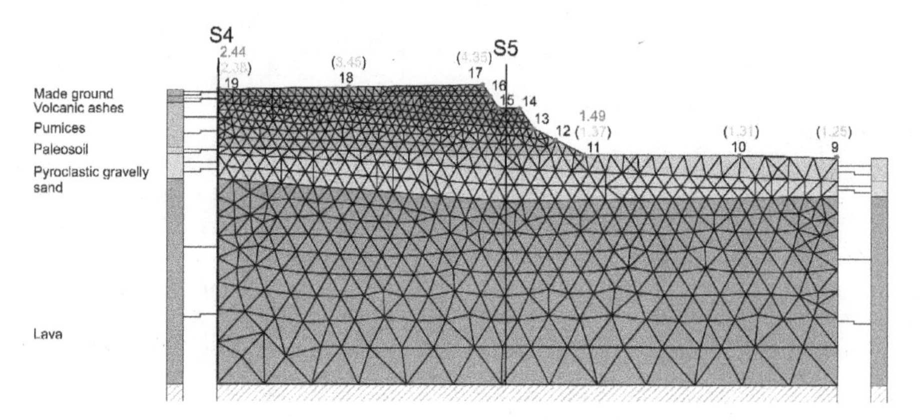

Figure 12. Section B, amplification function in 1D (out of brackets) and 2D analyses (in brackets).

The amount of amplification at some locations are shown in both the Figures 11 and 12 as well. The peak acceleration has to be intended as the mean value of the maximum ground acceleration for the selected earthquakes. The outcome of one dimensional (1D) ground response analysis is also shown for comparison at the toe and behind the crest of the slope. Such data have been gathered with the soil stratigraphy of the vertical line containing the selected point, so as to decouple soil (1D) effects from surface (topography) and morphology (valley) amplification. For cross section AA, the amplification factor is 5.42 at the crest (point 26), 1.85 at the toe (point 16) and 2.68 on the boundary (point 27). The effect at the crest is worthy of note and, what is more, far greater than the amplification factor based on conventional subsoil classification (1.539). The situation for cross section BB is similar; in this case the amplification factor is 4.32 at the crest (point 17), 1.37 at the toe (point 11) and 2.38 on the boundary (point 19). This outcome indicates that the method based on conventional subsoil classification fails completely the prediction of amplification effects behind the slope. The overall effect in cross section AA is even greater than that expected for cross section BB because of the additional amplification owing to the local morphological depression of the lava layer. Following Ashford et al. (1997) and Bouckovals & Papadimitriou (2005), topography and valley effects may be expressed through the 'aggravation' factors, A_{crest} and A_{toe}, defined as:

$$A_{crest} = \frac{a_{max}}{a_{crest}} \qquad A_{toe} = \frac{a_{max}}{a_{toe}} \tag{4}$$

where a_{max} is the maximum acceleration of the reference point at ground surface from 2D analysis, a_{toe} the maximum acceleration at the base of the slope and a_{crest} the one occurring at the crest with the last two quantities evaluated in free–field. The aggravation factor A_{crest} is 1.75 for cross section AA and 1.78 for cross section BB. According to Bouckovalas & Papadimitriou (2005), who provided a deep insight into the phenomenon of aggravation in proximity of a slope in homogeneous soil conditions, this factor lies in between 1.15 and 1.45. They also suggest to take a unit aggravation factor at the toe. This recommendation is fully consistent with the value of 0.92 coming from 2D analysis. The very pronounced peak of ground acceleration at the crest gathered from FE analysis is due not only to surface (topography) and morphology (valley) effects, but also to the intense interaction between topographic amplification and soil effects originating from the very shallow compliant rock. As a final comment, a proper modelling of local soil condition and geomorphological properties is mandatory for a reliable estimation of the seismic demand in the excavation fronts.

Figure 13 illustrates the profile of maximum acceleration at ground surface from the selected earthquakes scaled to 0.168 g. The leap of this function on the crest is certainly remarkable. Yet the profile is almost constant on the side of the covering, suggesting that a 1D response analysis may be acceptable to evaluate the seismic demand in this structure.

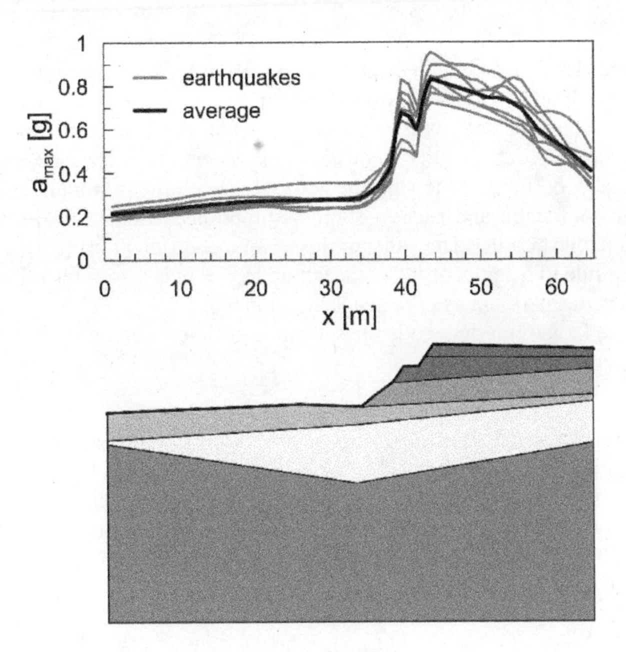

Figure 13. Profile of maximum acceleration at ground surface.

4.3 *New covering system: one-dimensional wave propagation analysis*

The local stratigraphy corresponding to the position of borehole S2 is first adopted for 1D ground response analysis. Figure 14a shows the response spectra of the selected earthquake at ground surface, the mean spectrum and the reference one based on conventional subsoil classification. Compared to this last function, the reduction of spectral acceleration at the very large period of the isolated structure is worthy of note. By contrast, at the very low structural periods, the mean function outstrips the code–spectrum, but this does not matter for the new covering. The change of mean spectra occurring from one alley to another is also investigated. Figure 14b shows the mean function corresponding to point 11 in Figure 12, on the opposite alley. It is argued from the comparison between these two figures that the local stratigraphy has a very little influence on the seismic demand for the covering system.

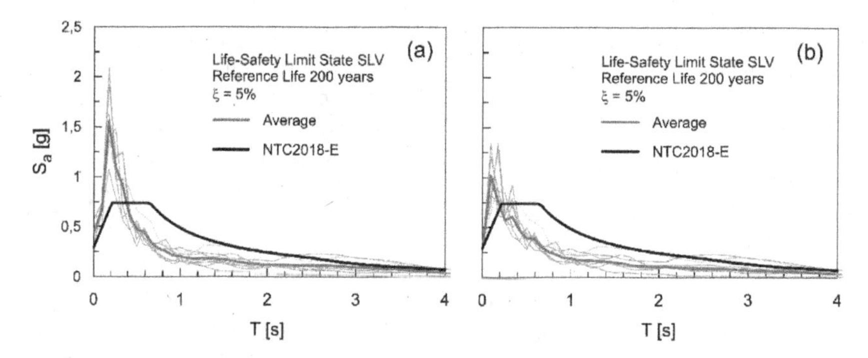

Figure 14. Response spectra from 1D analyses for cross sections AA (a) and BB (b).

5 EXCAVATION FRONTS

5.1 *General*

The most widespread approach in geotechnical engineering for slope stability analysis is the limit equilibrium method. It allows to identify the factor of safety (SF) of any potential failure surface. Many works in the literature deal with this approach. In all these methods, the earthquake action is idealized by a seismic coefficient (k_h) constant in space and time (pseudo–static analysis). The critical slip surface is defined as the slip line associated with the minimum safety factor. It may be defined under both static and pseudo–static assumptions, with the two critical surfaces not necessarily equal to one another. The value of the seismic coefficient bringing the soil mass above the critical sliding line to a status of incipient failure (SF = 1) is also referred to as 'critical' (k_{hc}) (and corresponding acceleration as critical acceleration a_c).

According to the Italian building code the seismic coefficient can be evaluated as:

$$k_h = \alpha \frac{a_{max}}{g} \tag{5}$$

where a_{max} is the maximum acceleration within the potentially unstable soil mass and α is a reduction factor accounting for the amount of tolerable displacement of the slope. In case of excavation fronts, for the earthquake action of the life–safety limit state the above reduction factor is equal to 0.38. Therefore, evaluating the maximum acceleration at the soil surface through the conventional subsoil classification would lead to $k_h = 0.10$. According to EN–1998–5 (CEN 250 2019), the equivalent seismic coefficient shall be taken as:

$$k_h = \frac{\beta}{\chi} \frac{a_{max}}{g} \tag{6}$$

where β is a reduction coefficient reflecting the ground motion asynchrony and χ a reduction coefficient accounting for the amplitude of the acceptable residual displacement for the considered limit state. As a matter of fact, χ is equivalent to the inverse of reduction factor α in Equation (5). For the life–safety limit state this coefficient shall be taken equal to 2. The reduction factor reflecting the spatial variability of ground motion may be evaluated as:

$$\begin{aligned} &\beta = 1\lambda \leq 0.05 \\ &\beta = 1 - \tfrac{0.65}{0.35}(\lambda - 0.05) \; 0.05 < \lambda \leq 0.4 \\ &\beta = 0.35\lambda > 0.4 \\ &\lambda = \tfrac{H}{T_m V_s} \end{aligned} \tag{7}$$

where H is the height of the slope, T_m the average period of the seismic action (Rathje et al. 1998) and V_s the mobilized shear wave velocity over the height of the slope. This simplified approach yields $\beta = 0.85$ for cross section AA and $\beta = 0.83$ for cross section BB, indicating about an additional 16% reduction due to the asynchrony of the motion as an average. Therefore, according to the above draft, the equivalent seismic coefficient would be about 0.11 for both the examined cross sections.

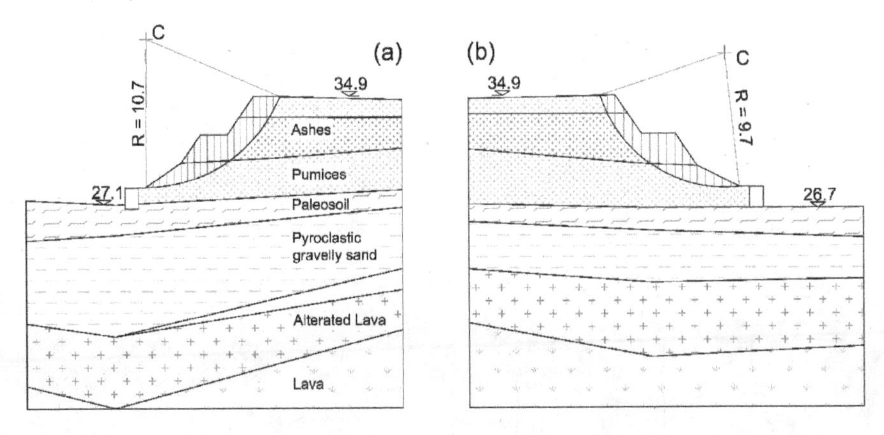

Figure 15. Critical surfaces for cross sections AA (a) and BB (b). Distances and elevation above sea level are in m.

The stability analyses are preliminary carried out by the pseudo–static approach following the method of Sarma (1975), which allows to evaluate the seismic coefficient corresponding to any prescribed value of the safety factor and vice versa. The selected method of analysis can be also applied under static conditions, by setting the seismic coefficient equal to zero. In this case, the critical slip lines (chosen as circular) represented in Figure 15 are obtained. The corresponding factor of safety are 1.342 and 1.463 for cross sections AA and BB, respectively. According to NTC (2018), under static conditions the partial safety factor for soil parameters (γ_M) is equal to 1.25, while the partial factor for resistance (γ_R) is 1.1. Since the slopes under examination are not subjected to variable load, the minimum allowable safety factor is $SF_{min} = \gamma_R \cdot \gamma_M = 1.375$. Although the SF of cross section AA is slightly smaller than SF_{min}, the excavation fronts can be considered stable enough under static conditions.

5.2 *Pseudo–static analysis*

A problem arises from the definition of the seismic coefficient equivalent to the earthquake induced lateral forces, variable both in space and time. Notably, while there is – at any instant of time – a remarkable variation of the calculated acceleration within the soil domain along the vertical direction, the variation in the horizontal direction is almost negligible. Based on this outcome, the asynchronism of ground motion is taken into account in a simplified manner. First, reference is only made to points at ground surface from the toe to the crest of the slope. Second, for each input signal, the evolution by time of the acceleration profile corresponding to points at ground surface on the slope is evaluated and the instant of time at which the integral of this profile is maximum (or minimum) is identified. The critical profiles of the selected earthquakes obtained by projecting horizontally target points at the ground surface on the y–axis are plotted in Figure 16. The resulting mean profile is also shown for both sections. This procedure leads to an average acceleration of 0.4 g for section A and 0.37 g for section B. The corresponding seismic coefficients (k_h), evaluated by taking the above equivalent accelerations in conjunction with a reduction factor $\alpha = 0.38$, are 0.152 and 0.140. They include soil surface (topography), morphology and ground motion asynchrony effects. As a noteworthy point, they are much greater than that gathered through the simplified approach by either the Italian code ($k_h = 0.10$) or EN–1998–8 revision ($k_h = 0.11$). The effect of the spatial variability of the ground motion may be roughly estimated by relating the average acceleration within the height of the slope to that expected at the crest. The ratio of these two quantities, which is equivalent to reduction factor β in Equation (5), is equal to 0.44 for Section A and 0.51 for Section B. Therefore, the simplified approach recommended by draft of EN–1998–5 revision underestimates the effect of ground motion asynchrony in a remarkable way.

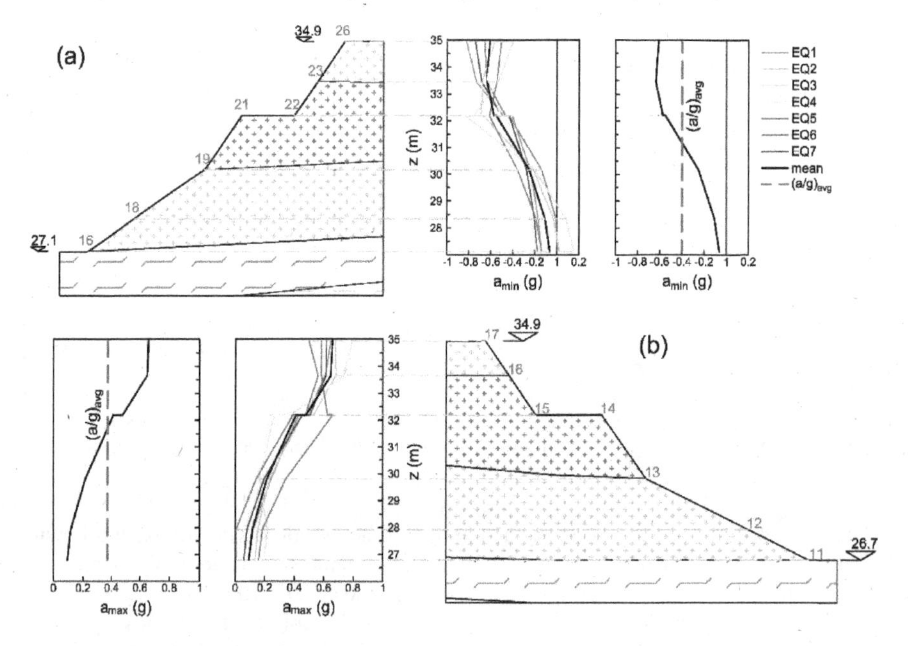

Figure 16. Acceleration profiles for cross section AA (a) and cross section BB (b).

The critical slip lines under pseudo–static assumption are identical to those identified in static conditions (Figure 15). The safety factors obtained by this approach for cross sections AA and BB are 1.05 and 1.17, respectively. Both are below the lowest allowable value prescribed by the Italian code for excavation fronts under earthquake actions ($SF_{min} = 1.2$). The distance from this value is particularly relevant for cross section AA, characterized by the morphological depression of the lava layer. As a final outcome, the slopes would be not stable enough according to the pseudo–static approach. Yet they could be re–profiled as shown in Figure 17. However, this expanded excavation would invade the archaeological remnants still buried. Furthermore, a flatter slope akin that in Figure 17 would imply far greater rainfall infiltration, a reduction of water suction and, consequently, a decrease of the 'apparent' cohesion. This design option is therefore in contrast with the need of preserving the archaeological ruins.

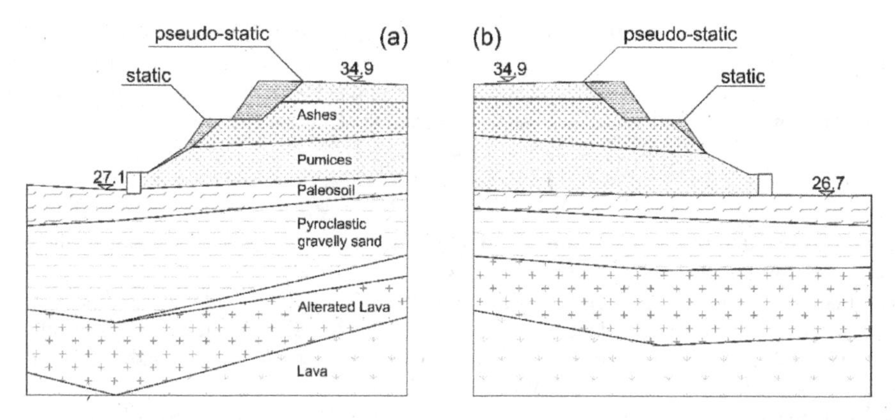

Figure 17. Slope profiles after conventional pseudo-static analysis for cross sections AA (a) and BB (b).

5.3 *Dynamic analysis by rigid block theory*

The stability of engineered earth slopes may be evaluated in several ways. The pseudo–static approach can only check whether or not slope instability will occur by judging the safety factor. An alternative, more elegant, option is the dynamic analysis by the rigid block theory (Newmark 1965), in which the landslide is modelled as a rigid–plastic friction block having a known critical acceleration. This method permits to evaluate the permanent displacements of the block subjected to an earthquake; the significance of the accumulated displacement is then judged. Compared to more sophisticated finite element calculations, the Newmark's model is a workable means, yielding much more useful information than the conventional pseudo–static approach (Jibson 1993). Permanent displacement analysis begins exactly when pseudo–static analysis ends, at the point where the critical acceleration, a_c, is exceeded. The block continues to move until the relative velocity between the soil mass and the base reaches zero and will slip again if the acceleration exceeds one more a_c. The process continues until the relative velocity drops to zero for the last time. While the classical Newmark theory was developed for translational sliding mechanism, the critical surface of the examined slopes has a circular shape. In this case, the damage or failure mechanism is represented by the accumulation of permanent rotations. A modification of the classical Newmark theory was therefore developed in this work to calculate the rotational displacement of the slopes. A similar approach has been proposed for instance by Zeng & He (2013).

The first step is the evaluation of the critical acceleration of the unstable soil mass. For the problem at hand, the pseudo–static analysis yields a critical seismic coefficient of 0.18 and 0.24 for cross sections AA and BB, respectively. The second step is the selection of ground input motion. In this case, an artificial earthquake is generated from each input signal, by averaging at any instant of time the acceleration profile within the height of the slope gathered from 2D analysis. Therefore, an artificial set of recordings is defined which encompasses the asynchronism of ground motion. Unlike the classical Newmark theory, reference is made to the rotation θ of the rigid friction block. When the seismic coefficient exceeds the critical one, the equation of dynamic rotational equilibrium of the sliding mass yields:

$$I_p\ddot{\theta} = (k_h - k_{hc}) \cdot W \cdot y_G \tag{8}$$

where I_p is the polar moment of inertia of the unstable soil mass, W is the weight of the same mass and y_G the lever arm of horizontal inertial action from the centre of rotation. Cumulated rotations are then calculated by double integrating those parts of the artificial ground motion lying above the critical acceleration using the rigorous algorithm of Wilson & Keefer (1983), taking into account explicitly the asymmetrical resistance of downslope and upslope sliding. A specific Matlab code was developed for the purpose of this work in which upslope rotations are prohibited. This is a reasonable assumption in many engineering problems as the critical acceleration in the upslope direction is usually much greater than the peak ground acceleration. The time histories of rotation evaluated by this approach are plotted in Figure 18. In line with what recommended by Italian code, the mean effect can be chosen to evaluate the seismic performance of the slope. The average permanent rotations yield to the following permanent displacements, w:

$$\begin{aligned} w_{pA} &= R_A \cdot \theta_{pA} = 33 \text{ mm} \\ w_{pB} &= R_B \cdot \theta_{pB} = 8 \text{ mm} \end{aligned} \tag{9}$$

where θ_{pA} and θ_{pB} are the average accumulated rotations, while R_A and R_B are the radii of the critical surfaces for cross sections AA and BB, respectively.

The main practical problem related to the application of the sliding block method is how to define tolerable permanent displacements. The limits on calculated values could be related to (Matasovic 1991): (a) functionality of structures on the crest or at the toe of the slope; (b) the post–earthquake stability of the slope. With regard to the last point, the main problem related to excessive slope movements comes from earthquake–triggered macroscopic ground cracking in that water percolation in earthquake opened cracks can significantly reduce the static stability. Ideally, the dilemma about whether displacement can be tolerated or not may be overcome through database in which observed earthquake–triggered slope movements are correlated to measures of damage. The first attempt set for slopes in this direction is the one by the *State of Alaska's Geotechnical Evaluate Criteria Committee* (Idriss 1985). According to this criterion, for earthquake events with a probability of exceedance of 10% in 100 years as those considered in this work, earthquake triggered movement must not exceed 12 inches. Wilson & Keefer (1983) held a value of 100 mm as a conservative estimate corresponding to the occurrence of macroscopic ground failure. Jibson & Keefer (1995) used the (5, 10) cm range as the level of critical displacement leading to ground cracking and general failure of landslides in Mississippi Valley. The *Guidelines for Analyzing and Mitigating Landslide Hazard in California* (Blake et al. 2002) claim that the median displacement should be kept less than 50 mm to avoid damage in buildings that might be potentially affected by the critical line. Finally, the *Guidelines for Evaluating and Mitigating Seismic Hazard in California* (California Geological Survey 2008) recommend that calculated displacements do not exceed 100 mm. Based on this review about tolerable earthquake–triggered slope movements, the calculated displacements for the problem under examination can be considered compatible with the life–safety limit state. Therefore, the profiles labelled 'static' in Figure 17 can be finally adopted for the stabilization of the excavation fronts.

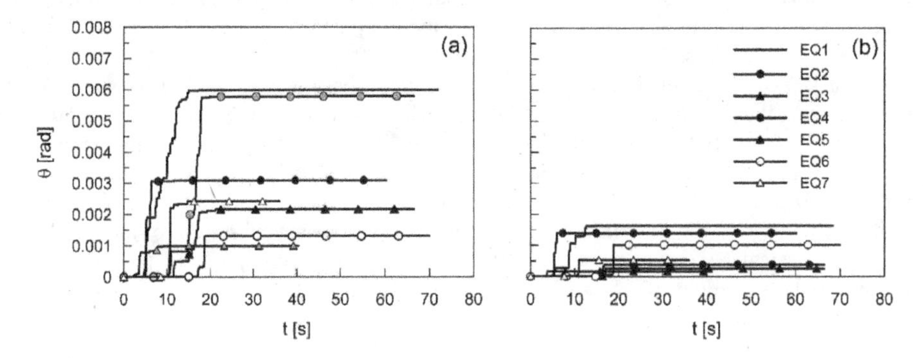

Figure 18. Permanent rotations of unstable soil mass by the sliding block theory.

6 FOUNDATIONS OF THE NEW COVERING SYSTEM

As mentioned before, the archaeological excavation of the *Insula* took place in more phases, simultaneously with the development of a disarticulated and uneven roofing system made up of corrugated aluminium sheets, steel pipes and joints, which were connected to a number of tubular steel frames protecting the archaeological ruins. The design of restoration interventions included the replacement of this roofing system with a single–span steel truss covering, as shown in Figure 19. The new single–span covering, supported by two alignments of heavy steel columns distributed along the lateral alleys, is equipped with classical double–concave friction–pendulum seismic isolators. Such devices were put at the top of the steel columns, in order to facilitate their maintenance and, what is of the utmost importance, to create a unique isolation plane.

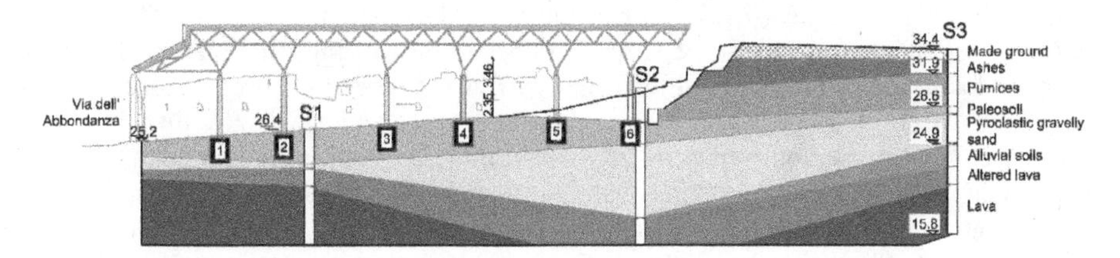

Figure 19. New covering system in place of the actual roof of steel pipes.

Concerning the foundation system, the key question to be answered is: what is the most appropriate choice in such difficult conditions? The large, multi–component loads transmitted by the columns are a critical issue, with the most severe eccentricity coming from the wind action (the effect of earthquake is of little importance due to the isolation devices). The design of the foundation system was inspired by the principles of the Venice Charter (1964) and the Cracovia Charter (2000), which suggest the choice of removable elements and the limitation of any visual impact. A weave of reinforced concrete, isolated caissons 2.1 m in diameter (D), 2.5 m high (h) and an embedment of 2.75 m, as shown in the cross section of Figure 20, has been adopted. It must be considered that this choice is mandatory to avoid the failure in uplift of the foundations under the leeward wind action. Yet it leads to undeniable advantages in comparison with shallow foundations. First of all, the irregular morphology of the lava layer is an inherent source of differential settlements which in turn may cause operational damages, while foundation settlements are minimized with such a choice. Second, the action of fluid concrete during installation on the adjacent ancient walls, if any, is of little importance due to arching effect in the soil. Third, the bearing behaviour of any caisson is independent of the wind direction. An effort is needed so as to minimize the impact of the circular foundations in the archaeological site by reducing their volume as much as possible.

Due to their geometric configuration, caissons are not easily treatable as shallow or deep foundations. From a conceptual point of view, the classical methods dealing with the calculation of bearing capacity of shallow foundations (e.g. Brinch Hansen 1970) may be indeed adapted to foundation caissons. Nevertheless, this method would neglect the tensile resistance and the shaft contribution to the ultimate capacity. An alternative approach is to use failure envelopes or more frequently, with respect to granular materials under drained conditions, yield envelopes, pointing out that the stress dependency of shear strength for granular materials leads to gradual yield and hardening with increased displacement in this case (Randolph & Gouvernec, 2011). Early studies on the yield surface of shallow foundations in sand were based on data collected by single gravity (1g) experiments (Butterfield & Ticof 1979, Dean et al. 1992, Georgiadis 1993, Gottardi & Butterfield 1993, Gottardi et al. 1999, Martin 1994, Nova & Montrasio 1991, Tan 1990, Ticof 1978), whose results indicated that the yield surface of such kind of foundation can be described by a rotated parabolic ellipsoid in (Q, H, M) force space, that is a rotated ellipse in the (M, H) plane and a parabola in planes along the Q axis at constant M/H ratio. A number of works has then focused on the response of caissons foundation for offshore wind turbines to combined loads (Achmus et al. 2013, Ahlinhan et al. 2019, Byrne & Houlsby 1999, Byrne & Houlsby 2001, Li et al. 2015, Villalobos 2006, Villalobos et al. 2009, Zafeirakos & Gerolymos 2016). Byrne & Houlby (2001) have extended the failure envelope equation by Gottardi et al. (1999) to include the effect to the foundation embedment ratio. Model tests on skirted foundations carried out in loose sand have highlighted also the existence of non-zero horizontal and moment capacity in the tensile range of vertical loads (Villalobos 2006). In order to accommodate these experimentally observed behaviours, Villalobos et al. (2009) have proposed the yield surface equation:

$$\left(\frac{M/D}{m_0 Q_0}\right)^2 + \left(\frac{H}{h_0 Q_0}\right)^2 - 2a\frac{HM/D}{m_0 h_0 Q_0^2} - \beta_{12}^2\left(\frac{Q}{Q_0}+t_0\right)^{2\beta_2}\left(1-\frac{Q}{Q_0}\right)^{2\beta_2} = 0$$
$$\beta_{12} = \frac{(\beta_1+\beta_2)^{(\beta_1+\beta_2)}}{\beta_1^{\beta_1}+\beta_2^{\beta_2}} \tag{10}$$

where Q_0 is the uniaxial vertical yield load, a accounts for the rotation of the elliptical cross section in the $(M/D, H)$ plane, β_1 and β_2 are shape parameters influencing where the peak horizontal and moment loads occur under vertical load, the coefficients (m_0, h_0) control the size of the yield surface in the (H, M) load plane, Q_t is the pullout resistance and $t_0 = Q_t/Q_0$. In this study, the Brinch Hansen (1970) equation is first adapted to determine the axial capacity in compression. The tensile capacity, arising only from friction on the sides of the caisson, is then calculated. The values of m_0 in Equation 9 is a function of the caisson aspect ratio h/D and, for the case at hand, it can be assumed equal to 0.1 (Villalobos 2006). In this case, the lateral load transmitted by the column under the wind action is only a few percentage of the axial load, so that, for the purposes of this study, reference can be made to the problem of a caisson foundation under vertical and eccentric load. Figure 20 shows the failure envelope in the (Q, M) plane evaluated through Equation (10) for caisssons No. 2 and 5 (see Figure 19) and the load paths expected under the upward and leeward wind action, referred to as path 1 and 2, respectively. Notably, the latter action yields to a remarkable reduction of the axial load upon the caisson foundation. But, what is more, the current state (C) along with the pertinent load path (ACC') is far closer to the failure envelope. In a radial mapping criterion (Butterfield 2006), the partial resistance factor relies upon the direction of the load path (= AB'/AB or AC'/AC). The minimum value is 2.71 for caisson No. 2 and 2.44 for caisson No. 5. They are both consistent with that prescribed by the Italian building code ($\gamma_R = 2.3$). It is clear from this example that the contribution from the embedded shaft of the caisson to the ultimate moment is crucial for a sustainable design of the foundation layout. A proper consideration of the load path direction is essential as well.

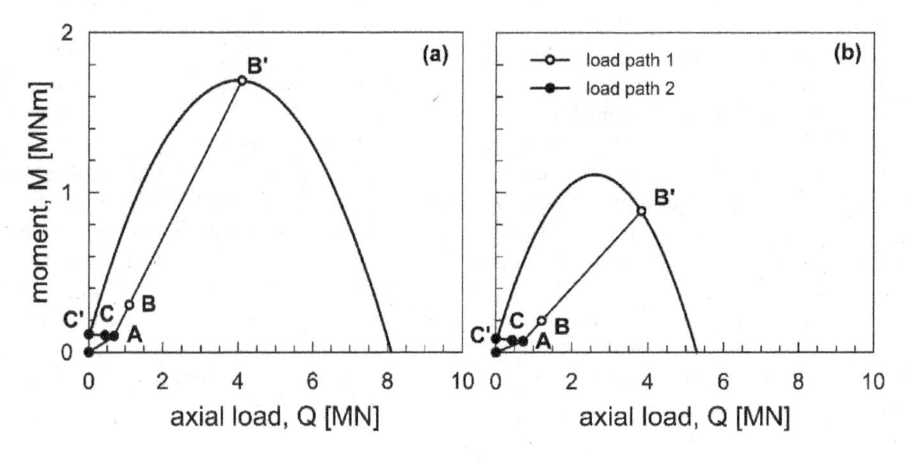

Figure 20. Interaction diagrams and load paths under the wind action for caissons (a) No. 2 and (b) No. 5.

7 DISCUSSION AND CONCLUSIONS

In this work the case study of the *Insula dei Casti Amanti* has been introduced as a ploy to discuss about the main geotechnical issues for the preservation of the ancient city of Pompeii. The restoration works included the re–profiling of the artificial slopes, the replacement of the actual roofing system with a single–span truss covering and the restoration and protection of the archaeological ruins.

The seismic demand in the artificial slope was evaluated by means of two–dimensional analysis of wave propagation using the equivalent linear approach. A remarkable and unexpected aggravation was identified at the crest of the slope owing to the intense interference between soil effect and the very shallow bedrock. The asynchronism of ground motion was tackled by averaging at any instant of time the spatial distribution over the slope height, turning out a seismic coefficient far greater than that evaluated with the conventional subsoil classification. Thus, a proper modelling of wave propagation is crucial for a safe and reliable prediction of the seismic demand in the artificial slopes. Re–profiling action based on the classical pseudo–static approach would have not been compatible with the need of preserving the archaeological remnants surrounding the *Insula*. By contrast, the alternative, dynamic analysis by the rigid block theory was proved to be a sustainable approach for slope analysis in the archaeological site.

The foundation of the new covering system consisted of fully embedded circular caissons. As the new roof is equipped with seismic isolators, the effect of the earthquake action was of minor concern. By contrast, large and multi–component loads transmitted by the columns under the wind action are a critical issue, with the leeward pressure on the exposed face of the roof being the most severe load combination. In this case, a proper consideration of the load path direction and the use of the interaction diagram concept are fundamental to minimize the impact of the foundation system in the *Insula*.

Notably, the very pronounced peak in the average spectrum for seismic analysis of the covering system corresponds about to the fundamental frequency interval of the subsoil, as it was argued from amplification function calculated with the Equivalent Linear analysis. In all likelihood, the fundamental periods of the ancient masonry textures fall within the same critical interval. Should this be the case, a double–resonance phenomenon could occur, with potential destructive effects for the archaeological site. As a matter of fact, after the Irpinia earthquake in 1980, a number of *domus* in *Pompeii* suffered severe damages, especially those without roofs. The major collapses occurred in *Regio* VII and cracking patterns appeared elsewhere in a very widespread manner, such that repair works of the earthquake damages started in 1982 and lasted about 10 years. It is difficult to establish what really happened then in terms of soil–structure dynamic interaction for the archaeological ruins, and this is beyond of the scope of the present work. This is indeed the aim of a specific research agreement between the Universities of Naples Parthenope and Federico II with the *Archaeological Park of Pompeii* defined at the end of 2021.

As a last comment, the geotechnical issues have been tackled through a strong interaction with a multitude of disciplines, with very different sensibilities, with the common aim of the preservation and enhancement of the archaeological site of Pompeii. The restoration works, in particular those related to geotechnical engineering, were not based on the principles of efficiency, economy, symmetry or uniformity of the behaviour, as it is usual in civil engineering, but rather on the principles of adaptation, removability and reversibility, which, in turn, led to solutions characterized by asymmetry and unevenness of the behaviour.

AKNOWLEDGMENTS

This work has been carried out under a three year research agreement between the Universities of Naples Federico II and Parthenope and the *Archaelogical Park of Pompeii*. The Authors are grateful to Dr. Vincenzo Calvanese from the Archaeological Park for his invaluable support.

REFERENCES

Achmus, M., Akdag, C.T. & Thieken, K. 2013. Load-bearing behavior of suction bucket foundations in sand. *Applied Ocean Research*, 43, 157–165.

Ahlinhan, M.F., Adjovi, E.C., Doko, V. & Tigri, H.N. 2019. Numerical analysis of the behaviour of a large-diameter monopile for offshore wind turbines. *Acta Geotechnica Slovenica* 16(1), 53–69.

Ashford, S., Sitar, N., Lysmer, J. & Deng, N. 1997. Topographic effects on the seismic response of seismic slopes. *Bulletin of the Seismological Society of America* 87(3), 701–709

Aversa, S. 2007. Preserving cities and monuments. *Proc. of Geotechnical Engineering in Urban Environments*. Madrid, 24–27 sept., vol. 5, Millpress, 453–462.

Blake, T.F., D'Antonio, R., Earnest, J., Gharib, F., Hollingsworth, R.A., Horsman, L., Hsu, D., Kupferman, S., Masuda, R., Pradel, D., Real, C., Reeder, W., Sathialingam, N., Simantob, E. & Stewart, J.P. 2002. Guidelines for analyzing and mitigating landslide hazards in California, Recommended procedures for implementation of DMG special publication 117. Southern California Earthquake Center pub., Los Angeles, CA, Blake, Hollingsworth & Stewart eds., 127 pp.

Bouckovals, G.D. & Papadimitriou, A.G. 2005. Numerical evaluation of slope topography effects on seismic ground motion. *Soil Dynamics and Earthquake Engineering* 25(7–10), 47–558.

Hansen, J.B. 1970. A revised and extended formula for bearing capacity.

Burland, J.B. & Standing, J. R. 1997. Geotechnical monitoring of historic monuments. In *Geotechnical engineering for the preservation of monuments and historic sites* (pp. 321–341).

Butterfield, R. & Ticof, J. 1979. Design parameters for granular soils (discussion contribution). In *Proc.* 7th *International Conference Soil Mechanics & Foundation Engineering*, 259–261.

Butterfield, R. 2006. On shallow pad-foundations for four-legged platforms. *Soils and Foundations* 46(5), 427–435.

Byrne, B.W. & Houlsby, G.T. 1999. Drained behaviour of suction caisson foundations on very dense sand. In Offshore Technology Conference. OnePetro.

Byrne, B. W. & Houlsby, G.T. 2001. Observations of footing behaviour on loose carbonate sands. *Géotechnique*, 51(6), 463–466.

Calabresi, G., & D'Agostino, S. 1997. Monuments and historic sites: Intervention techniques. *Proc. Arrigo Croce Memorial Symposium – Geotechnical Engineering for the Preservation of Monuments and Historic Sites*, Napoli, Viggiani ed., Balkema, 409–425.

California Geological Survey. 2008. Guidelines for Evaluating and Mitigating Seismic Hazards in California. *California Geological Survey Special Publication* 117A, 98 pp.

Carey, S. & Sigurdsson, H. 1987. Temporal variations in column height and magma discharge rate during the 79 AD eruption of Vesuvius. *Geological Society of America Bulletin* 99(2), 303–314.

Croce, A. (1985) Old monuments and cities. Research and preservation. Geotechnical Engineering in Italy: An overview, Special volume for ISSMFE – *Golden Jubilee, Associazione Geotecnica Italiana* ed., 361–415.
de Sanctis, L., Iovino, M., Maiorano, R.M.S. & Aversa, S. 2020. Seismic stability of the excavation fronts in the ancient Roman city of Pompeii. *Soils and Foundations* 60(5), 856–870.
de Sanctis, L., Maiorano, R.M.S., Brancaccio, U. & Aversa, S. 2019. Geotechnical aspects in the restoration of Insula dei Casti Amanti in Pompeii. *Proceedings of the Institution of Civil Engineers: Geotechnical Engineering* 172(2), 121–130.

Dean, E.T.R., James, R.G., Schofield, A.N., Tan, F.S.C. & Tsukamoto, Y. 1992. The bearing capacity of conical footings on sand in relation to the behaviour of spudcan footings of jackups. In Predictive soil mechanics: Proceedings of the Wroth Memorial Symposium held at St Catherine's College, Oxford, 27–29 July 1992, 230–253, Thomas Telford Publishing.

Ebrahimian, H., Jalayer, F., Forte, G., Convertito, V., Licata, V., d'Onofrio, A. & Manfredi, G. 2019. Site-specific probabilistic seismic hazard analysis for the western area of Naples, Italy. *Bulletin of earthquake engineering* 17(10), 4743–4796.

EN–1998–5 (CEN 250 2019). (pr)EN–1998–5:2019.2. Eurocode 8: Earthquake resistance design of structures – Part 5: Geotechnical aspects, Foundations, Retaining and Underground structures. European Committee for Standardization TC 250, Brussels, Belgium

Flora, A. 2013. Monuments, historic sites and case histories. General Report. *Proc. 18*th *International Conference on Soil Mechanics and Geotechnical Engineering*, Paris 2013, 3087–3094.

Georgiadis, M. 1993. Settlement and rotation of footings embedded in sand. *Soils and Foundations* 33(1), 169–175.

Gottardi, G. & Butterfield, R. 1993. On the bearing capacity of surface footings on sand under general planar loads. *Soils and Foundations* 33(3), 68–79.

Gottardi, G., Houlsby, G.T. & Butterfield, R. 1999. Plastic response of circular footings on sand under general planar loading. *Géotechnique* 49(5), 453–469.

Hudson, M.B., Idriss, I.M. & Beikae, M. 1993. QUAD4M – A computer program for evaluating the seismic response of soil structures by variable damping finite element procedures. Center for Geotechnical

Modeling, Department of Civil and Environmental Engineering, University of California at Davis, CA, USA.

Idriss, I.M. 1985. Evaluating the seismic risk in Engineering Practice. *Proc. 11th International Conference on Soil Mechanics and Foundation Engineering*, San Francisco, August 12–16, vol. 1, 255–320.

Jappelli, R. 1997. An integrated approach to the safeguard of monuments: The contribution of Arrigo Croce. *Proc. Arrigo Croce Memorial Symposium – Geotechnical Engineering for the Preservation of Monuments and Historic Sites*, Viggiani ed., A.A. Balkema, 11–27.

Jibson, R.W. & Keefer, D.K. 1993. Analysis of the Seismic Origin of Landslides: Examples from the New Madrid Seismic Zone. *Geological Society of America Bulletin* 105(5), 521–536.

Jibson, R.W. 1993. Predicting Earthquake–Induced Landslide Displacement Using Newmark's Sliding Block Analysis. *Transportation Research Record* 1411, 9–17.

Li, D., Zhang, Y., Feng, L. & Gao, Y. 2015. Capacity of modified suction caissons in marine sand under static horizontal loading. *Ocean Engineering* 102, 1–16.

Licata, V., Forte, G., d'Onofrio, A., Santo, A. & Silvestri, F. 2019. A multi–level study for the seismic microzonation of the Western area of Naples (Italy). *Bulletin of Earthquake Engineering* 17(10), 4711–4741.

Luongo, G., Perrotta, A., Scarpati, C., De Carolis, E., Patricelli, G. & Ciarallo, A. 2003. Impact of the AD 79 explosive eruption on Pompeii, II. Causes of death of the inhabitants inferred by stratigraphic analysis and areal distribution of the human casualties. *Journal of Volcanology and Geothermal Research* 126(3–4), 169–200.

Martin, C.M. 1994. Physical and numerical modelling of offshore foundations under combined loads. PhD thesis, University of Oxford, UK.

Matasovic, N. 1991. Selection of Method for Seismic Slope Stability Analysis. *Proc. 2nd International Conference on Recent Advances in Geotechnical Earthquake Engineering and Soil*, March 11–15, St. Louis, Missouri, Paper No. 7.20.

Meletti, C., Galadini, F., Valensise, G., Stucchi, M., Basili, R., Barba, S., Vannucci, G. & Boschi, E. 2008. A seismic source zone model for the seismic hazard assessment of the Italian territory. *Tectonophysics* 450, 85–108.

Newmark, N.W. 1965. Effects of earthquakes on dam and embankments. *Géotechnique* 15(2), 139–159.

Nova, R. & Montrasio, L. 1991. Settlements of shallow foundations on sand. *Géotechnique* 41(2), 243–256.

Rathje, E.M., Abrahamson, N.A., & Bray, J.D. 1998. Simplified frequency content estimates of earthquake ground motions. *Journal of Geotechnical and Geoenvironmental Engineering* 124(2), 150–159.

Randolph, M. & Gourvenec, S. 2017. Offshore geotechnical engineering. CRC press.

Rovida, A., Locati, M., Camassi, R., Lolli, B. & Gasperini, P. 2016. CPTI15, the 2015 version of the parametric catalogue of Italian earthquakes. Istituto Nazionale di Geofisica e Vulcanologia, doi: org/10.6092/INGV.IT–CPTI1 5.

Sarma, S.K. 1975. Seismic stability of earth dams and embankments. *Géotechnique* 25(5), 743–761.

Sigurdsson, H., Carey, S., Cornell, W. & Pescatore, T. 1985. The eruption of Vesuvius in A.D. 79. *National Geographic Research* 1(3), 332–387.

Tan, F.S.C. 1990. Centrifuge and theoretical modelling of conical footings on sand. PhD thesis, University of Cambridge, UK.

Ticof, J. 1978. Surface Footings on Sand Under General Planar Loads. PhD thesis, University of Southampton, UK.

Viggiani, C. 2013. Cultural heritage and geotechnical engineering: An introduction. *Proc. 2nd International Symposium on Geotechnical Engineering for the Preservation of Monuments and Historic Sites*, Napoli, Italy, CRC Press/Balkema, 3–12.

Villalobos, F.A. 2006. Model testing of foundations for offshore wind turbines. PhD thesis, University of Oxford, UK.

Villalobos, F.A., Byrne, B.W. & Houlsby, G.T. 2009. An experimental study of the drained capacity of suction caisson foundations under monotonic loading for offshore applications. *Soils and Foundations* 49(3), 477–488.

Wilson, R.C. & Keefer, D.K. 1983. Dynamic analysis of a slope failure from the 6 August 1979 Coyote Lake, California, Earthquake. *Bulletin of the Seismological Society of America* 73(3), 863–877.

Zafeirakos, A. & Gerolymos, N. 2016. Bearing strength surface for bridge caisson foundations in frictional soil under combined loading. *Acta Geotechnica* 11(5), 1189–1208.

Geotechnical Engineering for the Preservation of Monuments and Historic Sites III – Lancellotta, Viggiani, Flora, de Silva & Mele (Eds)
© 2022 Copyright the Author(s), ISBN 978-1-032-35998-4

Long term strategies for monuments care: The importance of monitoring and of a proper diagnosis

G. Russo
Department of Civil, Environmental and Geotechnical Engineering, University of Napoli Federico II

ABSTRACT: This paper is dedicated to highlight the importance of monitoring to determine a diagnosis. Two case studies are presented. The former case deals with the Benedictine Basilica of S. Angelo in Formis constructed in the XI century. It contains an outstanding cycle of medieval frescoes. Monitoring was carried out with the combined use of terrestrial survey and of satellite data. The Incurabili complex in Napoli with its Pharmacy (XVI – XVIII century) is the second case presented in the paper. The rooms of the Pharmacy with its paintings, its ceramics and its decorated floor is a marvelous monument located in the historical center of the ancient greek Neapolis. In this case movements were detected too and a monitoring plan was organized and carried out. The diagnosis was easier than in the first case and the remedial works have been fully designed.

1 INTRODUCTION

Italy is one of the countries with the highest concentration of monuments and historical sites in the world. The cities with their historical-architectural environment constitute a treasure of inestimable cultural, historical and artistic value. Most of the italian cities, even the smallest, hides art treasures that have always attracted a large number of visitors since the eighteenth century, when Italy was an obligatory stop on the "grand tour". As well evidenced by the various theories of restoration starting from the well-known Brandi's treatise (1963) the restoration after the damage is occurred is not the only possible one. There is also a preventive restoration well explained by Brandi himself (1963) which is concretely reflected in the prevention of the degradation process through the control of the artistic, historical asset to be protected but also through the control of the environment in which the asset is inserted or of the surrounding territory. On the basis of the knowledge acquired on the asset and on the environment in which the asset is inserted, it is possible to schedule a maintenance program that contributes to preventive restoration. Monitoring (Russo & Viggiani 2000) and analysis (Russo et al. 2009) with appropriate tools are quantitative indispensable steps to define preventive restoration programs. In this report two case studies are presented. They relate to monuments subjected to monitoring (Candela et al. 1997) and diagnostics operations for the subsequent development of hypotheses of intervention in order to stop a current degradation process. As it will be cleared in the present report, the diagnosis is not always simple and the interventions cannot always be said to be decisive. Obviously, in the protection of monuments and historical sites, other concepts such as vulnerability analysis and risk reduction with respect to extreme events (Russo et al. 2017) are becoming increasingly important but they will not be dealt with in this report for space reasons.

2 THE BASILICA OF SANT'ANGELO IN FORMIS

A temple dedicated to Diana Tifatina had been erected in V century BC at the base of the South West side of the Tifata calcareous mountains in Southern Italy (De Franciscis 1956). In AD 595

 DOI 10.1201/9781003329756-13

a Longobard church dedicated to the Archangel Michael was constructed over the ruins of the Roman temple, as reported in the "Regesto" of S. Angelo in Formis", in the library of the great Montecassino Abbey. In 1072 the Norman Richard the 1st, prince of Capua and count of Aversa, presented the church to the Abbott Desiderio, rector of Montecassino (Jacobitti, Abita 1992). At the site of the ancient church Desiderio erected the present Basilica whose façade is reported in the picture on the left of Figure 1.

Figure 1. The Benedictine Basilica of S. Angelo in Formis: on the left the present façade and on the right a painting representing an old version of the main façade.

On the right of the same Figure 1 a detail of the internal paintings is reported for comparison and is rather evident the difference between the two porticos: the present portico is characterized by 5 pointed arches while the one depicted in the painting show 5 round arches. This is in agreement with the findings occurred during relatively recent archeological excavations which provides the proof of the collapse of the original portico.

The floor is still lined with the white marble mosaic of the stylobates of the Diana temple, dating back to 150 – 170 BC. Several archeological excavations were carried out to determine the relationship in plan and in elevation between the old temple and the medieval basilica. In Figure 2 the picture on the right represents the old basement walls existing immediately below the current floor with holes excavated in different ages and for different purposes. Capitals, columns and marbles were also reused for the construction. The Basilica, in a simple and appealing Romanesque style, contains an outstanding cycle of frescos over the walls of its three naves.

Figure 2. The blessing Christ. Fresco in the central apse and the old structures of the temple to Diana.

They represent stories of the Old Testament in the lateral naves, and of the New Testament in the central one; a blessing Christ surrounded by the symbols of the four Evangelists is painted on the central apse (Figure 2 left), while the Final Judgment is painted on the opposite wall of the counterfaçade (Figure 3, right).

The frescoes may be dated back to 1087, i.e. to the time of construction of the Basilica; they are inspired by the iconography of the byzantine art and are believed to have been painted by at least five different artists (Wettstein, 1960). After the destruction of the Montecassino Abbey during the World War II, the S. Angelo frescoes are probably the most important document of the medieval painting in Southern Italy; they have been saved in relatively good conditions, probably because they have been covered by whitewash in XVIII and XIX centuries. The church has a plan divided in three naves, each one ending in semicircular apse (Figure 3, left). The columns that divide the naves are made of different materials (granite, white marble and green marble) and have original Corinthian capitals from the temple of Diana. The façade is graced with a delicate portico or narthex of five arches, upheld by four Corinthian columns (Figure 1 left).

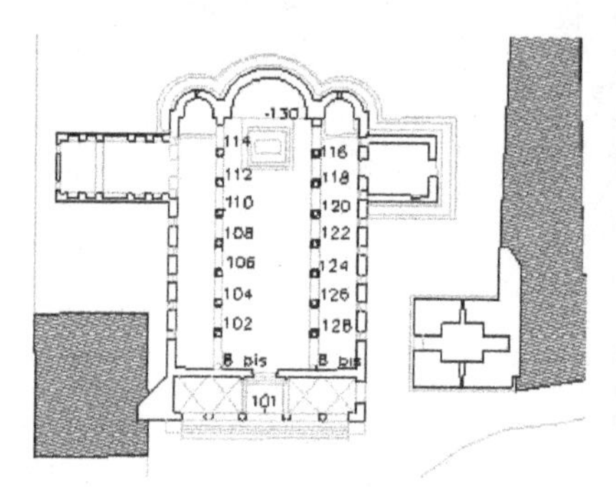

Figure 3. The Benedictine Basilica of S. Angelo in Formis: schematic geometrical plan (left) and the Western internal façade with a portion of the left colonnade (right).

2.1 *The subsoil of the Basilica and its geological setting*

The slope (Figures 4 and 5) is composed by limestone and dolomite belonging to Trias – Cretacic, heavily fractured, somewhere brecciated and even mylonitized, interested by diffuse karst phenomena. The limestone is superimposed to Oligocene – Miocene deposits, composed by sandstones and variegated shaley clays with a chaotic structure. Limestone slipped over the Miocene formation, and such a tectonic superposition generated the fracture system affecting it. The rock formation is crossed by significant fracture lines in the North – South direction and become thinner towards the plain located in front of the basilica.

Later on (30,000 to 35,000 years B.P.) the plain and the slopes have been covered by the grey Campanian tuff erupted by the Phlegrean volcanoes. Since then, the morphological and structural pattern of the area remained unchanged, except for some anthropic actions. Among them the retaining walls and the fill (Figure 6, left) to realize, in different stages, the courtyard of the Basilica, and the opening of stone quarries. At least three of such quarries are located within a few hundred meters from the Basilica; they have been cultivated till the 1980s with extensive use of explosives.

The water table is generally found at the boundary of the limestone, very permeable by fracturation and the Boreholes and geophysical investigations allowed a detailed reconstruction of the subsoil of the zone surrounding the Basilica, which includes three horizons. The upper one is composed by made ground, for a thickness ranging from a few decimeters to some meters. Below the made ground a layer of fractured rock is found with a thickness variable between 15 and 30 m; it includes dolomite, slightly cemented dolomitic limestone and intensely fractured dolomite, besides cemented calcareous debris. Finally, the base formation of sandstones and variegated shaly clays is found. The geological sketch in Figure 6 shows clearly that the Basilica is located across a stratigraphic discontinuity, with the apses and the backward part of the naves founded on rock, and the front on the debris cover or even on the made ground.

Figure 4. The basilica in the calcareous dolomitic slope.

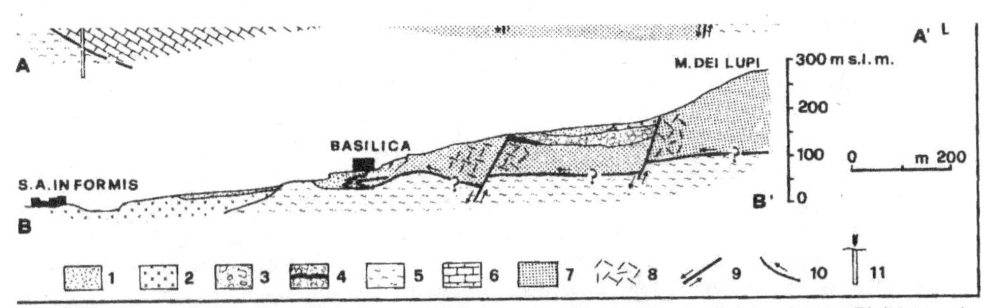

Recent slope debris (upper Pleistocene-Holocen); 2. Campanian grey tuff Auct. (upper Pleistocene); 3. Alluvial deposits (upper Pleistocene); 4.Ancient cemented slope breccias with basal pyroclastic level (lower-middle Pleistocene); 5. Sandy-clayey flysch (Caiazzo Sandstone-Tortonian); 6. Limestones (Cretaceous); 7. Dolomitic limestones and dolomite (Triassic-Jurassic); 8. rock; 9. Fault (dashed if supposed) ; 10. Thrust; 11. Boreholes

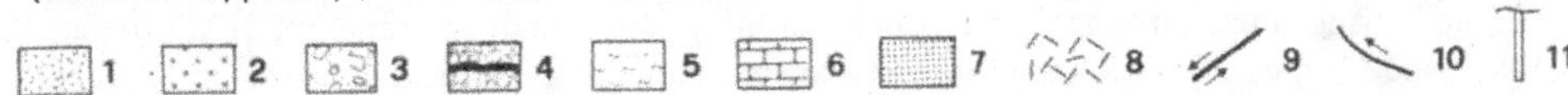

Figure 5. Geological section of the slope; A) North – South; B) East -West (after Corniello & Santo 1995).

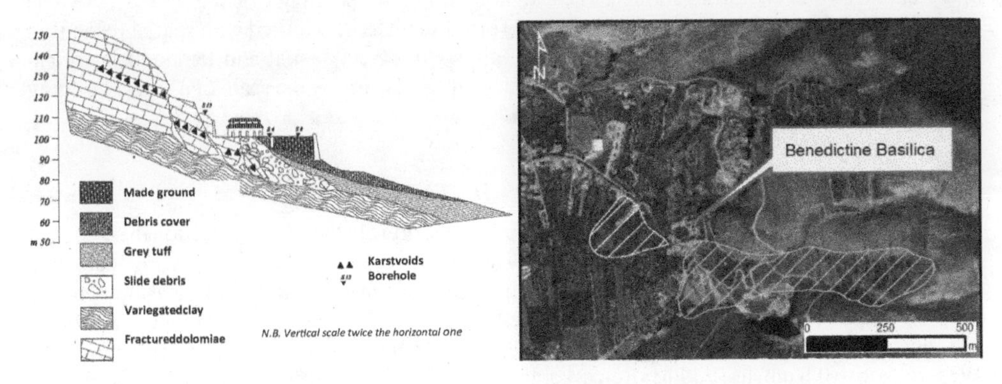

Figure 6. Geological section of the area (left) and landslide inventory map (right) within the area of Basilica (after ISPRA 2008).

As regards the slope instability phenomena, "IFFI" project, dedicated to landslide inventory complex, respectively (ISPRA 2008) (Figure 6 right) while no landslides were detected as involving the site of the Basilica.

2.2 *Previous restoration works and damages*

A stone in the right apse of the Basilica refers to repair works carried out in 1732 under the Cardinal Giuseppe Renato. Little is known about the damage requiring such interventions; it is likely they were consequence of an earthquake occurred in the same year with an MCS intensity equal to X in Irpinia. In 1930 a similar earthquake produced the collapse of the wooden roof. A number of fissures in the walls were repaired and the frescoes restored "with the help of German experts". As far as known however further earthquakes registered in 1962, 1970 and 1980 produced only minor effects.

In 1969 some fissures appeared in the walls above the columns of the central nave; in the following years the fissures gradually opened and extended to other parts of the Basilica, with a concentration in the right side of the façade and in the right nave near the façade. The occurrence of fissures was interpreted as the consequence of distortions caused by differential settlement at the foundation level and monitoring activities started in 1972/73. At the end of the 1970s, following repeated alarms on the safety of the Basilica, a Committee formed by a geotechnical engineer, a structural engineer and a geologist was installed, promoting a geological survey of the area, some further subsoil investigations, a geodetic monitoring of a number of points both inside the Basilica and outside and some further measurements of the fissures opening. Results of investigation and monitoring are discussed in Adriani et al. (1980) and in Pellegrino and Pescatore (1981) documents. The Committee conclusions related the damage mainly to the activity of the nearby quarries, where explosives were systematically used. Accordingly, the quarrying has been completely stopped since 1981. Neverthless, fissures in the Basilica went on increasing, endangering the integrity of the frescoes; as consequence, a new cycle of investigation and monitoring has thus been started in 2012, with the aim of planning remedial and consolidation works. Some of the monitoring data have been reported in a previous paper by Cammarota et al. (2013) together with the description of remedial works. Here new data are presented and discussed. A recent comparison with data obtained via satellite is also reviewed with the aim to fix at least some conclusions on the observed type and rate of movement. In the last decade, a growing interest has been directed towards the exploitation of remote sensing approach as a tool for displacements monitoring of monuments (Di Martire et al., 2016; Tapete & Cigna, 2019; Tomás et al., 2012): in particular, Differential Interferometry Synthetic Aperture Radar (DInSAR), based on processing of multi-temporal repeat-pass SAR images, has demonstrated to be a powerful technique to measure displacement annual rates and evolution of scatterers identified on Earth's surface. The availability of non-operating (ERS, ENVISAT-ASAR) and operating (COSMO-SkyMed, SENTINEL-1) satellite images, acquired with spatial resolution up to 3×3 m and revisiting time up to 6 days, allows to detect spatial and temporal historical movements since 1992, from large to local-site investigations, also at the scale of a single building (Infante et al., 2019; Zhou et al., 2015).

2.3 *Monitoring data: summary of the available experimental evidence*

Apart from some historical information on repair works reported in the previous section the evidence of movements and their records is not so large. The earliest detected and systematically collected data are those coming from the early '70 of the previous century. Two series of measurements of width of cracks in the walls are available: the first one in 1974/75 and the second one in 1980/81. The increase in the width of fissures was up to 0.5 mm, in nearly two years period spanning over 1974/75, and to 0.8 mm in 1980/81. Remarkable was the fact that the trend of the fissures measured in 1974/75 was compatible with a hogging deformation of the church. Furthermore, in January 1980, the offset of the columns, measured with reference to the vertical direction, was accurately determined. The offsets ranged from 10 to 89 mm depending on the columns location. These offsets were obtained via topographic technique and were indeed more compatible with a sagging mode of deformations. Indeed, the recorded offsets kept constant over a range of 1 year and were judged as small enough to be attributed to possible original construction defects.

As to regards the monitoring of displacements, both geodetic and satellite data are available.

In the periods February 1980-March 1981, May 2012-January 2013 and 2018/2021 vertical displacements of some points inside and outside the Basilica (Figure 7) have been measured. The benchmarks inside the church are better represented in the zoom of Figure 3 while in the Figure 7 the local geomorphology of the site and the outside benchmarks CS1 and CS2 are represented.

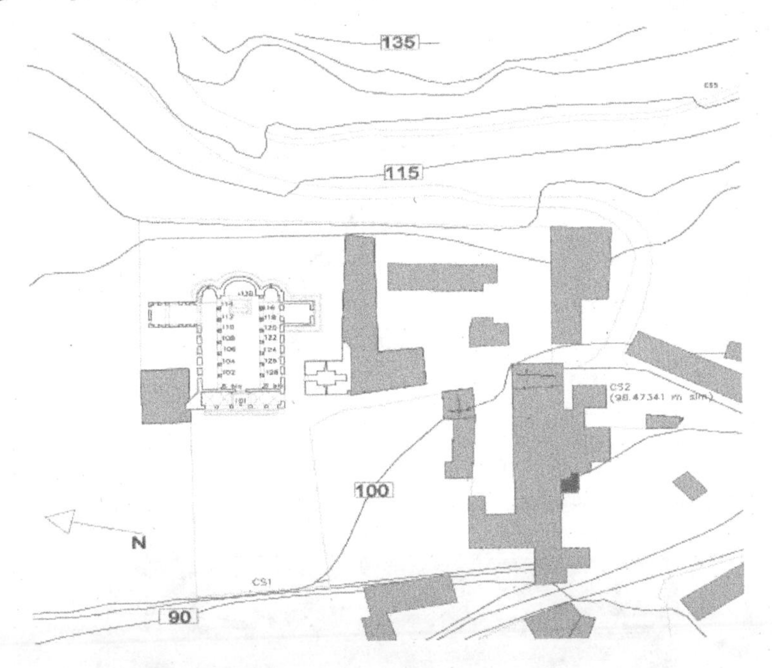

Figure 7. Plan of benchmarks used for topographic survey.

The survey was carried out with optical levels at the beginning and now is going on with electronic level using an invar bar based on the technology of the bar code has a reference system for the electronic level. The benchmarks CS1 and CS2 are substantially stable and relative movements in the monitoring period have been always fluctuating within the range of a couple of mm. Furthermore, since the 2018 the GPS is also used to independently check the possible movements of the references CS1 and CS2.

In Figure 8 it can be appreciated that in the first period (1980-1981) the maximum settlement was about 3.5 mm and occurred near the front façade and in the columns close to the three apses confirming an hogging mode of deformation. Forty years later with a few intermediate surveys carried out in the last decade a maximum value of nearly 80 mm is recorded at the benchmark 130 which is on the floor of the central apse which underwent significant remedial works in the 18th century when also the overall cycles of frescoes were cleaned by white cover applied at the beginning of the same century. The hogging mode of deformation is clearly confirmed, and the average value of the distortions applied to the masonry above the columns in the central nave are respectively $0.8x10^{-3}$ on the left and $1.5x10^{-3}$ on the right. In terms of velocity the phenomenon going on, whatever it is, is producing nearly 2mm/year increase in the settlement of the benchmarks close to the central apse. For the front façade the velocity is larger than 1 mm/year and in the central zone of the church the slowest rate of movement is recorded in the range between 0.5 mm/year and 1 mm/year. The above rates are only confirmed by the comparisons between sets of measurements taken at rather long-time interval.

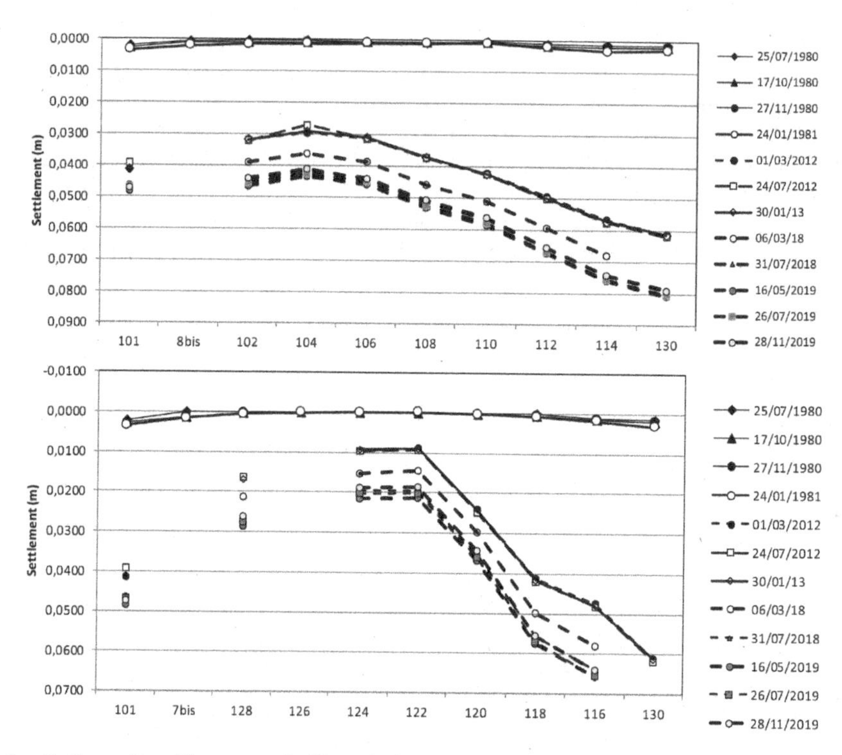

Figure 8. Settlement profile measured with optical survey at the base of the columns.

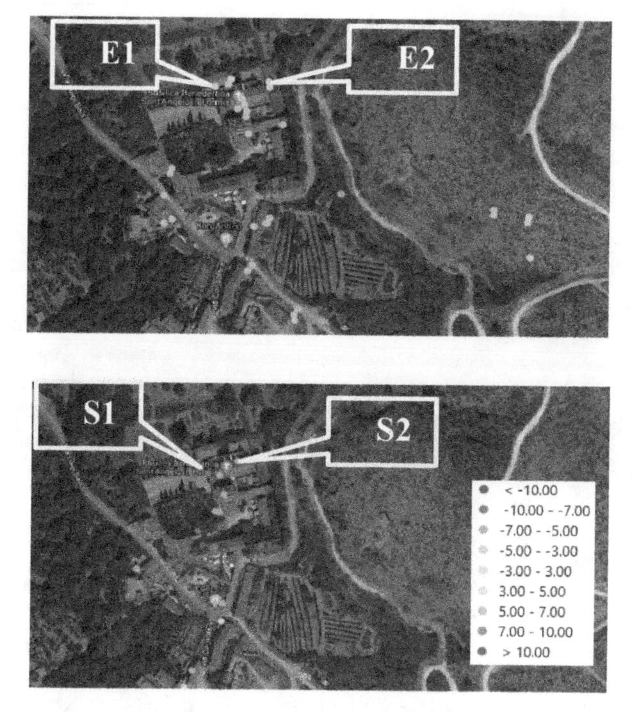

Figure 9. “LoS” displacement velocity map (mm/yr) of targets identified by descending ENVISAT ASAR (a) and SENTINEL-1 (b) interferometric products.

To further check the outlined trend, C-band interferometric data have been used: firstly, ENVISAT ASAR descending products, just available on the Geoportal of the Italian Environmental Ministry in the period June 2003-June 2010, have been analyzed. Subsequently, 91 SENTINEL-1 descending images, acquired in the period January 2017- January 2020 have been processed by means of Coherent Pixels Technique (CPT, Mora et al. 2003). As a result, for each dataset, displacement mean velocity maps of identified targets, measured along Line of Sight (LoS), have been obtained (Figure 9). Moreover, starting from 'LoS' displacement value, vertical component of movement has been evaluated according to Di Martire et al. (2013), as function of satellite incidence angle in descending geometry.

Such analysis allows to obtain settlements time series of some targets identified on the Basilica, as indicator of structure deformations (Figure 10). Maximum value of measured settlement is about 25 mm in the time span June 2003-June 2010 and about 15 mm in the time span January 2017-January 2020. As it can be seen by the proposed simple linear fitting in the period 2003-2010 the selected points E1 and E2 which are located on two opposite sides of the church are showing similar scatter and, above all, similar velocities (i.e. 3.5 mm/year). This rate was already recorded in the early years of optical survey (1980–1981).

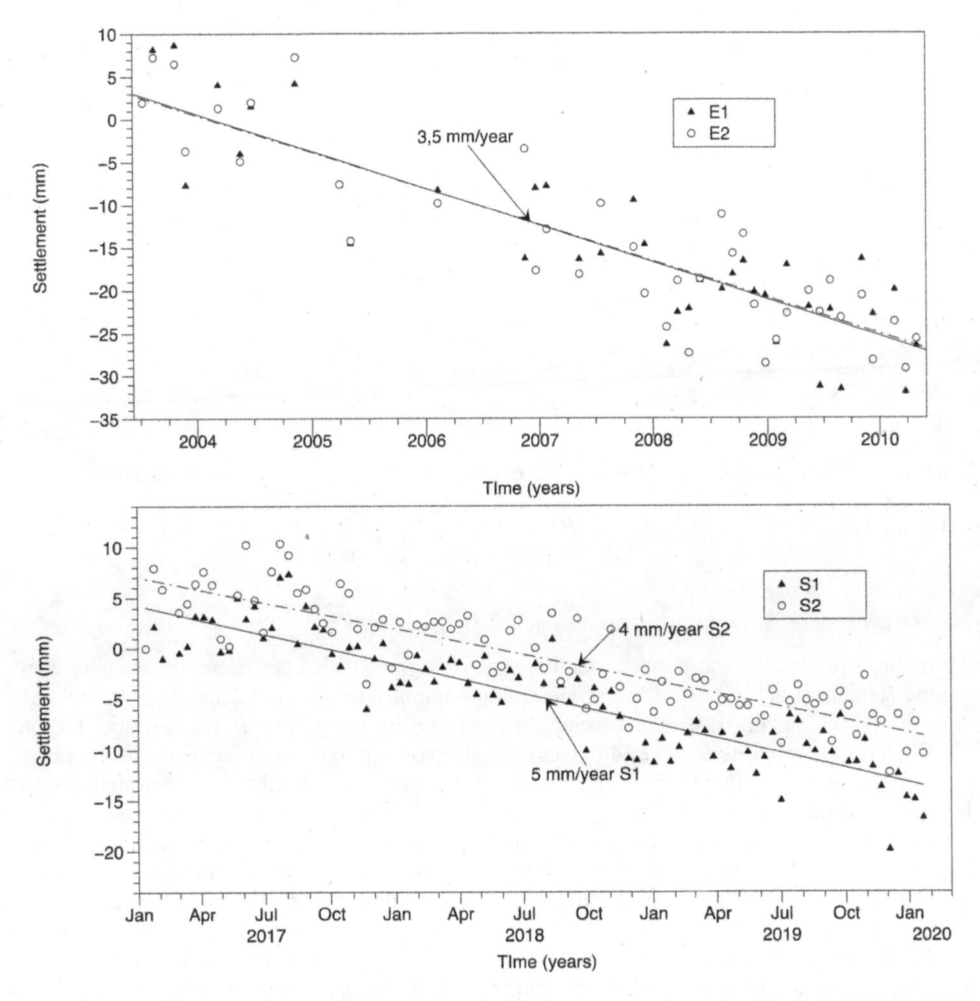

Figure 10. Settlements time series of some targets identified on the Benedictine Basilica by ENVISAT ASAR (a) and SENTINEL-1 (b) interferometric products.

In the lower part of Figure 10 the data from Sentinel Satellite in the three years period 2017–2020 show for the selected points S1 and S2 a slightly faster rate of movement which approaches on the average 4 to 5 mm/years. It is noteworthy that the rate of the point S2, approximately located in the middle of the church, is smaller than the rate of point S1 located on the front façade of the church confirming the hogging mode of deformation deduced by survey via electronic level.

Finally in Figure 11 the available data are plotted all together showing a rather clear trend. The plot has been obtained using as a reference the linear fitting trough the survey data (i.e. benchmark 130). The movement obtained by satellite data have been simply averaged on the two selected points and adapted to the trend provided by the terrestrial survey data.

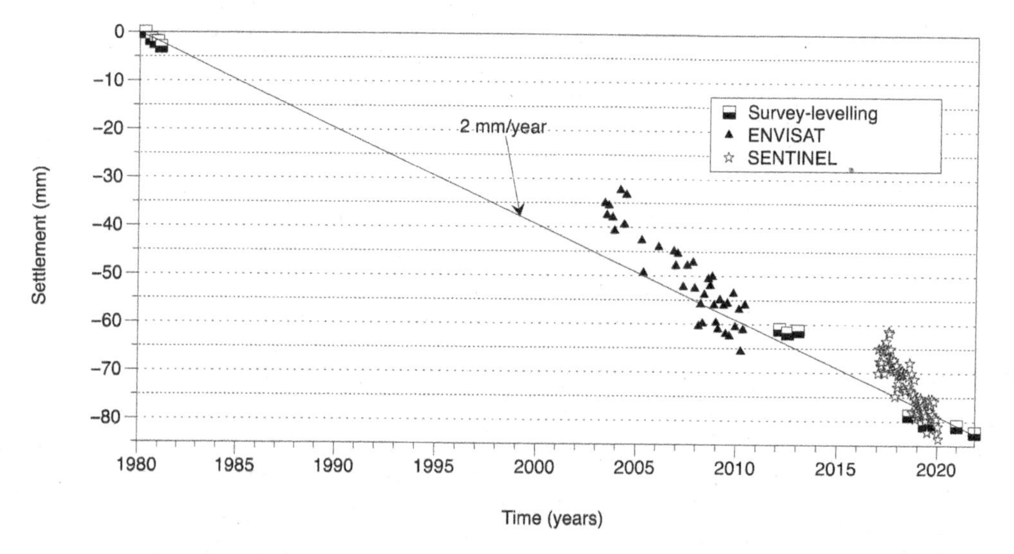

Figure 11. Settlement time series recorded in the last 40 years for the Basilica di Sant'Angelo in Formis.

2.4 *Discussion of the monitoring results and latest repair works*

All the data reported in the previous section show that the basilica is affected by a rather continuing settlement at least since 1980. The average initial rate was of 1.5 mm/year in 1980/81 with the marks moving faster approaching even a rate of 3.5 mm/year. The average rate on the whole observation period (i.e. 40 years) mainly based on terrestrial survey is in the range between 1.2 mm/year, for the right nave, and 1.5 mm/year for the left nave. The satellite data have contributed to throw light on periods where the survey was not available. They confirm the general trends commented before with slightly larger rate of movements obtained by fitting the typical scattered response. What maybe alarming is that the last three years covered by satellite have shown the largest movement rate ever detected in the range between 4 and 5 mm/year.

The deflection ratios calculated separately for the two naves and already introduced in the previous section are compatible with the occurrence of fissuring (Burland & Wroth 1975) which were clearly present in the church in the period 2012–2014. In those years some remedial works were decided, planned and carried out (Figure 12).

Figure 12. Recently carried out remedial works in the church.

The interventions could only cure the symptoms rather than removing the cause(s).

Interventions at the foundation level were difficult to conceive and to realize without interacting with the underlying structures of the ancient roman temple and with the floor which is still lined with marbles coming from the mosaic of the ancient temple. Without having yet a precise idea of the causes of the movements and considering the limitations described above any intervention at the foundation level was at least delayed.

On the other hand, as already mentioned at the beginning of the paper, the roof of the Basilica was completely rebuilt, after the collapse of the existing wooden roof due to the earthquake occurred in 1930. Large attention has been paid to this relatively new roof which was made by bricks and concrete and covered by wood thus resulting in a heavy load both on the side masonry and on the foundation.

This roof has been dismantled allowing first the construction of peripheral concrete beam at the top of the lateral masonry walls, to strengthen the structural box against seismic events. A new wooden roof, lighter than the existing one, has been finally installed. Further interventions have been dedicated to recover and protect the frescoes after appropriate and careful sealing of the existing cracks.

However, both satellite data and terrestrial survey data confirm that the movements at the foundation level are still going on and as matter of fact some new small cracks in the recently plastered walls are appearing.

On the other hand, it must be outlined that if the recorded rate of settlement had been active since the construction of the basilica, an average settlement between one meter and two meters should have accumulated; such figure is clearly impossible. It is evident that the observed movements initiated relatively recently, for some unknown reason; they have a rather continuous trend at least in the last 40 years with no traces of cycling.

In previous papers possible reasons that have been considered by various Authors are: (i) the extensive use of explosives in the nearby stone quarries; (ii) the effect at the surface of karst phenomena in the limestone; (iii) residual slow tectonic movements of the slope. None of these suggestions seems totally convincing. The quarrying activity which was considered one of the most likely causes has been completely stopped in the early '80's when the evidence of movements was first detected but the movements are continuing 40 years later. There are no evident reasons why the karst phenomena should begin at a time of some decades ago. The third hypothesis (Di Nocera

2013) is based on the possibility that a tectonic activity has been triggered either by recent seismic activity or by the explosions in the quarries. At the time being, no definite conclusion can be drawn from the available data; it is believed however that there is a clear need for keeping active a careful, long term and relatively frequent monitoring.

3 THE INCURABILI COMPLEX WITH ITS PHARMACY

The eighteenth-century pharmacy of the INCURABILI COMPLEX is the place where Art meets Science. Here the Rococo Baroque style, designed by Domenico Antonio Vaccaro, is combined with the enlightenment of the masters of Anatomy and Botany (Domenico Cotugno, Domenico Cirillo) in the most important Hospital of the Bourbon Kingdom. The alchemical preparations of the ancient tradition open the passage to medicinal chemistry in the place where skilled craftsmen of wood, gilding and ceramics prepared rooms where the voices of researchers who wrote the history of European thought resounded. In particular, the saplings, the hydras and the "riggiole" (tiles) of the Massa brothers represent an extraordinary unicum for the elegance of the colors, the refinement of the iconographic program aimed at the esoteric-masonic function of the Great Hall, where the uterine matrix operated is proudly displayed.

The Incurabili complex is quite articulated with its history from the beginning to the current situation covering a span of many centuries. In particular as reported also in a recent ph.d thesis (Micillo 2013), the main construction phases can be summarized as follows:

1. The foundation from 1519 to 1585;
2. The extension on the viceregal walls of 1729 operated by Alessandro Manni;
3. The first restoration from 1747 to 1751 directed by Bartolomeo Vecchione;
4. The incorporation of the conterminous monastic structures of the XIX century (1800 Santa Maria delle Grazie, 1836 Santa Maria della Consolazione);
5. Minor construction episodes and modifications of the XX century.

The whole hospital complex is delimited to the North, from East to West, by Maria Longo street, to the West (from North to South) by Madonne delle Grazie square; to the East (from North to South) by Consolazione alley, Domenico Capozzi and Luciano Armanni street, to the South by the monastic complex of Santa Maria delle Grazie and that of Santa Maria Regina Coeli.The investigations in the subsoil show clearly that if we include in the complex and in its monumental part its subsoil as it should always be done in highly stratified areas respecting the concept of the uniqueness of the ground-monument system the history that is covered extends over two thousand five hundred years, reaching the ancient Greek Neapolis with its surrounding walls still in place (Figure 13).

Maria Lorenza Longo founded the Incurabili hospital. Born in Barcelona with the original name of Lorenza Requenses, she arrived in Naples in 1506 with her husband Giovanni Lonc, minister of the king Ferdinand the Catholic of Spain, suffering from a severe form of rheumatoid arthritis. After the death of her husband, she went to the sanctuary of Loreto to ask for a cure, promising to devote herself to the care of others if received. And so it happened and Maria Lorenza Longo began her work in Naples by founding the Hospital of the Incurable in 1519. The history of the oldest part of the city of Naples had begun many centuries earlier. For example, the Greek wall of the fifth century BC arranged in an east-west direction and characterized by a double curtain made of regular large blocks of tuff and filled in Roman times with walls in opus reticulatum is still visible along Constantinopoli street and Foria street. In the picture in Figure 13 a section of the original greek wall (the so-called "second phase wall") of the fourth century BC with a scarp and a stepped foundations for an overall height of 9.20 m is visible. Traces of the greek walls were also found in the boreholes carried out for soil investigations purposes and summarized in the next paragraph. The Incurabili Hospital was from about 1525 until 1813 the only institution destined to the reception of the mentally ill people and only during the "french decade" in compliance with a law by Joachim Murat the sick were transferred to the asylum of Aversa.

Figure 13. Greek walls surrounding the original Greek city of Neapolis.

The hospital keeps substantially its original conformation from the end of the sixteenth century until the intervention of the engineer Alessandro Manni in the 1729 who executed the designs of D. A. Vaccaro. In this phase the hospital is enlarged, and the pharmacy is created with its current configuration. In Figure 14 the façade of the complex on the Consolazione alley is sketched as reconstituted on the basis of many ortophotos. By an historical point of view this is an interesting perspective because the initial period of the hospital started from the existing XVth century houses.

Figure 14. Façade on Consolazione alley with the back windows of the Incurabili Pharmacy.

In Figure 15 the marvelous grocery room of the Pharmacy with the roof frescoes by Pietro Bardellino and the monumental stairs in the courtyard of the Incurabili complex.

3.1 *The subsoil of the INCURABILI COMPLEX and its geological setting*

The foundation levels of the monumental buildings of the INCURABILI COMPLEX are placed at different elevations. The oldest part of the complex visible today and founded at lower levels is certainly the one on the Consolazione alley. The inspections in the existing houses and their maps allowed to determine the main levels of the foundation walls on that side of the complex. These walls are made by continuous masonry similar to the ones that continues in elevation with thicknesses often unchanged for several meters above the foundation level.

The soil layering as deduced by boreholes reported in Figure 16 can be summarized as follows:

Figure 15. The eighteenth-century pharmacy of the INCURABILI COMPLEX: internal picture of the main hall and external picture of the monumental stair.

a) topsoil with a thickness varying from a minimum of a few decimetres in the borehole S4 to less than a metre in soundings S2 and S3 and up to a little less than three metres in the soundings S1 (in the center of the courtyard) and S5, along the Consolazione alley;
b) under the topsoil (manmade), with the exception of survey S4, located near the access from Maria Longo street, which logged for about 10 m walls that are part of the ancient greek walls (Neapolis), in all cases there are alternations of sandy-silt cinerites, layers of volcanic pumices and layers of coarser sand sometimes mixed with lava slags; the alternating layers correspond to a well-known stratigraphic series linked to the activity of nearby volcanic complexes;
c) at the end of the series of alternating pyroclastic uncemented materials, in all the surveys Neapolitan yellow tuff, a well-known soft rock, is found with a top layer locally named "cappellaccio" and characterized by poorer mechanical properties and often lacking a significant degree of cementation.

The yellow Neapolitan tuff found in the final part of all the surveys is at a rather constant level above sea level ranging between +26 m and +30 m asl. The deepest point of the roof of the tuff is just that relative to the borehole S4 that in on the line of the ancient Greek walls (+26 m asl).

Taking into account the distance among the boreholes the variation of the tuff layer elevation is in the ordinary range.

Figure 16. Plan view with locations of boreholes Si and Dpsh tests.

In the area of the Incurabili Hospital during subsoil investigations several deep shafts have been detected. They connect the above buildings with a very deep network of rather small tunnels excavated in the tuff basement. The network is partially related to the famous ancient roman Bolla aqueduct. These vertical wells are 20 meters deep or more and generally end inside the tuff basement where a network of horizontal or sub-horizontal tunnels connects the most of them (Figure 17). The smaller wells were used to draw water inside the hospital complex and in the houses facing on the Consolazione alley. Many of these wells today are filled with debris and waste material. Some larger wells are empty, inspectable and located close to the foundations walls of the Incurabili Pharmacy on the edges of its plan area. In the recent past these shafts have been involved in problems with water infiltration from above produced by huge leaks caused by failures of the pipes of the internal white water drainage system. Some available video inspections carried out with the support of speleologists (Figure 17) show that the upper part of these wells, which is developed in loose and uncemented soils above the the water table, is lined with walls of tuff blocks that are not always in good conditions.

3.2 *Damages, monitoring data and diagnosis*

In its recent history the Incurabili monumental complex has been seriously affected by damages represented by crack openings in the walls, in the floors, sometimes also highlighting detachments between structurally independent bodies. To keep the monumental building under observation, considering both its value and the concerns that inevitably accompanied the sporadic appearance of new cracks, a monitoring system was implemented. At the present time several months (about 10

Figure 17. Shafts, cavities close to the Incurabili Pharmacy foundations.

months) of precision survey carried out by leveling on the benchmarks in the wall and on nails/bolts on the floor referred to fixed points in external areas close to the complex are available.

In this section these data referred to medium-term monitoring are presented and discussed aware that one or more full annual cycles are by far the best option to make a satisfactory interpretation of the phenomena going on and to filter off the thermal influence on the measurements if any. The Figure 18 reports the settlement of the external benchmarks R2 and R3 referred to the main reference R1 versus the temperature. In this case all the movements are very small and whitin +/− 0.5 mm without showing a particular trend with the temperature even in the rather large experienced range from 10°C to 32°C. In the same figure the temperature recorded is plotted vs. the time during the whole monitoring period.

On the other hand, Figure 19 reports plots of movements versus temperature to show how close correlations can be sometimes detected. In this case the correlations can be adversely affected by the fact that the temperature in the plots represent a unique record measured in the entrance area of the complex. Of course, a bad correlation may also simply represent the fact that observed movements are related to causes other than the simple temperature. Keeping this in mind it can be remarked that the marks on the Consolazione alley have movements strictly related to temperature with a R^2 value close to 0.8 and quite small value of the residuals in the whole range of records, while the marks inside the courtyard on the façade of the Pharmacy and on the hospital monumental chairs present much lower value of R^2.

The same can be clearly observed by direct comparison of the trend of the benchmarks movement with the time in Figure 20 and the temperature with the time in Figure 18. This finding can confirm the fact that real movements nonrelated to thermal response of the measuring systems and of the buildings are occurring mainly in the area inside the courtyard of the complex where the largest settlement with a lower correlation with temperature are clearly observed. Some benchmarks inside the courtyard have reached during the period of observation even values close to 5 mm, which although small in less than 1 year represent a significant value. To summarize the above findings in Figure 21 is plotted a 2D contours map of the measured subsidence, obtained filtering out, at the best possible, effects of temperature on the records. The maximum values corresponding to the colors of the central zone of the basin are between 2 and 3 mm. Therefore, net of thermal effects,

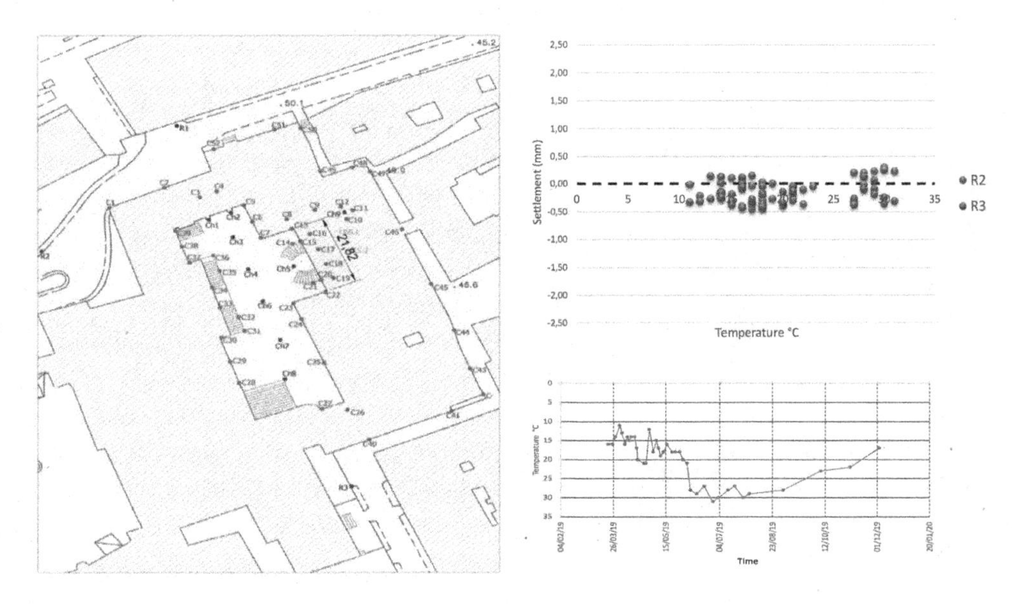

Figure 18. Settlement and temperature: Fixed benchmarks in the area outside the Incurabili Complex.

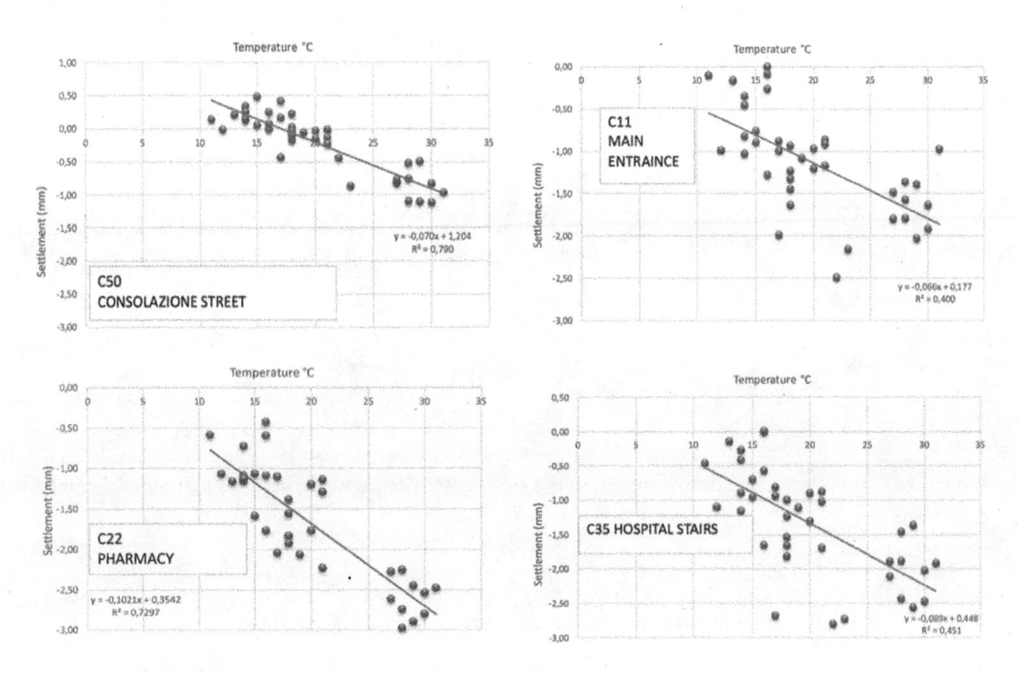

Figure 19. Settlement vs. temperature for different zones of the monument.

the values are rather small and tend to stabilize over time as can be seen from the diagrams in Figure 20. Obviously, the monitoring should continue for time longer than the exposed period and cover, as mentioned before, at least one or two full year cycles, but in this case, at least, the diagnosis was quite simple.

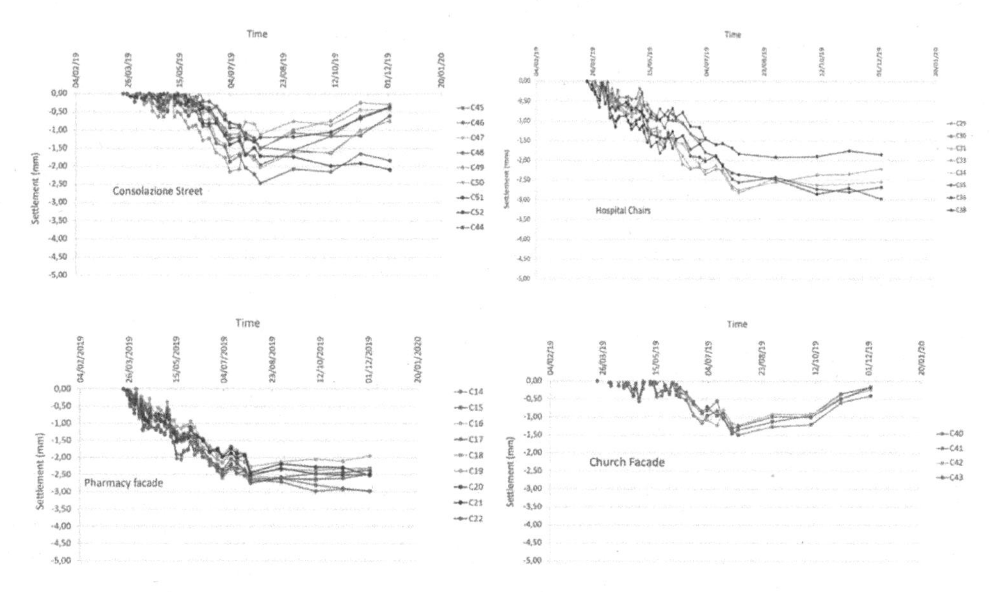

Figure 20. Settlement vs. time for different zones of the monument.

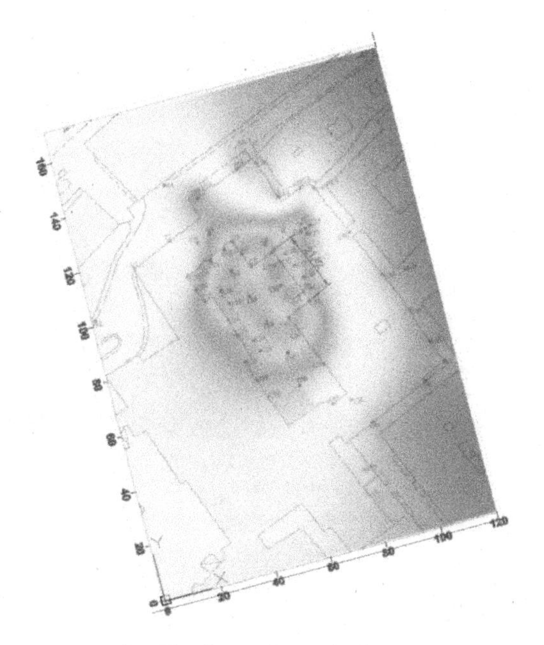

Figure 21. 2D settlement contours after filtering out thermal effects on measurements.

3.3 *Diagnosis and therapy*

The available evidence including the monitoring data reported in the previous paragraph and the survey of the cracks (Figure 22) that are not documented in detail to be brief outlines a fairly clear situation. Based on its history and of the available documents the Incurabili monumental complex has been certainly stable for a few centuries from its foundation. Apart from problems related to the ordinary maintenance of the structures no problems have been discovered looking at the history of the monument. A few years ago, some subsidence problems have appeared involving in different ways and with different degree of danger several parts of the monumental complex. Some problems which are not documented here for space reason have been clearly related to structural failures of weakened portion of the roofs or of the slabs. In the case of the Pharmacy the problems were clearly related to subsidence problems involving the bearing walls starting from the foundation level.

Figure 22. Cracks in the rooms and on the floor of the INCURABILI PHARMACY.

As previously mentioned in the months preceding the start of the monitoring operations and during the previous autumn and winter at the turn of 2018–2019 several reports had been made to the authorities in charge for sudden manifestations of water leaks in pipes present in the subsoil of the courtyard of the complex. Being the ancient monumental complex partially used as an hospital with all the needed sewers and aqueduct pipes it is not hard to guess that problems with the water system could have occurred even before. A careful program of inspection of the whitewater and the blackwater sewer systems was organized. In some cases, given the antiquity of the complex, it was not even possible to reconstruct exactly all the passages of the sewer systems even using video inspections and other modern techniques. However, it is a fact that, as for example reported in the picture of Figure 23, conspicuous water losses occurred at several times with consequent dragging of fine sandy material that sometimes was even delivered inside the deep wells inspected and described previously.

Once established a close relationship between the severe cracks in the monumental complex and the problems with water leakages in the sewer system, the patients of the hospital were quickly cleared out and the drains were quickly interrupted. It is to be underlined that also the inhabitants of the houses on the Consolazione alley immediately below the floor of the Pharmacy were for the first time in the history of the complex cleared out. The white-water network was sectioned at the base of the building and was delivered to the sewer with temporary by-passes. These decisions were promptly taken, and the actions consequently made just at the beginning of the monitoring period. Therefore, as anticipated above, it is extremely likely that the observed settlements or subsidence effects net of the thermal effects are simply the viscous tails of phenomena induced by erosion of

Figure 23. Evidence of damaged pipes of the sewer system of the Incurabili complex.

fine material and micro-collapse by saturation in limited areas of pozzolana above the free surface of the water table.

An overall project of recovery of the prestigious monument is in progress because the degradation induced by the crumbling networks of underground services was coupled with degradation by infiltration of rainwater from above through floors and waterproofing no longer suitable.

4 CONCLUSIONS

In the paper two examples of very valuable monuments in the city of Napoli and in its surroundings are presented to show how vulnerable may be such structures that have however a very long life and many lessons to give us.

In the case of the medieval basilica of Sant'Angelo in Formis even if the overall picture is not one of fear for its future it has to be recognized the substantial difficulty to understand the movements recorded in the last 40 years of its life. The use of satellite data has confirmed the findings obtained via terrestrial survey with more precise but less continuous in time methods. In such a case remedial works have been carried out even recently removing some effects or damages but, probably, non-removing the ultimate cause of the movements that is not yet been certainly identified.

The case of Incurabili Complex on the other hand represents a typical situation of the historical center of the city of Napoli where the monumental buildings are still used and sometimes their use is not at all compatible with the safety and mainly with the preservation of the artistic part. The movements recorded for about 1 year have clearly shown that some viscous effects were still taking place while the causes of the movements had been likely removed simply preventing additional water leakage occurring from an old and widely damaged sewage system.

REFERENCES

Adriani, L., Pellegrino, A., Bescatore, T.S. (1980). La Basilica Benedettina di S. Angelo in Formis: prima relazione. Indagini sui terreni e sulle strutture del monumento per lo studio dei dissesti. Unpublished report, Soprintendenza ai Monumenti della Campania, Napoli.

Brandi, C. (1963) Teoria del restauro, Ed. di Storia e Letteratura, Torino – pp. 1–158

Candela, M., Mandolini, A., Russo, G. (1997). Monitoring Castel dell'Ovo in Napoli – Preliminary results. Proc. *Geotech. Eng. For the Preservation of Monuments and Historical Sites*, Balkema, pp. 343–347.

Corniello, A., Santo, A. (1995). I dissesti della Basilica Benedettina (XI secolo) di Sant'Angelo in Formis (Capua) ed il complesso assetto geologico dell'area. *Geologia Applicata e Idrogeologia*, vol. XXX, I, 125–137

De Franciscis, A. (1956). Templum Dianae Tifatinae. Soc. Di Storia Patria di Terra di Lavoro, Caserta, 60 pp.

Di Martire, D., De Luca, G., Ramondini, M., Calcaterra, D., (2013). Landslide-related PS data interpretation by means of different techniques. In *Landslide science and practice* (pp. 347–355). Springer Berlin Heidelberg

Di Martire, D., Novellino, A., Ramondini, M., Calcaterra, D., (2016). A-Differential Synthetic Aperture Radar Interferometry analysis of a Deep Seated Gravitational Slope Deformation occurring at Bisaccia (Italy). *Science of The Total Environment*, Volume 550, 15 April 2016, Pages 556–573. doi:10.1016/j.scitotenv.2016.01.102

Di Nocera, S. (2013) Personal communication.

Jacobitti, G.M., Abita, S. (1992). La Basilica Benedettina di Sant'Angelo in Formis. ESI, Napoli, 97 pp

Infante, D., Di Martire, D., Confuorto, P., Tessitore, S., Tòmas, R., Calcaterra, D., Ramondini, M. (2019). Assessment of building behavior in slow-moving landslide-affected areas through DInSAR data and structural analysis. *Engineering Structures*, 199, 109638.

Italian National Institute for Environmental Protection and Research, ISPRA, 2008. Landslides in Italy. Special Report 2008. ISPRA, Rapporti, 83/2008.

Micillo, A. (2013). Il Complesso Ospedaliero di Santa Maria del Popolo degli Incurabilidi Napoli: evoluzione storico urbanistica. Ph.d. thesis XXV Ciclo, Dottorato di Ricerca in
Storia e Conservazione dei Beni Architettonici del Paesaggio, Tutor prof. L. Di Mauro

Mora, O., Mallorquí, J.J., Broquetas, A., (2003). Linear and nonlinear terrain deformation maps from a reduced set of interferometric SAR images. *IEEE Trans. Geosci. Remote Sens.* 41, 2243–2253. http://dx.doi.org/10.1109/TGRS.2003.814657.

Pellegrino, A., Pescatore, T.S. (1981). La Basilica di S. Angelo in Formis. Indagini sui terreni di fondazione ed analisi dei dissesti: relazione finale. Unpublished report, Soprintendenza per i Beni Ambientali e Architettonici della Campania, Napoli.

Russo, G., Viggiani C. (2000) The stability of monuments over coastal cliffs in the bay of Napoli. Proc. International Millenium Conference on Safeguarding of our Cultural Heritage Vol. (U) pp.10 held by UNESCO & ICOMOS in Bethlehem (Palestine), Oct. 2000.

Russo, G., D'Agostino, S., Lombardi, S., Viggiani C.(2009) Structural engineering and geology applied to the static problems of the Etruscan "Tomba dell'Orco" (Tarquinia, Central Italy). Journal of Cultural Heritage ISSN 1296-2074, vol. 11(2010), available on line since 2009, pp. 107–113 – doi 10.1016/j.culher.2009.11.001

Russo G., Viggiani C., Cammarota A., Candela M. (2013) The Benedictine Basilica of S. Angelo in Formis (Southern Italy): a therapy without diagnosis?. pp.225–232. In Geotechnical Engineering for the Preservation of Monuments and Historic Site (2013) vol. 1 Int. Conf. TC. 301

Russo G., Alterio, L. Silvestri, F. (2017) Seismic Vulnerability Reduction for Historical Buildings with Non-Invasive Subsoil Treatments: The Case Study of the Mosaics Palace at Herculaneum. International Journal of Architectural Heritage, 11 (3), pp. 382–398. DOI: 10.1080/15583058.2016.1238969

Tapete, D., & Cigna, F. (2019). COSMO-SkyMed SAR for detection and monitoring of archaeological and cultural heritage sites. *Remote Sensing*, 11(11), 1326.

Tomás, R., Garcia-Barba, J., Cano, M., Sanabria, M. P., Ivorra, S., Duro, J., & Herrera, G. (2012). Subsidence damage assessment of a gothic church using Differential Interferometry and field data. Structural Health Monitoring, 11(6), 751–762.

Zhou, W., Chen, F.L., Guo, H.D. (2015). Differential radar interferometry for structural and ground deformation monitoring: A new tool for the conservation and sustainability of cultural heritage sites. *Sustainability*, 7, 1712–1729.

Wettstein J. (1960) Sant'Angelo in Formis et la peinture médiévale en Campanie. Geneve

Geotechnical Engineering for the Preservation of Monuments and Historic Sites III – Lancellotta, Viggiani, Flora, de Silva & Mele (Eds)
 ISBN 978-1-032-35998-4

The Grand Canal at Versailles: Geotechnical investigation, II

J.-D. Vernhes
UniLaSalle, Beauvais, France

P. Saulet
ESRI, Meudon, France

A. Heitzmann
Etablissement Public du Château, du Musée et du Domaine National de Versailles, Versailles, France

ABSTRACT: The park of the Palace of Versailles required major geotechnical developments whose center-piece is called the Grand Canal, designed by André Le Nôtre and dug in the 1670s on behalf of King Louis XIV. This exceptional work was the subject of an archival study, the main results of which were communicated on the occasion of the ECSMGE in Reykjavik in September 2019. The article proposed for the 3rd TC301 International Symposium reports on the in situ investigations and numerical field models carried out, with UniLaSalle students, over a period of two years, allowing in principle to answer questions that remained unaddressed.

The construction of the current relief map and of the supposed relief map before work makes it possible to estimate a volume of excavated material and embankments of nearly 700,000 m^3 and 500,000 m^3 respectively, realistic orders of magnitude. The apparent unbalance has no justification other than the current uncertainties in topographic models. The question of the canal watertightness and its hydraulic relationship to the surrounding ground, given the available field data, paradoxically appears less clear today than in the conclusions of the previous article. The real nature of the basin deposit, apparently colluvial to significant depths, complicates attempts to interpret both Le Nôtre's project and the current functioning of his work.

1 MAIN ELEMENTS OF THE WORK FROM THE FIRST ARTICLE

For a general presentation of the Grand Canal of Versailles today, we refer the reader to the first article dedicated to its geotechnical analysis (Vernhes & Heitzmann 2019).

The scientific contribution of this article consisted of a detailed analysis of the '*Comptes des bâtiments du roi sous le règne de Louis XIV*' in their simplified form, published in five successive volumes by Jules Guiffrey at the end of the nineteenth century (Guiffrey 1881-1901). These volumes have been reviewed for all that concerns the works related to the Grand Canal of Versailles, for a total of 137 articles explicitly related to this work. The construction of the water body was carried out over a period of nearly 14 years, from 1667 to 1680. Here is the main information resulting from this survey.

1.1 *Phasing of the work*

The irregularity of expenditure over the 14-year period made it possible to identify four phases of work. The first (1668-1669) and the third (1671-1672) are intense periods of activity. They see respectively the construction of most of the East-West and North-South branches of the cruciform canal. It represents 200,000 m^2, nearly 85% of the total surface of the built basin. This area does not include peripheral earthworks, which double the area of natural land likely to have been topographically modified, slightly or strongly – in particular at the western end of the canal (Figure 1).

DOI 10.1201/9781003329756-14

Figure 1. View to the West of the Grand Canal of the park of the Palace of Versailles taken from the Latona basin (Sept. 2018). Below the Green Carpet and slightly above the canal, the Basin of Apollo and the sculptures of its fountain. At the other end of the water body, the Basin of Gally, that is not very visible on this picture.

An important aspect of these results for further earthworks analysis is that a coherent scenario of land movement should take into account this phasing: the balance of movements at the end of a phase should be in equilibrium.

1.2 *Earthworks volumes*

For a basin area of about 230,000 m^2 and an excavated depth rounded to 2 m, we obtain an earthworks volume of about 460,000 m^3. This first figure makes it possible to fix the order of magnitude of a minimum volume of the lands moved and deposited – because the draught of the canal is two meters and the excavation depth is therefore a little higher – assuming that the Grand Canal would have been built entirely out of excavated material.

1.3 *Draining nature of the Grand Canal in question*

The Grand Canal is, in terms of altimetry, the lower basin of the complex of basins and fountains of the park of the Palace of Versailles, therefore its first outlet, before the waters return to the natural environment in the Gally creek by means of drain plugs.

As we see in *Les Comptes* (‘*The Accounts*’), the canal, navigable and intended for a ceremonial fleet, had been designed during the first phase of work as a water reserve for the upstream of the park thanks to return mills. Furthermore, it is reported that the large and very flat area of the park where the canal is now largely situated, with a historically marshy nature, presented development difficulties during the attempts of sanitation by drainage. This would have led the designer of the project, André Le Nôtre, to suggest and carry out a permanent draining work (Baridon 2013).

2 FIELD SURVEYS IN 2018 AND 2019

During the academic years 2018–19 and 2019–20, field surveys were carried out on the site of the Grand Canal in Versailles, thanks to the work of geological engineering students. The types and quantities of surveys carried out are given in Table 1 below, located on the map in Figure 2 and detailed in the following paragraphs.

Table 1. In situ surveys carried out in 2018 and 2019 (modified according to Cordier et al. 2019).

Type	Quantities
Manual augers, number	122
Manual augers, linear (m)	76.4
Manual penetrometer, number	60
Seismic refraction array, number	7
Seismic refraction array, linear (m)	161
Electrical panels, number	11
Electrical panels, linear (m)	1072
Topographic profiles (levelling X,Z), number	43
Topographic profiles (levelling X,Z), points X,Z	1602
Topographic profiles (leveling X,Z), linear (m)	5072

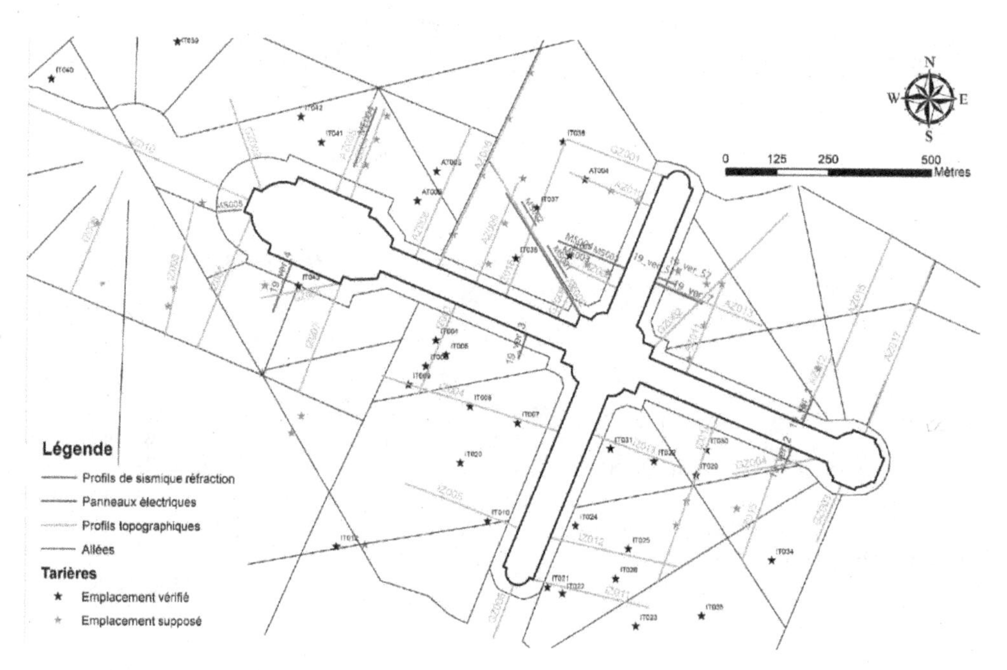

Figure 2. Map schematizing the in situ implementation of the measurements carried out around the Grand Canal in 2018 and 2019 (Cordier et al. 2019, p. 20).

2.1 *Geotechnical measurements*

2.1.1 *Manual auger*

The purpose of manual auger samplings was to highlight lithological variations in the first meter of soil around the Grand Canal. The agreed idea was to associate them with the stratigraphic changes visible on the geological map (Figure 5) but also with the "lithological anomalies" linked to earthworks.

2.1.2 *Portable dynamic penetrometer*

The purpose of indirect soil compactness measurements using the portable dynamic penetrometer was similar: to detect, at a depth limited to 1 m, a possible global variation in soil response depending on its geotechnical state: in place, excavated or backfilled.

2.1.3 *Assessment of geotechnical measurements*

Even taking into account the presence of a recent surface cover rich in organic matter, which can locally be several decimeters thick, auger and penetrometer survey data showed a wide dispersion of the results. These results proved to be refractory for a direct interpretation according to the criteria initially planned, which were intended to highlight a difference distinguishing between in place and remoulded soil, in other words between natural and excavated material on the one hand, embankment on the other hand.

We thought at some stage that the wide dispersion of the results was related to the natural variability of tertiary rocks at advanced stages of alteration in place. In a geostatistical language, it would then be a "nugget effect" problem, a problem that can be solved by a better distribution and densification of the sampling. The discussion presented in Part 3 sheds significant enlightenment on these difficulties.

2.2 *Geophysical measurements*

2.2.1 *Seismic refraction*

The purpose of the seismic refraction measurements (Figure 3) was comparable to the purpose of penetrometric tests. It has indeed been implemented in order to identify variations in the state of

Figure 3. Installation of a base of seismic refraction measurements (Sept. 2018). The device is installed near the Gally basin. To the east, above the other end of the canal, we can see the Palace.

compactness of the soil by correlation with the seismic velocities of the P wave in the ground. The depth of investigation of the method, with the adopted equipment and parameters, was in the order of 5 to 10 m.

2.2.2 *Electrical panels*

The purpose of electrical panel measurements, through a resistivity imaging, was always to look for signs of variation in the compactness of soil in relation to the effects of earthworks. This approach previously required hypotheses on lithology but also on the water content of soils, to which electrical resistivity measurements are particularly sensitive. The depth of the investigation of the method, with the equipment and settings used, was in the order of 15 m.

2.2.3 *Assessment of geophysical measurements*

As a synthesis of the findings, geophysical measurements have everywhere highlighted a stratified structure of the soils. This is an unsurprising result in seismic tests since the method is firstly adapted to highlight velocities variations of the P wave (v_P) on both sides of interfaces more along horizontal than vertical direction. The observation stated above is less clear with electrical panels and, in fact, this method is not supposed to favor one direction over another in geometric field modelling.

In seismic refraction, a "seismic bedrock" in a range of velocities v_P from 1700 to 2000 m/s is encountered at moderate depths from three to five meters. It is systematically surmounted by a complex of two "slower" horizons, the most superficial from one to two meters thick for $v_P \approx 300$ m/s, the deepest from one to three meters thick for $v_P \approx 1000$ m/s.

In electrical panel, it is more convenient to describe the general situation through the following observation. Soils are characterized by low to moderately low resistivities, from 10 to 50 Ohm.m, except, where appropriate, near the surface. In these cases, the resistivities of the soil layers are typically in the range of 50 to 300 Ohm.m on very variable thicknesses from a few decimeters to a few meters.

Whatever the method, and as in geotechnics, difficulties of interpretation emerged by establishing a link with the indications of the geological map and the categories of remoulded / in place soil. This is discussed in Part 3. However, general reasonings makes it possible to take advantage of the information provided by geophysical prospection. For example, it is very likely that the soil is water saturated at few meters depth. This hypothesis can be supported on two points: the general weakness of the resistivities measured on site below a possible resistant layer, fresh water having a resistivity in the order of 40 Ohm.m and significantly influencing the resistivity of the soils it soaks; on the other hand on the velocities v_P, which overcome in depth the threshold of 1500 m/s, speed of compression-al waves in the water. A persistent problem is that this saturation is not necessarily synonymous with superficial phreatic level since phenomena of capillarity may very well pro-duce the same geophysical results.

2.3 *Topographic measurements*

2.3.1 *Airborne photogrammetry by drone*

As part of the students' work, a drone photogrammetry campaign (with a three-color RGB sensor) took place in December 2018. It aimed to set up a Digital Elevation Model (DEM) for the entire perimeter of the Grand Canal. This campaign gave a result difficult to exploit for the geotechnical survey. We have chosen not to include the data in the final DEM.

Anticipating the problem, we were going to encounter with the canopy of woods arranged around the Grand Canal, we chose the beginning of the winter season to carry out the campaign. The characteristic shaving lighting of this season at the latitude of Versailles, producing ample drop shadows, added to a particularly dense twigs vegetation, disrupted the process that should have allowed the photographic cover to produce reliable and continuous altimetry data. We now consider that only a relatively expensive LIDAR-type sensor would allow to achieve the expected result.

2.3.2 *Levelling (X,Z) with level and target*

More modest means of topographical measurements were implemented by the teams of students, a levelling type according to a profile determined in advance (Figure 4). These profiles consist of points known by their horizontal distance to a point of origin (X) and by their relative altitude to that point (Z) measured with a level and a target. The resulting accuracy is in the order of one centimeter over aiming distances of a few tens of meters, very sufficient for a geotechnical objective. The advantage of these measures in the context of the site is that they have not been prevented, but only complicated, by vegetation. All the gathered data were then linked to a fixed reference, the limestone margin of the canal, established at 109 m NGF (General Levelling of France). The resolution (the measuring step) of the measured profiles is variable, the accuracy obtained is in the order of 1 to 10 cm in altimetry and from 10 cm to 1 m in planimetry.

The obtained result, combined with a public data source from the National Geographical Institute, is presented and discussed in parts 4.2 and the following parts.

Figure 4. Levelling measures (Sept. 2019). View taken to the East, the visible part of the Grand Canal is the Gally Basin.

3 A SUBSURFACE GEOLOGY PROBLEM

3.1 *Data from the geological map of France*

The geological map of France at 1:50,000, Versailles sheet (n° 182) was published in 1967 under the authority of Jean Goguel (Figure 5). It has not been updated since then. It highlights a system of Tertiary formations frequently encountered in the Paris Basin. These are the sandy, clayey, marly or limestone ensembles, from the Stampian to the Lutetian, aged about 30 to 45 million years. The principle of drawing the cartographic boundaries of these strata is to associate with their sub-horizontal character the pattern of the relief, with the result of relating approximately these boundaries to the topographical level lines. The Grand Canal was mainly developed in the topographic depression of the ru de Gally, within which the geological map indicates as supragypsous Marls of the Ludien (code e7), level dated on average 35 million years, the most recent of the Eocene in the Paris Basin.

Figure 5. Extract from the geological map of France at 1:50,000, Versailles sheet (GOGUEL, 1967), cen-tered on the Grand Canal. The castle is about 1 km east of the canal.

The 2018 and 2019 field measurements were based on the principle that, apart from a first decimetric layer of "topsoil" covering the entire site, the lithology encountered on site had to correspond to the stratigraphic data of this map or be affected by changes related to earthworks. These modifications were not meant to be purely random but to be unified by simple principles of excavated areas and deposit areas, in connection with the Grand Canal construction project of course.

Initial difficulties in interpreting the survey results presented in Part 2 of this article are now explained.

3.2 *Taking into account the data of surveys carried out in 1997 and consequences*

In the period of exploitation of the field surveys of 2018 and 2019, we have accessed an internal report of the Palace of Versailles issued in December 1997 by the Chief Architect of Historic Monuments at that time (Lablaude 1997), dealing with the restoration project of the banks and margins of the Grand Canal. This project resulted in the installation of four piezometric tubes in boreholes carried out with the helical auger Φ 80, at depths of about 8 meters, positioned around the perimeter of the water body at the cross of the North-South and East-West branches, as well as in the vicinity of its western end.

These four boreholes all show the presence of a high thickness of colluvium from the Fontainebleau sands and quaternary spreading silt (Lablaude 1997, p. 254), up to depths of 5 to 8 m. These materials are described as incoherent fine sand interspersed with clay lenses, very different from the expected plastic and homogeneous marl-clay facies, lithological translation of the supragypsous Marls of the Ludien. At the scale of the seventeenth century earthworks, this means that a large part of the displaced materials do not correspond to the indications of the geological map, which primarily describes the tertiary or anterior geological substratum and commonly excludes surface formations. For topographical reasons, this condition probably does not apply to the North and especially South end of the North-South arm of the Grand Canal, in areas where the project approached the outcropping sands of Fontainebleau. However, the idea that the earthworks under Le Nôtre's remoulded materials that had been in place for several tens of millions of years, turned out to be false. Colluviums, materials already reworked during the quaternary, cannot present a geotechnical facies as different before and after earthworks as was initially expected.

This evidence explains the difficulties of correlations between the indications of the geological map and the results of the geophysical and geotechnical investigations on all the sectors investigated in 2018 and 2019. Under these conditions and at the stage of the current results, possible criteria for differentiating the soils in place from those remoulded by means of intrusive or indirect investigation are unfortunately too ambiguous to allow for a conclusion. The analysis of the topographic models is the only tool to make advancements in the problem.

4 EARTHWORKS: SEARCH FOR A BALANCE OF MOVEMENTS

4.1 *Status of the issue in 1997*

The map in Figure 6 appears in the report of the late Pierre-André Lablaude presented above. On the occasion of his general diagnosis before the planned works, the architect wonders about what he calls the "anthropogenic topography", resulting from the lands reprofiling by earthworks. His analysis led him to use the level line at 110 m altitude NGF to distinguish two opposite situations: the one, in general case, where the canal is built in a "natural recess", and the one where this situation is in "artificial trench", which according to him corresponds to the particular cases "of the western ends (Gally basin), North (Horseshoe basin) and South (Menagerie basin)" (Lablaude) 1997, p. 241). Its conclusion is mapped by a modification of the route of the 110 m NGF level line at these three ends of the Grand Canal.

Our own analysis of his work is divided in two stages. Firstly, we note the ambiguity of his geotechnical vocabulary. Is it necessary to understand that the canal originally under 110 m was built in an embankment, or in a levelled surface, and excavated above this height? The equidistance of the level lines drawn on his maps being 5 m, he does not seem to seek to give details on this issue, especially in cases where he considers that the canal was developed under the natural altitude 110 NGF. Secondly, by interpolating this level line as he does on his interpretative map, he assumes that the canal axes at the three North, West, and South ends were almost always orthogonal to the

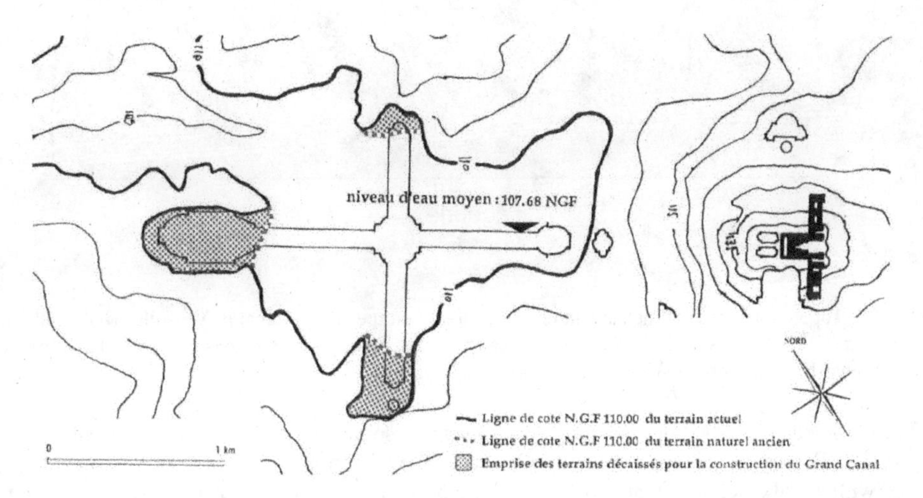

Figure 6. Grounds altimetry plan with location of the NGF 110 m altimetric line, extract from the map p. 242 (Lablaude 1997).

level line. In doing so, he considers that these three ends, and particularly the western end, caused only excavated earthworks. The problem of their deposit does not arouse his attention. These two re-marks about Lablaude's work are extended in the following paragraphs.

4.2 *Current topography*

The map in Figure 7 was developed from two data sources: the synthesis of on-site leveling profiles made by the students in 2018 and 2019 (see section 2.3.2), and the 5 m resolution Digital Elevation Model (DEM) of the National Geographical Institute (IGN). It should be noted that we have accessed the IGN 1 m resolution DEM but to note that, for the park of the Palace of Versailles, it has not a real gain in resolution. All the difficulty of the work then focused on the problem of harmonizing two sources of topographic information with their own characteristics, on the one hand the profiles (X, Z) precise but with a very heterogeneous planimetric arrangement, therefore very artefactual in a DEM, and on the other hand the DEM of the IGN, with a very homogeneous dot pattern, but insufficient resolution/accuracy for the geotechnical survey conducted.

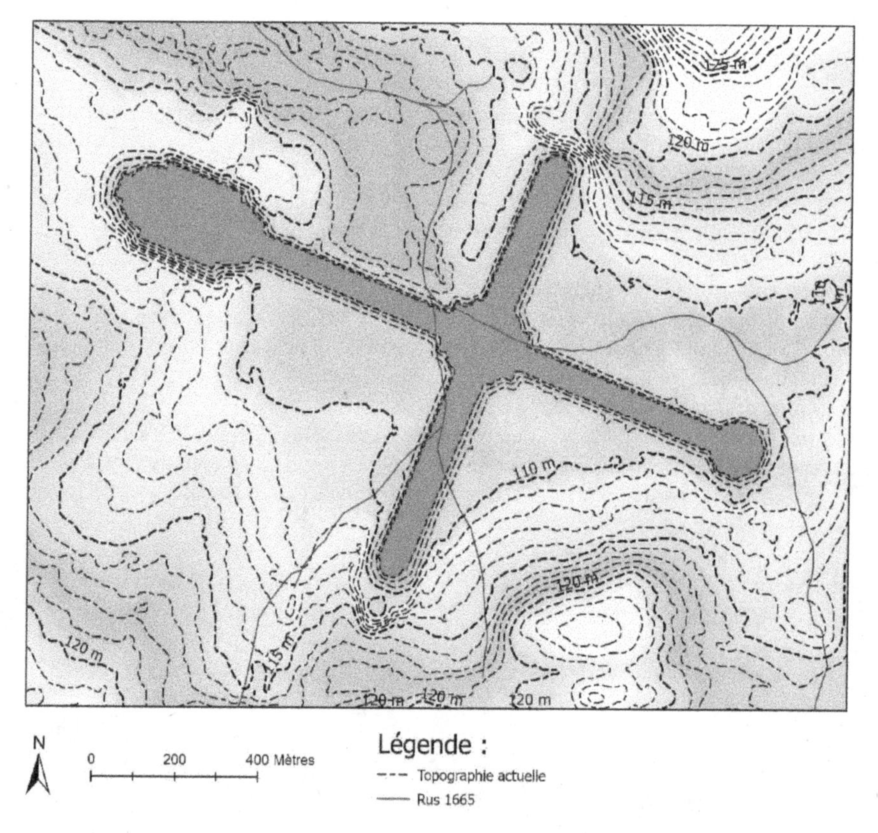

Figure 7. Topographic map of the current terrain centered on the Grand Canal of Versailles (IGN and UniLaSalle data). Equidistance of 1 m level lines. Location of the water streams according to the plan attributed to Delapointe 1664/65, Bibliothèque nationale de France, Va 448 f.

The obtained map is the most accurate available at present and therefore makes it possible to lay a basis for the analysis of earthworks by observing the terrain model. However, this map has its limits, with a real resolution that is between 1 and 5 m and not 1 m everywhere, and furthermore, it suffers from visible artifacts, for example the effect of steep slope instead of a vertical at the place where the peripheral edge is located. This situation is taken into account in the analysis proposed at the stage of this article. This map makes it possible to find the first-order facts already known, for example the role of the level line at 110 m NGF, but it provides everywhere decisive precisions. We georeferenced and then superimposed on this map the system of streams prior to the development of the canal (named "Rus 1665") in order to use these archival hydrographic data to support morphological reasoning. These arguments are presented in the next chapter.

4.3 *Topography before works ("1665")*

The map in Figure 8 was constructed on the one hand by means of a naturalistic reasoning, based on the observation of the natural environment, and on the other hand on the integration of the logic of earthworks, including their phasing in four stages. The naturalistic reasoning is based on the fact that the lines of levels distant from the Grand Canal, those which are considered to be unchanged by other sites in the park, are shaped by a principle of hydrographic erosion following lines of weakness in the geological substratum, and those at several scales nested in each other. We therefore analyze the location of high points and thalwegs, considering the gravity trajectory of the flows of water or mate-rial transported by water. According to this principle, it is noted that the streams are logically located in the thalwegs or more generally following the lines of topographical low points.

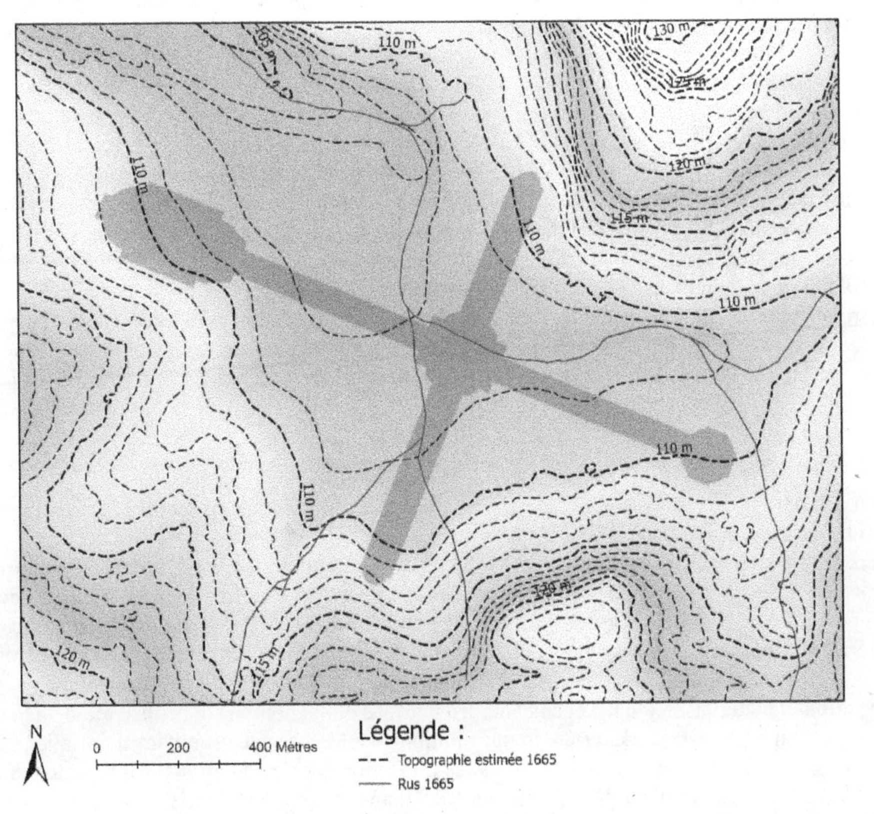

Figure 8. Topographic map of the land before work (symbolic date: 1665) centered on the Grand Canal of Versailles (IGN and UniLaSalle data). Equidistance of 1 m level lines.

Thus, the topography of 1665 (Figure 8) in the large central sector of the canal appears very horizontal, while a light steep difference in altitude is observed in the area of land bounded by the North and West arms of the cross. The level lines have thus been rede-signed with an assumption of greater progressivity of the difference in altitude throughout the area in question. On the other hand, two sectors were identified since the first article as embankment deposits, delimited by steep embankments in comparison with all the slopes observed elsewhere in the park. They are particularly visible thanks to the leveling profiles measured on the field. This concerns the area west of the north arm of the canal and the area north of the Gally basin. These artificial mounds have therefore been "erased" in the map in Figure 8. According to the morphological principles set out above, the level lines at the northern and southern ends of the Grand Canal, but also on the side of the Apollo Basin to the east, have been smoothed to obtain a more regular relief. Finally, to complete the description of the main changes made to the current level lines, we corrected in the field area between the west and south arms of the canal the concavity reversal of the level line at 110 m NGF. This correction consisted in "erasing" a kind of earth platform with no apparent link to the system of relatively marked thalwegs that go up to the south, and where two arms of one of the streams that existed there before 1665 flowed.

4.4 *Deduction of a cut and fills Map*

The map in Figure 9 was obtained by subtracting the previous two maps. This subtraction had to be delimited by a boundary beyond which we considered the impact of the works – especially the embankment depositing activities – as 'negligible'. This hypothesis was made necessary by the observation that large quantities of embankments can, by calculation, be generated by an infra-decimetric difference in altitude between the map 1665 and the current, therefore by the minimum of uncertainty on the topographic data. On the other hand, in anticipation of future improvements, areas related to the four phases of work have been demarcated on this map, in order to extract detailed cubatures phase by phase – for the moment not exploited. These internal limitations, however, pose other problems because it is likely that the areas thus associated with a single stage of construction may have been affected by several successive phases of work.

These intuitions have remained qualitative until then, and now they are visualized and become accessible to quantification. This is particularly the case of the large deposits West of the northern arm of the canal (height up to 2.5 m) and North of the Gally basin (height up to 4 or even 5 m). With regard to the embankment of the Gally basin, André Le Nôtre's skill is to have used a topography angled to the main axis of the canal to manufacture from scratch, by an earthmoving process, a symmetry of the banks on both sides of the basin, with all the appearance of a natural theater.

4.5 *Resultant from the movement of land: an unbalance in the current state of knowledge*

We recall here the result of part 1.2 where we put forward the initial figure of excavated material produced by the work of the Grand Canal in the order of 460,000 m^3.

From the work on Digital Elevation Model presented above, there are nearly 705,000 m^3 of excavated material for about 515,000 m^3 of embankments. There should not be this considerable difference. First, it is not reported in *Les Comptes* (Guiffrey 1881-1901) to evacuate excavated material at a distance from the park, none of the excavated materials having any real economic value and the principle of the lower energy consumption prevailing at that time as today. Then, the deposition of materials that have been damaged because of phenomena of proliferation and natural recompaction over several centuries, in our opinion should not cause significant variations in the volumes initially occupied, therefore those resulting from our calculations. This means that our map of the current terrain combined with the 1665 map hypothesis and the resulting embankment-excavated volumes, although representing an improvement over the previous state of the issue, is still unsatisfactory. As an indication and subject to all the constraints expressed above, details of the calculations phase by phase are provided in Table 2.

As discussed in Section 2.3.1, significant progress is likely to occur if a more accurate/resolution DEM can be exploited, for example obtained by a LIDAR coverage of the land. It would remove uncertainties that are not definitive, which would justify a more detailed work on the original topography of the site.

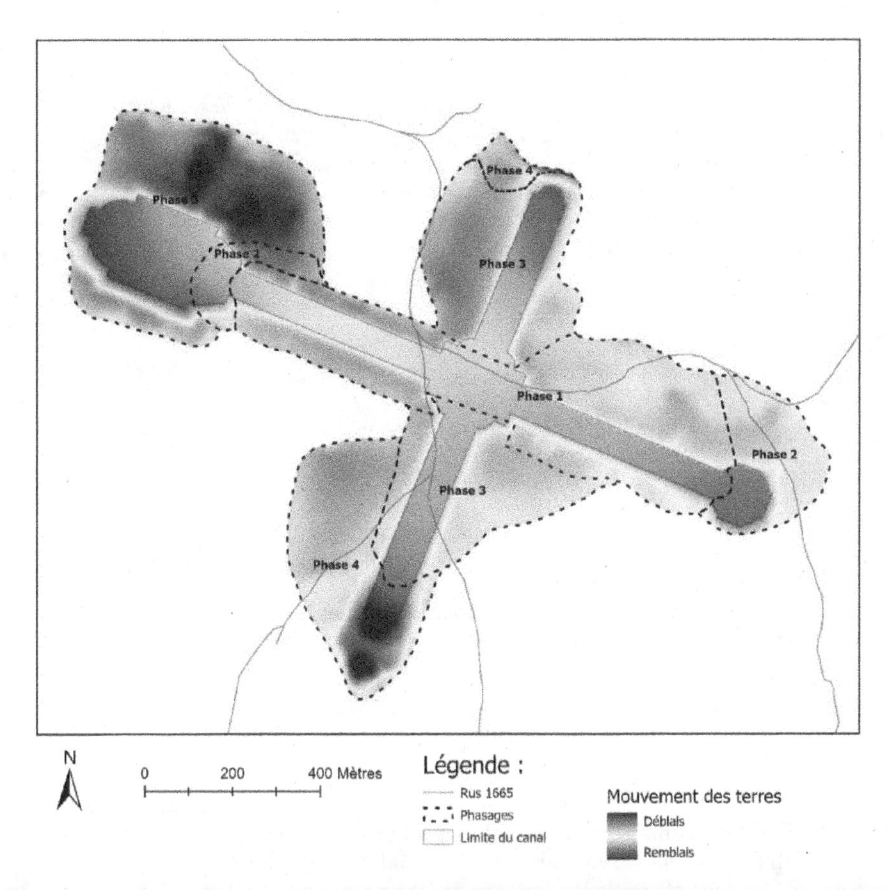

Figure 9. First interpretative cuts and fills map resulting from the digging work of the Grand Canal of Versailles.

Table 2. Results of embankment-excavated volumes by phase in 2021.

Construction phase.	Excavated material (m^3)	Embankment (m^3)
1	165	85
2	75	25
3	350	340
4	115	65
Total 1667–1680	705	515

5 WATERTIGHTNESS OF THE CANAL

5.1 *Two opposing points of view*

In part 1.3 we highlighted an ambiguity about the problem of the watertightness of the Grand Canal. The first functions assigned to it (navigable basin, reservoir for the fountains of the park) suggest that it had to be sealed to guarantee a water level and therefore a minimum available volume at any time of the year. But the analysis of André Le Nôtre reported by Michel Baridon seems on the contrary to indicate that there should have been no perimeter wall or watertight invert between the basin and its casing so that the canal can fulfill its function of drainage and sanitation of the old swamp, by catchment or permanent drawdown of water levels in the surrounding land.

This ambiguity is not clearly removed by the analysis of the *Comptes*, and on this point, we qualify the opinion expressed in our first article (Vernhes & Heitzmann 2019) where we defend the thesis of generalized watertightness, a thesis translated by a schematic section showing a continuous lateral and bottom processing by application of waterproof clay. It has certainly been noted that resources have been devoted to this processing of the water body under construction, but without knowing precisely where (among the spaces between temporary embankments and peripheral wall, at the bottom of the excavation), under what conditions and how systematically this was done throughout the years of construction.

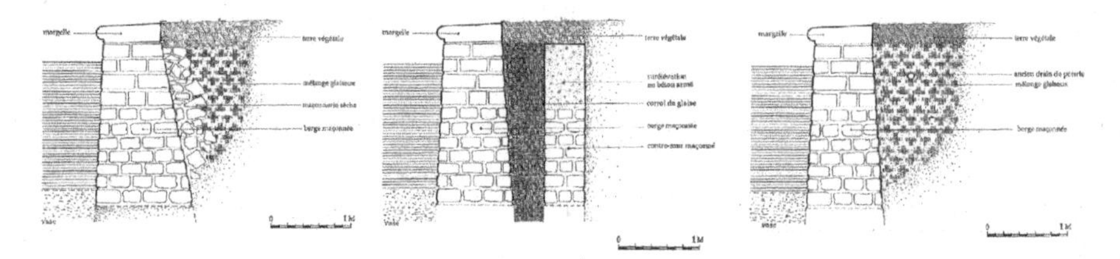

Figure 10. Interpretative diagrams of the results of the three borehole surveys of the peripheral masonry wall of the Grand Canal (June 1997), carried out with a manual shovel, in Lablaude 1997 (pp. 280, 282 and 284). They represent, from left to right, a variety of situations at the northern, southern and eastern branch of the Grand Canal.

5.2 *Observations and analyses in 1997*

The Lablaude 1997 report, already cited, provides interesting hydrogeological information on the Grand Canal site. One of the architect's goals was to restore at least part of the clay core that he assumes he will find behind the entire perimeter retaining wall of the Grand Canal. To justify the reconstruction, he must affirm its importance for the proper maintenance of water levels in the basin.

The report is based on piezometric levels measured in the terrains surrounding the canal in four points. These four points are not precisely identified on the map of the survey performer (Sol Progrès company, Figure 11a) while their altitude is curiously of great precision and stability in the documents of exploitation of the piezometric data provided by the architect, always at 108.24 NGF. This elevation is also given in the document as the nominal altitude of the edge capping the retaining wall. We express a doubt in particular about the real altitude of the head of PZ3 (Figure 11b), which seems to us possibly higher and therefore far from the piezometric surface.

Lablaude notes that the water level in the surrounding soil of the canal is lower or almost equal to the level observed at the same time in the Grand Canal. The height differences are from a few centimeters to one meter in PZ1, PZ2 and PZ4. They are more marked, from one to two meters, in the PZ3, without any explanation of Lablaude. The architect highlights that the phreatic water sits in a formation of sandy-silt colluvium and that the muddy level of the bottom of the basin is on average two meters below the margin. He does not consider the colluvium or the muddy bottom as impermeable.

What is surprising in the rest of his analysis is that he believes that the water in the field feeds the canal as soon as its static level is higher than the level of the bottom of the canal and not that of the water in the canal, which contradicts the principle of communicating vessels. He therefore believes that, except for the PZ3, all levels recorded in the autumn of 1997 attest that the canal is primarily an outlet for the ground waters sur-rounding it, from the bottom. In doing so, Lablaude admits both that the bottom of the canal has no sealing function, that it would therefore not have been wrought, and that work to repair the wrought on the peripheral wall is of interest since it would maintain the operating regime of this wall originally planned.

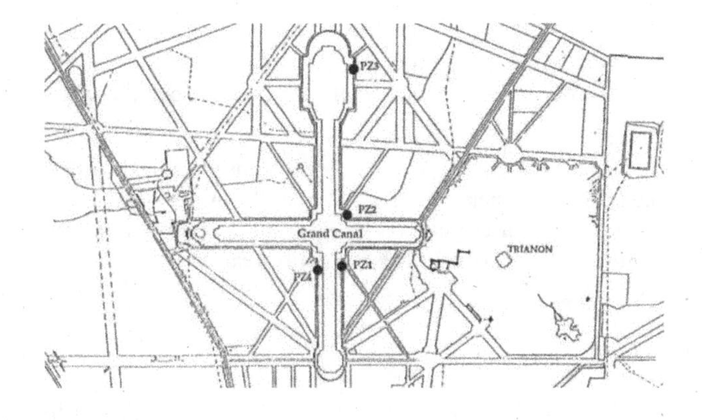

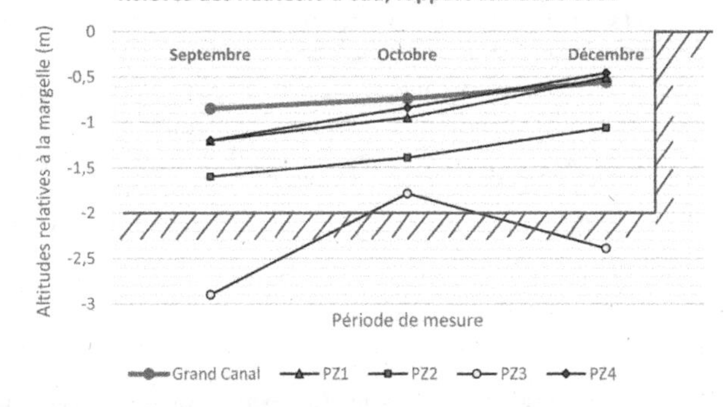

Figure 11. (a) map of the 1997 piezometric boreholes provided in the Sol Progrès report and reproduced on p. 252 of Lablaude 1997. (b) Graphical translation of the point water levels recorded in the autumn of 1997 in the Grand Canal and in the piezometers, with casing showing the level of the edge and the average level of the muddy bottom. Water data, op. cit. cit., pp. 268, 270, 272.

The Lablaude reasoning we tried to understand is therefore ultimately difficult to follow, but it illustrates the difficulty that the question of the sealing of the Grand Canal represents.

5.3 *Current hydraulic analysis: three possible regimes*

Assuming that the decimetric variations in the water level observed in the Grand Canal are naturally stabilized over the long term around an average satisfactory for its functions as a navigable reservoir basin, three possible solutions could concurrently explain such a condition:

a. Effective watertightness everywhere at the basin/terrain interface and therefore independence of water levels in the ground (piezometrics) and in the canal (limnimetrics); this hypothesis is associated with the generalized presence of a wrought several decimeters thick with an effective permeability in the order of 10^{-6} m/s or less. This hypothesis seems to be contradicted by the condition of the foot of the perimeter wall on the landward side on excavation surveys with shovels (Figure 10), according to Lablaude,1997.
b. It is the low permeability of the surrounding soil that actually ensures the watertightness of the Grand Canal. Even when the water level in the basin is higher than the water level in the field, the flows transmitted to the ground are negligible over time, especially compared to other loss phenomena such as evaporation. Under these conditions, the discontinuous nature of the sealing works and a certain heterogeneity of soils at the basin/terrain interface are no longer at the heart of the hydraulic analysis of the Grand Canal and can be relegated to the rank of second-order causes.

c. The phreatic level is stable during the year. In this case, regardless of the permeability of the soil, the water levels between the land and the Grand Canal are slowly or rapidly balanced and no problem arises if the natural water levels in the land are close to those required for the water body.

With the evolution of our own thinking, in the end only solutions b. and c. are retained as possible, and can even complement each other. One way to check the preponderance of one over the other would be to look at the degree of correlation between piezometric levels and water level in the canal by continuous measurements over at least an entire year.

6 CONCLUSIONS

This article draws a picture of the knowledge acquired on the Grand Canal from a geotechnical perspective through archival studies, field investigations conducted in 2018 and 2019, and through the exploitation of a Digital Elevation Model initiated in 2019 and recast in 2021. The main advance is the one obtained by the creation of a DEM leading to a first hypothesis of the volume of the earthworks by excavated material and embankments. The corresponding figures are, respectively, 705 000 m^3 and 515 000 m^3. These figures, which will have to be even better placed, show the order of magnitude of the works carried out, very comparable to those currently assumed in the period 1679-1686 for the earthworks in the park of Marly (Vernhes, 2017), which followed almost immediately the last finishes for the Grand Canal.

ACKNOWLEDGMENTS

I wish to acknowledge the students of UniLaSalle (Beauvais, France) who dedicated their Thesis or Collective Research Project in Geoscience to the Grand Canal in Versailles: Abdel Hussein Cissé, Guillaume Le Motheux du Plessis, Maxime Peron, Corentin Rubio, Ghislain Cordier, Félix Fleuriot, Pauline Lesecq, Yesmine Meslamani, Guillaume Meyfroidt, Matthieu Poulain, Michael Vanhee, Nicolas Dubeau and Elodie Ricard. Our thanks also go to Gilles Bultez, Head of the Fountains Department of the Parc deVersailles, Sandrine Pallandre, Head of the State Resources Department and Daniella Manar, historian of the Fountains Service, for their scientific support.

REFERENCES

Baridon, M. 2013. "Les jardins de Le Nôtre et d'Hardouin-Mansart", dans Arizzoli-Clementel P. (dir.), Versailles, Paris, Citadelles et Mazenod.

Cordier, G., Fleuriot, F., Lesecq, P., Meslamani, Y., Meyfroidt, G., Poulain, M. & Vanhee, M. 2019. Rapport de Project Collectif Géotechnique 5A sur le Grand Canal de Versailles, 80 p. Non publié.

Guiffrey, J. (éd.) 1881-1901 (5 tomes), *Comptes des bâtiments du roi sous le règne de Louis XIV* (1664-1715). Source gallica.bnf.fr / Bibliothèque nationale de France.

Goguel, J. 1967. Versailles, Carte géologique de la France à 1/50000, n° 182, Orléans, Bureau de recherches géologiques et minières.

Lablaude, P.-A. 1997. Etude préalable à la restauration des berges et margelles du grand canal, rapport destiné à l'Etablissement Public du Musée et du Domaine National de Versailles, 145 p. Non publié.

Le Guillou, J.-C. 2000. Le domaine de Versailles de l'Aube à l'Aurore du Roi Soleil (1643-1663), in Versalia n° 5. Reproduction en p. 65 d'un plan attribué à Delapointe 1664/65, Bibliothèque nationale de France, Va 448 f.

Vernhes, J.-D. 2017. Grands travaux d'aménagement du vallon de Marly. Bulletin du Centre de recherche du château de Versailles. DOI : https://doi.org/10.4000/crcv.14555

Vernhes, J.-D. & Heitzmann, A. 2019. Le Grand Canal à Versailles : enquête géotechnique. XVII European Conference on Soil Mechanics and Geotechnical Engineering. 2019, Reykjavik, Iceland. ISBN 978-9935-9436-1-3

Geotechnical Engineering for the Preservation of Monuments and Historic Sites III – Lancellotta, Viggiani, Flora, de Silva & Mele (Eds)
 ISBN 978-1-032-35998-4

Geotechnical studies to optimize the protection measures against flooding of St. Mark square (Venice, IT)

P. Simonini & F. Ceccato
Università degli Studi di Padova, Padova, Italy

ABSTRACT: The famous St. Mark square, located on an island characterized by the lowest gap between ground and sea level, is currently flooded during high tide, even though the MOSE barrier system is operative. To design cost-effective interventions to safeguard this historical heritage, a understanding of flooding mechanisms and the relationship between groundwater pressure and tidal oscillations was necessary. This paper presents the results of a recent monitoring campaign carried out at St. Mark square as well as a discussion on the selected interventions to protect the square against flooding.

1 INTRODUCTION

The historic city of Venice, located on several islands in the middle of a lagoon, and the surrounding environment are characterized by a rather precarious equilibrium, with the safety margin reducing at an accelerated rate. The rate of deterioration is due to the increasing flooding frequency of the historic city, that is caused by the natural eustatic rise of the sea level, by the natural subsidence and by the regional man-induced subsidence, which was significant between the '40s and '70s of the last century.

The city of Venice is located in the middle of the omonimous lagoon, whose origin is traced around 6000 years ago, when the sea water diffused into a preexisting lacustrine basin during the deglaciation period. The tide flows in and out from three lagoon inlets; its normal excursion is around 60–70 cm and lasts approximately 6 hours.

Since the gap between the ground level of Venice islands and the sea (lagoon) water level is small, at tides exceeding +0.8 m above mean sea level (referred to as mean sea level at the *Basilica di Punta della Salute* - s.l.P.S.), a few of island pavements begin to be flooded. At tide levels exceeding 1.1 m., occurring under low atmospheric pressure, strong winds blowing from Adriatic Sea and enduring rain, the flooding of a great part of the city is observed.

To protect the city of Venice against this recurrent flooding, several projects have been undertaken, the most important being the design and construction of movable gates located at the three lagoon inlets, namely the MOSE barriers (MOSE is acronym for *Modulo Sperimentale Elettromeccanico*: Experimental Electromechanical Module). These gates, controlling the tidal flow, temporarily separate the lagoon from the sea at the occurrence of high tides exceeding 1.1 m above s.l.P.S..

Other relevant interventions consisted in erosion mitigation of marshes and wetlands, morphological cost-line restoration, renovation of the existing jetties at the inlets, fishing farms reopening and, as far as the historic city concerns, raising the elevation of banks, pavements and sidewalks in some selected areas to prevent the floods due to tide below +1.1 m.

DOI 10.1201/9781003329756-15

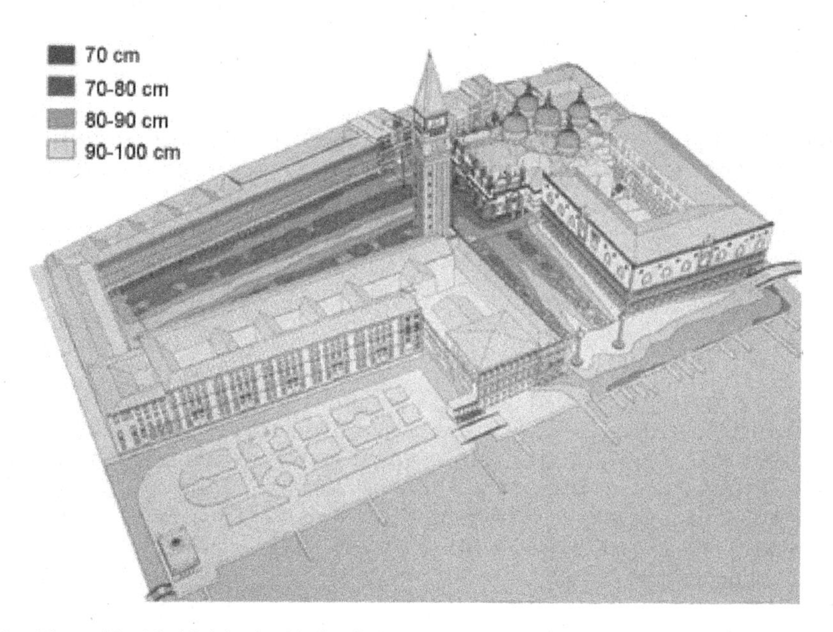

Figure 1. View of St. Mark's island with flooded area at corresponding values of tidal level (cm) (adapted from www.mosevenezia.eu).

The St. Mark's square is the lowest part of the historic center of Venice and is characterized by an elevation between +0.80 m and +0.90 m above s.l.P.S. More particularly the area facing the St. Mark's Cathedral is located at an even lower level, approaching +0.60 m above s.l.P.S.

The increasing frequency of flooding of the St. Mark's square induces considerable deterioration of masonry walls, foundations and decorative architectural elements of very important ancient buildings (Bettiol et al. 2015; Ceccato et al. 2014; Fletcher & Spencer 2005), jeopardizing the historical heritage of city and especially of that facing the St. Mark's inner square. When the water level reaches +0.60 m above s.l.P.S., a small zone in front of the Basilica and its narthex begin to be flooded. At +0.90 m, approximately 65% of the square is covered by water and at 1.15 m it is fully submerged. A tide exceeding +0.90 m is currently recorded at around one per cent of the time on the annual basis, but this percentage will rise considering a small effect of natural subsidence (0.5 mm/year) and sea level rise due to climate change.

The activation of the MOSE gates, when +1.10 m above s.l.P.S. (or higher) is forecasted, does not prevent the historical area of St. Mark's square to be flooded. Specific protection countermeasures are therefore needed, but selecting optimal solutions in this special context was not straightforward as it might be respectful of the historical heritage, compatible with the touristic activities as well as cost-effective.

Figure 2 shows a cross-section of a typical quay-wall facing the surrounding canals or the St. Mark's basin with an indication of different flow paths potentially concurring to flooding:

1 Back-flow through the existing drainage system;
2 Overtopping, from the St. Marks' basin;
3 Heavy rainfalls;
4 Seepage through the soil.

As shown in Figure 2, seepage flow may currently occur through or below the quay-walls, through the ancient and pervious drainage network, the open joints between the stones forming the pavement, which are in hydraulic connection with the water basin through the soil.

To protect the square against flooding a preliminary project was proposed in 1998, in which several countermeasures have been planned, the most important being the construction of vertical cut-off continuous sheet pile walls around the perimeter of St. Mark's island to avoid water infiltrating through and below the quay walls and the installation of an impermeable membrane below the stone pavement, to prevent seepage flow through the permeable joints between the stone elements of the pavement. Some doubts of the need and effectiveness of such huge interventions arise in the late '10 s and therefore different studies have been undertaken. More particularly, mechanisms 1–3 have been investigated in maritime and hydraulic studies (Ruol et al. 2020; P. Salandin 2020;) while mechanism 4 was investigated by specific geotechnical investigations presented and discussed in the following.

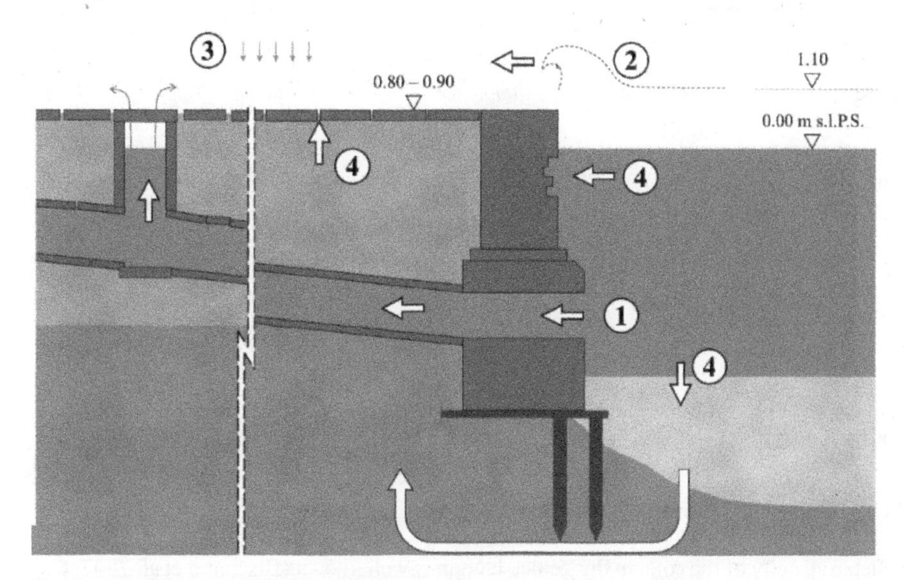

Figure 2. Main flooding mechanisms of St. Mark's Square: 1. Back-flow through the drainage system, 2. Overtopping, 3. Heavy rainfalls, 4. Seepage through the soil.

2 THE OLD DRAINAGE NETWORK

The still active ancient drainage system of the St. Marks' square, formed by a network of masonry tunnels (referred to as *gatoli* in Venetian language) collecting both rainfall and wastewater, was mostly constructed in the 18th century by the Republic of Venice.

The typical sections of these *gatoli* (Figure 3a) are characterized by rectangular sections between 0.4 m and 0.8m wide and between 0.6 and 0.9 m high. The vertical masonry walls are about 0.30 m-thick; the *gatoli* are closed above by a massive calcareous block, called *stelere*. For larger *gatoli*, an arch vault reaching a width of 1.50 m and a height of 2 m is used (Figure 3b). As a consequence of the construction technique, they are relatively pervious and seepage water can flow through the fissures between the *stelere* and the lateral walls, and, for the most deteriorated ones, between the brick elements forming the lateral walls.

Figure ...shows an inner view of one conduct in good conditions and one in worst conditions.

The embedded drainage elements are gently sloping toward the canals or the St. Mark's basin. The water discharge rate is therefore very slow, thus fostering sediment deposition, which can reduce over time their cross section to 50–75% or, completely obstructing it in some cases (Volpato 2019). Hydraulic measurements and models (Volpato 2019) showed that the hydraulic head inside the *gatoli* coincides with the lagoon water level, i.e. the system is fully connected to the lagoon such as water can flow in and out according to tide oscillation.

Due to the perviousness of the *gatoli*, the water flowing into drainage system, is in direct connection with the surrounding shallowest soil forming the foundation ground of the entire square.

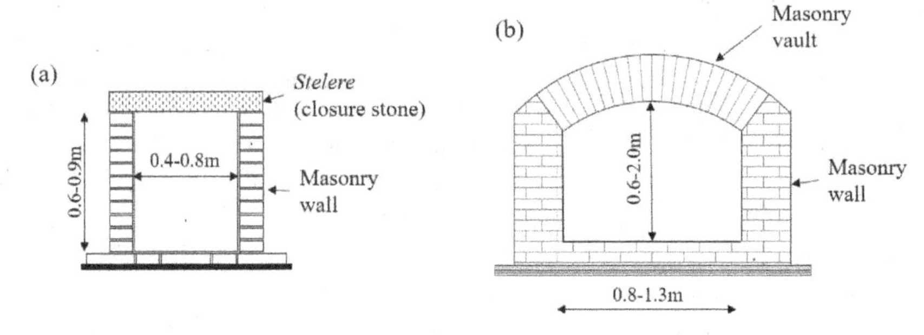

Figure 3. Typical cross section of Venetian *gatoli.*

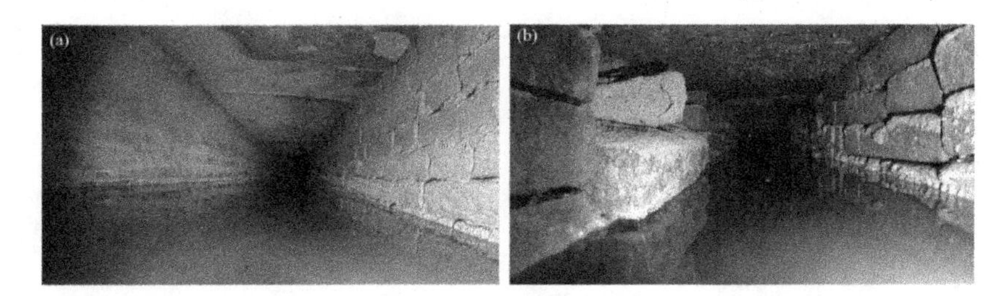

Figure 4. Inner view of two drainage conducts in S. Mark's square: (a) gatolo in good conditions, (b) gatolo in bad conditions.

3 GEOTECHNICAL SITE INVESTIGATIONS

The heterogeneity of the soils in the Venice lagoon is well known (Biscontin et al. 2007; Cola et al. 2008; Ricceri 2007; Simonini et al. 2007) and the island of St. Mark is no exception. Given the complexity of the system, three main geotechnical campaigns have been carried out in St. Mark square in order to define the geotechnical model of the subsoil. The first set of site investigations was commissioned by CVN in 1993 and included 6 geotechnical boreholes (16 m-deep) with collection of undisturbed samples, 10 standard penetration tests (SPT), 6 Lefranc permeability tests, 3 Piezocone Cone Pentration tests (CPTU) with dissipation tests. In 1997–1998, 6 geotechnical boreholes up to 20 or 32 m depth (with undisturbed sample collection), 9 standard penetration tests (SPT), 8 Piezocone Cone Pentration tests (CPTU), Boutwell in situ permeability tests were conducted. The third campaign was carried out in 2019 and consisted in 10 geotechnical boreholes up to a depth of 20 m, with collection of undisturbed samples; 13 CPTU, 5 seismic piezocone tests (SCPTU), 5 Dilatometer tests (DTM); 5 Seismic Dilatometer tests (SDTM) driven 20m-deep. Geotechnical laboratory tests included classification tests (grain size distribution, specific gravity, water content, Atterberg limits, etc.), permeability tests in triaxial cells, with constant head and variable head permeameters and one-dimensional oedeometric compression tests. The locations of the tests are shown in Figure 4.

The results of these site investigation campaigns allow us to identify a superficial anthropic layer (Unit 1/1A) with a thickness of 3.0 to 4.5 m. This layer is extremely heterogeneous because it experienced several different anthropic actions along the centuries and it hosts the drainage network (Bortoletto 2019). In particular, the permeability varies along 7 order of magnitudes, depending on the type of test, the depth, and the location. Higher values, up to 1.210^{-3} m/s, are measured for Boutwell permeability tests at very shallow depth; lower values, up to 210^{-10} m/s, are obtained from CPTU dissipation tests and oedometric tests.

A low permeability layer (Unit 2/2A) lies below Unit 1 and it has a thickness between 2.0 m and 7.0 m. The material ranges from sandy silt (2) to clayey silt (2A). The permeability varies between 110^{-10} m/s and 110^{-7} m/s. Beneath this layer there is a more permeable sandy formation, characterized by uniform fine sand (Unit 3) and silty sand (Unit 3A). The thickness varies from

0.50 m to approximately 8.0 m. From the depth of −9.80 m below s.l.P.S., up to the maximum sounding depth, there is the typical alternation of prevalently clayey layers (4A) with lenses of moderately silty sand (4). A typical stratigraphy is shown in Figure 6.

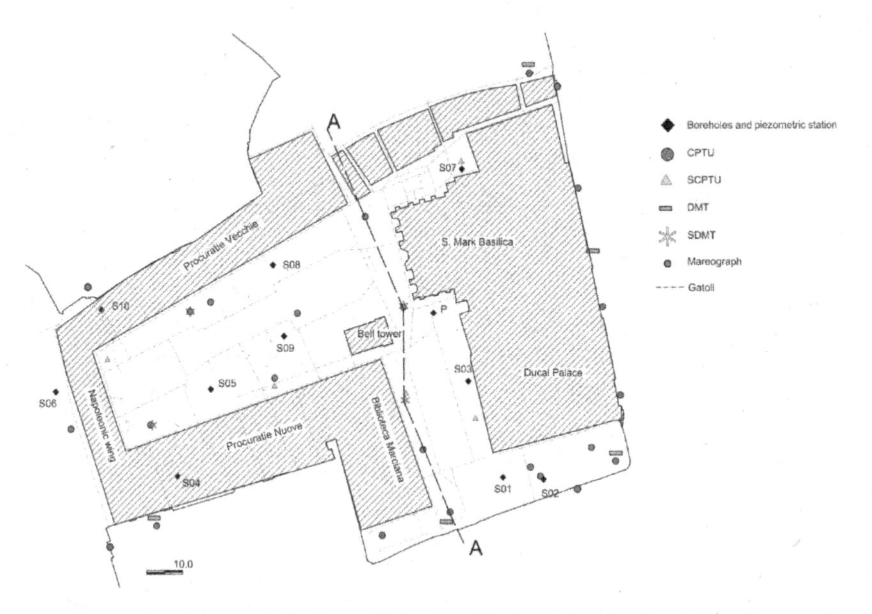

Figure 5. Position of in-situ geotechnical tests (CPTU, SCPTU, DMT, SDMT), boreholes, piezometric stations and old drainage pipes (*gatoli*).

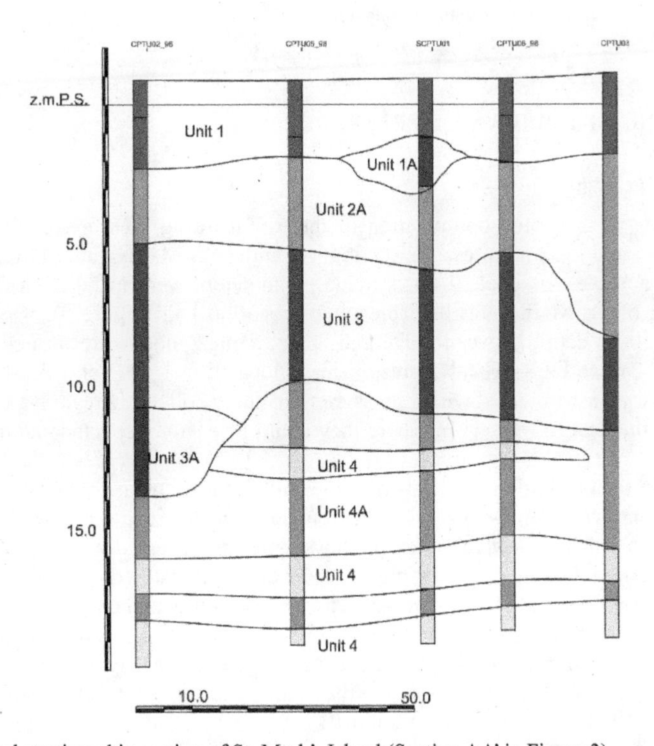

Figure 6. Typical stratigraphic section of St. Mark's Island (Section AA' in Figure 3).

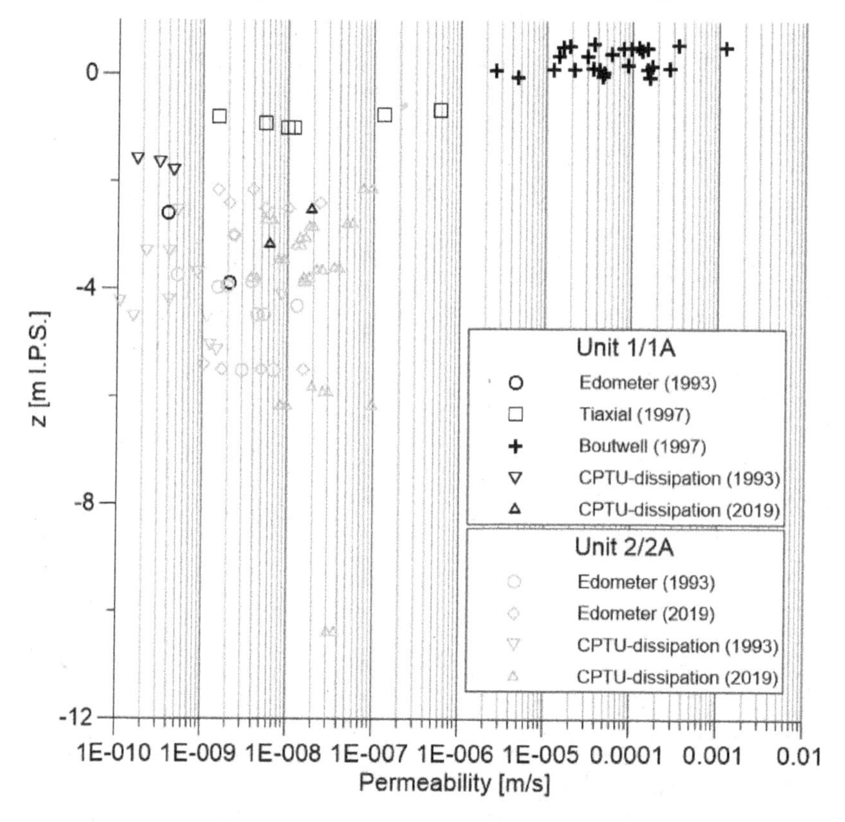

Figure 7. Results of permeability measures along depth.

4 MONITORING OF PORE WATER PRESSURES

4.1 *Pore-water pressure readings*

One of the key aspects for the optimization of the safeguarding measures is understanding of the seepage flows and the pore pressure distribution under St. Mark square in response of tidal oscillations. To achieve this goal, 9 Casagrande piezometers were installed at 2018 in a pilot site at the south of St. Mark's Basilica (piezometric station P in Figure 5). 4 piezometers are located in Unit 1 at a depth between 2.0 m and 2.3 m, 4 piezometers are located in Unit 2 at a depth between 3.2 m and 3.8 m, and 1 piezometer is located in Unit 3 at a depth of 7.6 m. The Casagrande cell was instrumented with a pressure transducer collecting readings every 6 minutes and transferring the data to the server, where they could be easily inspected and downloaded for elaboration.

Figure 8 shows the oscillation of the average piezometric level measured in soil units 1, 2, and 3 and the sea water level from October 16 to November 15th, 2018. In this period, the water level increased above ground level several times and the square was repeatedly flooded for a few hours. Moreover, a 156 cm (above s.l.P.S.) very high tidal level was registered on October 29th due to a storm surge. The amplitude of the pore pressure oscillation is reduced compared to the tidal wave in all layers, but the reduction is minimum for soil Unit 1, which reaches pressure levels closer to sea level when the floor of the square is flooded because water can easily enter from the permeable pavement. During this observation period, the maximum piezometric level measured in Unit 1 is 129 cm, in Unit 2 it is 106 cm and in Unit 3 it is 91 cm.

During high tide, higher water levels are observed in Unit 1 compared to Unit 3, thus the hydraulic gradient is directed downward. This proves that during high tides seepage flow from deeper layers could be excluded. During low tide levels, lower groundwater levels are observed in Unit 1 compared to Unit 3 and, therefore, the hydraulic gradient is directed upwards.

The average piezometric level in Unit 3 is slightly higher than the average level in Unit 1, meaning that these two layers can be considered hydraulically disconnected and Unit 2 can effectively prevent the flooding mechanism 4.

Since local soil heterogeneities, distance from the perimeter quay walls and the *gatoli* network, as well as their geometric features and state of preservation, significantly influence the hydraulic response of the entire subsoil system, 25 new piezometers were installed in 2019 at 10 monitoring sites distributed throughout St. Mark's square. According to the local conditions, it was decided to use 2 or 3 piezometers for each specific site, measuring the absolute pore water pressure in the different soil units (1, 2 or 3).

The main findings of the pilot site highlighted above are confirmed; moreover, new interesting considerations can be drawn. Figure 9 shows the pore pressure response in units 1 and 3 as a function of sea level for piezometer S05. The minimum ground level near this piezometric vertical is +84 cm l.P.S. When the tide increases, water enters the *gatoli* and can infiltrate the soil through their permeable walls; thus, the piezometric level in unit 1 increases. When the paving becomes submerged, the pressure response in the superficial layer is very fast because water can infiltrate from the top boundary. The maximum piezometric level may eventually be higher than the maximum sea level, as a consequence of some other significant contributions, such as rainfall or anthropic sources. After the peak, while the tidal level decreases rapidly, a slower response is observed in Unit 1, and for a certain time the pressure can be higher than the sea level. This can be explained considering that for medium water levels, water drains out mainly from the walls of the *gatoli*, and then, for very low tide level, only from quay walls or very deep ducts, resulting in a slower pressure dissipation rate. In addition, part of the superficial layer may become unsaturated during low tide, thus reducing its permeability. Numerical analyses considering different boundary conditions and unsaturated material properties confirmed this explanation (Ceccato et al. 2021).

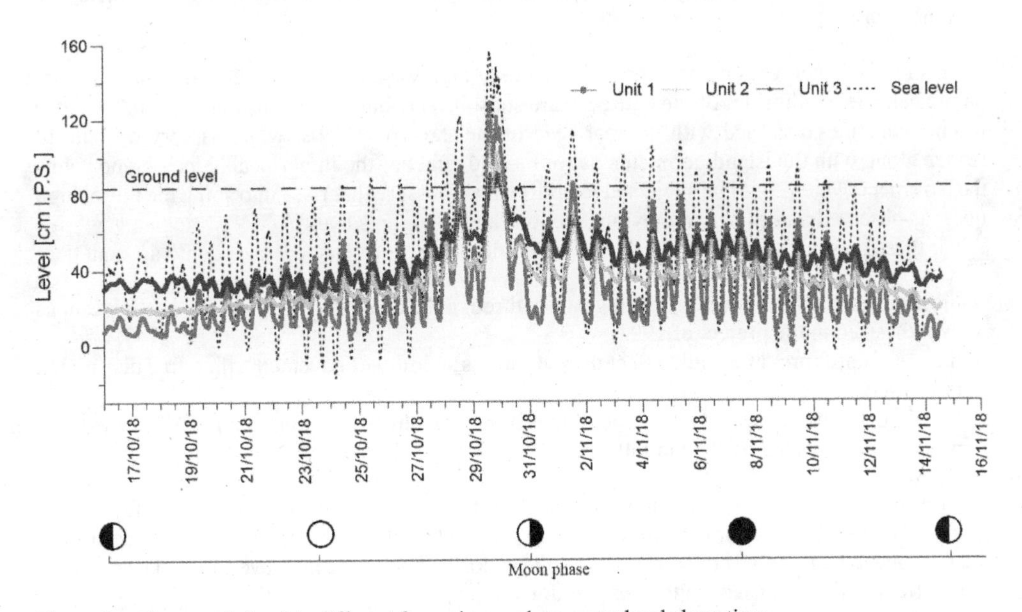

Figure 8. Piezometric level in different formations and sea water level along time.

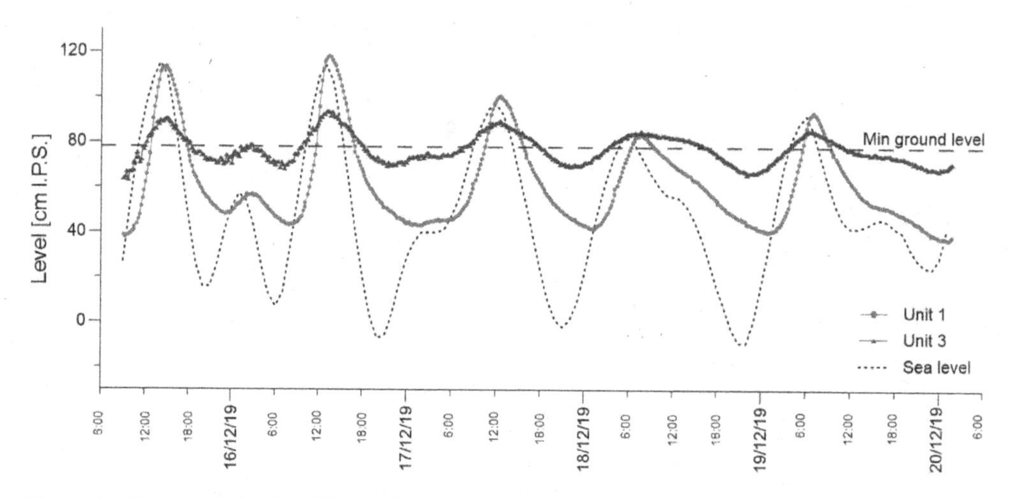

Figure 9. Piezometric level in different formations and sea water level along time in station S05.

5 OPTIMIZATION OF PROTECTION MEASURES OF ST. MARK'S SQUARE

As pointed out in Section 1, to protect the square against tides up to 1.1 m above s.l.P.S. a preliminary project was proposed in 1998, in which several countermeasures were planned such as:

- installation of vertical cut-off continuous sheet piling around the perimeter of St. Mark's island to avoid water infiltrating through and below the quay walls;
- installation of an impermeable membrane just below the stone pavement, to prevent seepage flow through the permeable joints between the stone elements;
- closure of the ancient drainage network by realizing a new one, collecting the rainfall water to be pumped and then discharged into a specific caisson realized close to the quay-wall facing St. Mark's basin.

This type of intervention appeared to be not only highly expensive, but also to impact heavily on the delicate equilibrium of the entire square subsoil. The main questions that arise in the more recent years are concerned with the real need of the sheet pile walls, which is very difficult to realize along with the island perimeter, as well as the effect of the impermeable membrane below the pavement, which could be subjected to uplift water pressure that rises into soil Unit 1 after high tides, as observed and shown in previous sections and in Figures 8 and 9.

For these reasons, a more gentle approach was discussed and selected, taking into account that:

- the seepage rate from the surrounding canals through the quay walls might be not so relevant as hypothesized in the project of 1998;
- the water uplift pressure could affect the long-terms stability of pavements if no free dissipation is allowed;
- the old drainage network system could be restored (as already done in the past) and used in a proper way to discharge the rainfall water.

To examine the effect of this updated approach to the design of the long-term safeguarding protection measures, a pilot test site was set up around the St. Mark's Basilica, whose narthex is the first one to be flooded, being its minimum level located at +0.62 m above s.l.P.S. The outcome of this field trial is discussed in the next section.

6 PROTECTING ST. MARK'S BASILICA

In 2018 specific protection measures for St. Mark's Basilica were realized to reduce the frequency of flooding of its nartex and surrounding areas that are the lowest of the entire island. The intervention consisted in the restoration of the old drainage pipes next to the Basilica with the installation of closing systems (valves) to avoid water backflow during high tides and a pumping station (Figure 10a).

The system is designed to operate for medium-high tides, i.e. between 62 cm above s.l.P.S. (the lowest level of the nartex floor) and 88 cm above s.l.P.S. (maximum ground level around the nartex). For sea levels lower than 62 cm, water can freely flow in and out of the system (Figure 10b). When the tidal level rises above 62 cm, the valves close and eventually the pumping station turns on, preventing the flooding of the nartex (Figure 10c). If the sea level rises above 88 cm, an overflow occurs, nartex flooding cannot be avoided and the pumping system is turned off (Figure 10d).

The MOSE system is currently operating to limit the maximum tidal level at 130 cm s.l.P.S. and in the next future this level will be progressively lowered to 110 cm s.l.P.S. The protection measures of the St. Mark Island is designed with this reference value and will be progressively realized in the coming years. Meanwhile, additional measures to prevent overflow are currently under discussion. The construction of a transparent and impermeable barrier of 1.20 m height around the basilica is being considered, among others.

The piezometric station at the south of St. Mark Basilica, that is, station P in Figure 5, is very close to the pumping system and these measurements offer information on the effects of the operating protection system on the pore water pressures in the subsoil.

Figure 11 shows the average piezometric level in units 1, 2, and 3 during the activation of the system. Only the piezometric level in Unit 1 appears to be influenced. When the sea level reaches 62 cm, the system switches on and the pressure immediately decreases. Afterward, it increases again following the tide, but the maximum piezometric level is lower compared to the case without an operating system. After the peak, when the tide decreases below 62 cm the system is deactivated and the pressure increases slightly, but it reduces rapidly following the descending sea water level.

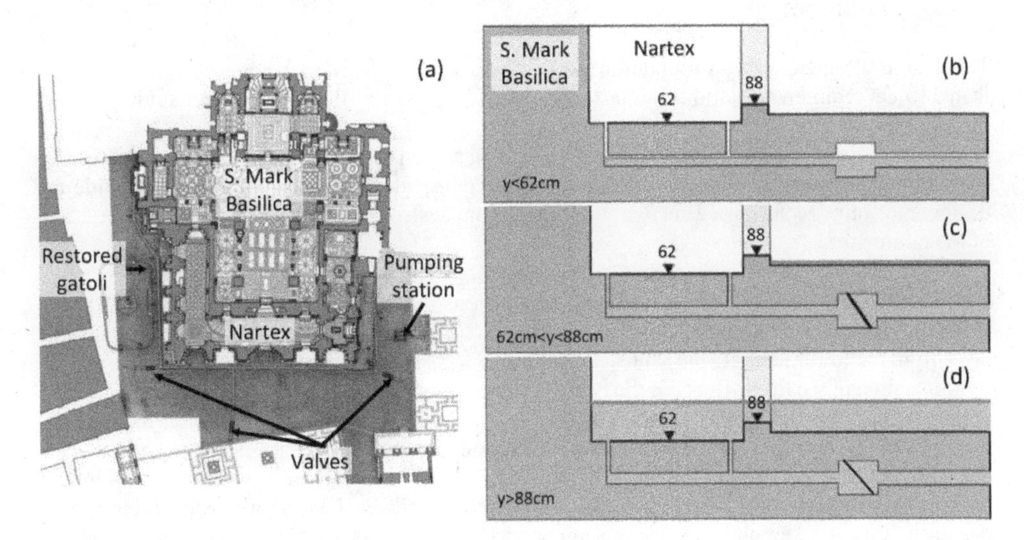

Figure 10. Protection measures of the St. Mark Basilica: (a) plan view of the improved drainage system, (b-d) simplified view of operating phases.

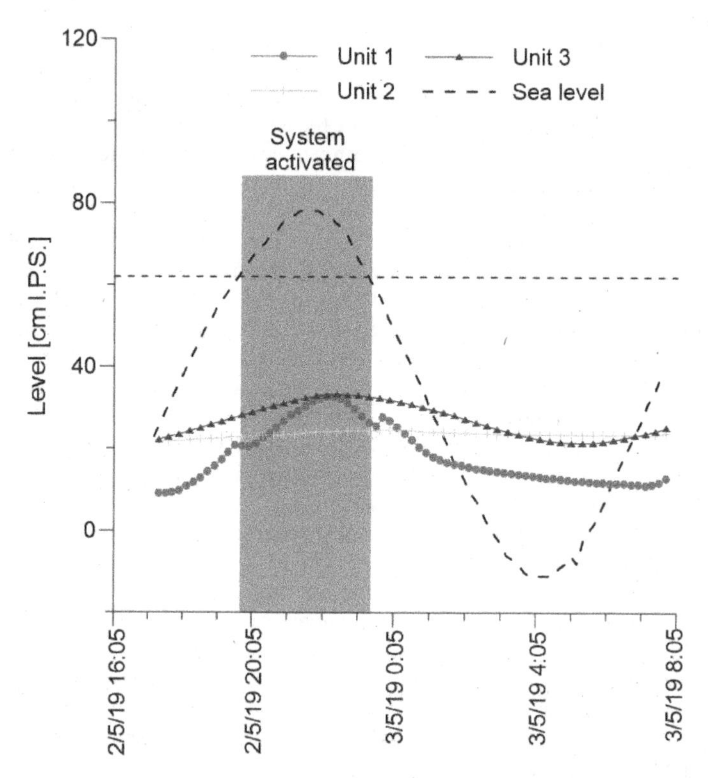

Figure 11. Average piezometric level in the ground measured in station P and sea water level during the activation of the protection measures.

7 CONCLUSIONS

The design of suitable flood mitigation measures to protect St. Mark's square against tide less than 110 cm required a multidisciplinary approach within which the geotechnical study played a fundamental role.

Within this frame, the results of geotechnical testing as well as the measurements of pore pressure oscillations in the soil of St. Mark's Square in Venice provided relevant information to guide the designer to optimize the most effective solutions that must also be respectful of the ancient situation of the entire area.

The local soil profile is characterized by a heterogeneous alternance of layers composed of a mixture of sand, silt, and clay in different proportions, but basically subdivided in three layers, whose intermediate one is composed of a fine-grained soil separating the upper permeable anthropic layer from the lower natural fine sands.

Fluctuation of the tide influences the pore pressure in both formations 1 and 3. The local vertical component of hydraulic gradient between these two layers is directed downward for higher tides and upward for the lower ones. Thus, seepage flow from deeper layer is impossible during high tide peaks. In the shallower layer, maximum water levels are close to lagoon level, whereas minimum ones are higher; moreover, the pore pressure may remain higher than the water level during tide decrease. This is a key phenomenon when carrying uplift analysis for the long term stability of paving especially in the more depressed area of St. Mark's Square.

In addition, it was noted that the seepage in the formation is characterized by very small flow rates, thus the actual infiltration flow to be drained is very low.

From all the observations and interpretation carried out so far, it was clear that the intervention such as the one selected for the area surrounding the Basilica and its narthex, could be extended to the whole square, showing that the previously hypothesized impermeabilization systems could not the proper solution to prevent the flooding due to seepage.

ACKNOWLEDGEMENTS

This work was supported by Consorzio Venezia Nuova [Prot. N. 14224 UGA/VA, 30/11/2018]. The authors thank the companies belonging to the Consortium Kostruttiva, in particular Mate Eng. and Thetis, and the colleagues of ICEA Department for the fruitful discussions regarding the flood protection of St. Mark's Square.

REFERENCES

Bettiol, G., Ceccato, F., Pigouni, A.E., Modena, C., Simonini, P., 2015. Effect on the structure in elevation of wood deterioration on small-pile foundation: Numerical analyses. *Int. J. Archit. Herit.* 3058. https://doi.org/10.1080/15583058.2014.951794.

Biscontin, G., Cola, S., Pestana, J.M., Simonini, P., 2007. Unified compression model for venice lagoon natural silts. *J. Geotech. Geoenvironmental Eng.* 133, 932–942. https://doi.org/10.1061/(ASCE)1090-0241(2007)133:8(932).

Bortoletto, M., 2019. Relazione archeologica - Interventi di salvaguardia dell'Insula di Piazza San Marco a Venezia.

Ceccato, F., Simonini, P., Koppl, C., Schweiger, H.F., Tschuchnigg, F., 2014. FE Analysis of degradation effects on the wooden foundations in Venice. *Riv. Ital. di Geotec.* 2, 27–37.

Ceccato, F., Simonini, P., Zarattini, F., 2021. Monitoring and Modeling Tidally Induced Pore-Pressure Oscillations in the Soil of St . Mark ' s Square in Venice , Italy. *J. Geotech. Geoenvironmental Eng.* 147, 1–14. https://doi.org/10.1061/(ASCE)GT.1943-5606.0002474.

Cola, S., Sanavia, L., Simonini, P., Schrefler, B. a., 2008. Coupled thermohydromechanical analysis of Venice lagoon salt marshes. *Water Resour. Res.* 44, n/a-n/a. https://doi.org/10.1029/2007WR006570.

Fletcher, C.A., Spencer, T. (Eds.), 2005. *Flooding and environmental challenges for Venice and its lagoon: state of knowledge.* Cambridge University Press.

P. Salandin, 2020. Evaluation of flooding of the Sankt Mark's Island in: 'Analysis of possible interventions to safeguard the Sankt Mark's Island from high tides' Part 3 (in italian). Padova.

Ricceri, G., 2007. *Il ruolo della geotecnica nella salvaguardia della città di Venezia e della sua laguna.*

Ruol, P., Favaretto, C., Volpato, M., Martinelli, L., 2020. Flooding of Piazza San Marco (Venice): Physical model tests to evaluate the overtopping discharge. *Water* (Switzerland) 12. https://doi.org/10.3390/w12020427.

Simonini, P., Ricceri, G., Cola, S., 2007. Geotechnical characterization and properties of Venice lagoon heterogeneous silts, in: Leroueil, Hight (Eds.), *Characterisation and Engineering Properties of Natural Soils.* Taylor & Francis Group, London, London, pp. 2289–2327. https://doi.org/10.1201/noe0415426916.ch17.

Volpato, M., 2019. Relazione idrologica e idraulica - Interventi di salvaguardia dell'Insula di Piazza San Marco a Venezia. Venice.

Geotechnical Engineering for the Preservation of Monuments and Historic Sites III – Lancellotta, Viggiani, Flora, de Silva & Mele (Eds)

Observed interaction between Line C of Roma underground and the *Cloaca Maxima*

G.M.B. Viggiani
Cambridge University Engineering Department, UK
Università di Roma Tor Vergata, Italy

N. Losacco
Politecnico di Bari, Italy

E. Romani
Metro C S.c.P.A., Rome, Italy

A. Sonnessa
Politecnico di Bari, Italy

ABSTRACT: Contract T3 of Metro C crosses the archaeological area of the historical centre of the city of Rome. In the final part of the contract, the running tunnels, which were excavated using two Earth Pressure Balance (EPB) shields, pass directly under one of the earliest sewage systems of the world, the *Cloaca Maxima*, functional to the present day. This ancient sewer consists of two arch shaped masonry and conglomerate tunnels, at a relatively close spacing. The monitoring data gathered during the passage of the TBMs underneath the ancient tunnel indicate that the ancient sewers are relatively flexible, substantially conforming to the greenfield displacements.

1 INTRODUCTION

Excavation of underground railway tunnels through the historic centre of big cities is increasingly common, given the growing urban population and the consequent need to reduce traffic and CO_2 emissions, while providing efficient and sustainable mobility systems. Although construction techniques have been developed to address minimisation of tunnelling-induced displacements, particularly with the adoption of state-of-the-art tunnel boring machines, the assessment of the effects induced by tunnel construction on existing structures is mandatory, especially when these are highly vulnerable and highly exposed. In this context, Rome represents an extreme example, as the excavation of the twin running tunnels of Contract T3 of the third line of Roma underground (Line C) was carried out underneath the archaeological area of the historic centre, crowded with both surface and buried archaeological remnants of outstanding documental, historical, monumental, and artistic value.

As shown in Figure 1, Contract T3 of Line C comprises approximately 3 km of running tunnels, two stations and two ventilation shafts. A third station, Piazza Venezia, located at the end of the contract, is currently under design. The tunnels run from San Giovanni station, at the boundary with Contract T4, where there is a connection with the existing Line A, to Piazza Venezia, delimiting the boundary with Contract T2. The tunnels were excavated mainly using two identical Earth Pressure Balance (EPB) shields, with a maximum diameter of 6.71 m at the cutting wheel and a length of approximately 11 m. The precast reinforced concrete segmental lining, consisting of 1.4 m long and 0.3 m thick rings with an internal diameter of 5.8 m, was erected inside the shield. In order to minimise induced ground displacements, pressurised backfill grouting of the tail void was carried out during each advancement stage of the EPB shield, using a bi-component fast-hardening grout.

In the last stretch of Contract T3, approximately halfway between the Basilica of Maxentius and Piazza Venezia, the tunnels were excavated at relatively short vertical distance beneath two

DOI 10.1201/9781003329756-16

ancient sewer tunnels of Roman age, belonging to the *Cloaca Maxima* sewage system, and a more recent sewer tunnel built in the Renaissance, the *Chiavicone della Suburra* (simply referred to as *Chiavicone* in the following). This paper describes the effects of the construction of the twin tunnels of Line C underneath the ancient sewer tunnels, in terms of measured settlements induced both in the soil – in quasi-greenfield conditions – and on the buried structures.

Figure 1. Aerial view of the historic centre of Rome showing the route of Contract T3 of Line C.

2 THE CLOACA MAXIMA AND THE CHIAVICONE DELLA SUBURRA

The *Cloaca Maxima* represents the most important sewage system of ancient Rome, dating to the end of the 6th century BC. Initially conceived as an open-air canal, supposedly by Tarquinius Priscus, it was subsequently covered and expanded into a sewer tunnel by Tarquinius Superbus. The sewage served as a drainage system for spring and rainfall waters coming from the valley between the Esquiline and Viminal hills, as well as for the frequent floods of the Tiber that made the Roman Forum and the *Argiletum* valleys swampy and unhealthy. During the Flavian Age, the *Cloaca Maxima* was truncated and diverted, first by Vespasian, to build the *Templum Pacis*, and later by Domitian, to construct the new Forum of Nerva (Antognoli & Bianchi 2009). This last intervention was carried out on the stretch of the *Cloaca Maxima* that today crosses via dei Fori Imperiali, beneath which the running tunnels of Line C were excavated, and left a deactivated branch that is the remnant of the western end of the Vespasian's route of the sewage.

Today, the active branch of the *Cloaca Maxima* channels the urban wastewater and storm water run-off coming from the Esquiline, the *Fora* valley, and the buildings located between the Roman Forum and the Velabrum, which then flows into the collector sewer built at the end of the nineteenth century under the Tiber Embankment (Corsetti 1925). In the area where intersection with the Line C running tunnels occurs, shown in plan in Figure 2(a), the active branch of the *Cloaca Maxima* runs approximately North-East to South-West; it bends some 20° northwards while crossing via dei Fori Imperiali, and then it recovers the initial direction, as it leaves the embankment of the main road. The deactivated branch extends approximately 36 m beyond the South-West retaining wall of the main road embankment. As shown in the cross-section in Figure 2(b), the tunnels of both the active and the deactivated branch are arch-shaped, with an average height of 3.5 m, a width of 3.0 to 3.5 m and cross-sectional areas of 10.5 m^2 and 9.2 m^2, for the active and deactivated branch, respectively. In the stretch in which they underpass via dei Fori Imperiali (with a ground level of approximately 22.0 m a.s.l.), the crowns of the two sewers are located at a depth of 8.0 to 8.5 m below the modern ground level.

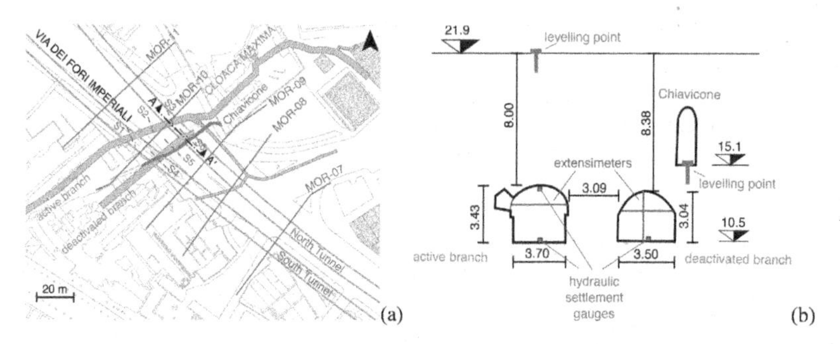

Figure 2. (a) Relative plan position of the ancient sewers and Metro C running tunnels and (b) cross section (AA') through the *Cloaca Maxima* and the *Chiavicone*.

In the context of the site investigation carried out by Metro C S.C.p.A. during the design stage of Contract T3, a detailed exploration of the active branch of the *Cloaca Maxima* was carried out, which included coring of the tunnel lining, digital photogrammetry, laser scanning and topographic surveys (Di Mucci & Miniero 2015). The results of these investigations showed that a variety of construction techniques had been employed for the ancient tunnel lining.

In the first part of the sewer, North-East of the via dei Fori Imperiali, and in the deactivated branch, both belonging to the sewer of Vespasian age, see Figure 3, the sewer tunnel rests on pozzolanic mortar conglomerate foundations (1) and the lining consists entirely of blocks of *Tufo Lionato* (2). The vault is made of 9-11 radial ashlars, while the abutments are made of three rows of superimposed blocks.

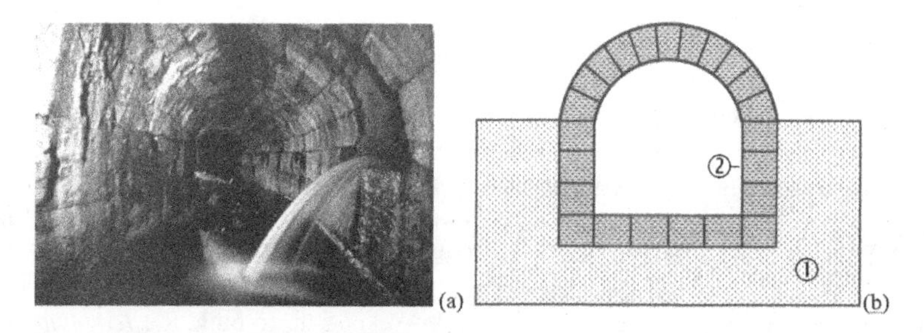

Figure 3. Lining of functional stretch and de-activated branch of Vespasian sewer: (a) photograph (E. Bianchi) (b) schematic drawing.

In the currently active branch, belonging to Domitian sewer, see Figure 4, the floor consists of travertine blocks (1) resting on a pozzolanic mortar conglomerate foundation slab (2). The abutments are made of two or more rarely three courses of large blocks of *Lapis Gabinus* (3), surmounted by *Opus Latericium* (4), *i.e.*, bricks with lime and pozzolana mortar, of varying height up to about 1 m, serving as a levelling layer on which a vault in cement conglomerate (5) is set. Despite the lack of homogeneity of the construction systems, the state of conservation of the sewer appears to be of an extraordinary level.

The *Chiavicone* was built at the end of the 16th century, to expand the existing ancient sewage system, as required by the creation of the Alessandrino district, between the Coliseum and the current location of Piazza Venezia (Antognoli 2015). As shown in Figure 2(a), starting from the Forum of Nerva, the *Chiavicone* follows the route of the deactivated branch of *Cloaca Maxima*, although at a shallower depth, as its floor is located at approximately 6.8 m below ground level, see Figure 2(b). The initial and final stretch of the sewer were demolished in the recent past – in the 1930s and during the archaeological surveys carried out in 1985-1986 and 1995-1997, respectively. The remaining part of the *Chiavicone* crosses via Alessandrina and via dei Fori Imperiali, extending some mere 4.0 m beyond the South-West retaining wall of the via dei Fori Imperiali. The tunnel is currently used as a pedestrian underpass walkway joining the two unburied parts of the Forum of Nerva. It is narrow and arch-shaped with an average height of 3.6 m and a width of 1.2 m, and the lining is made of reused material from Roman Age houses. The abutments consist of alternating courses of brickwork, tuff blocks, marble, and travertine fragments, grouted with hydraulic mortar. The barrel vault is made in *opus cementicium*; the floor is composed of two rows of stone slabs and is slightly concave at the centreline.

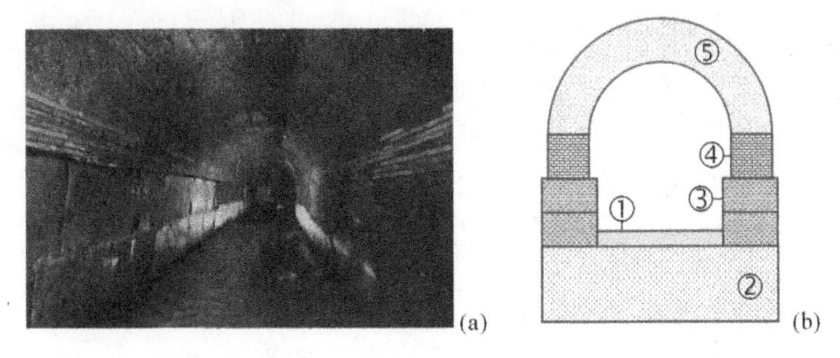

Figure 4. Lining of functional stretch of Domitian sewer: (a) photograph (E. Bianchi) (b) schematic drawing.

3 THE RUNNING TUNNELS OF METRO C UNDER THE CLOACA MAXIMA

3.1 *Layout of tunnels and monitoring sections*

As shown in Figure 2(a), the twin tunnels of Line C run almost parallel to one another under via dei Fori Imperiali and start diverging slightly as they approach Piazza Venezia just before passing under the *Cloaca Maxima*. The average distance between the tunnel axes beneath the ancient sewers is about 20 m. The tunnels cross the *Chiavicone* and the deactivated branch of the *Cloaca Maxima* almost at a right angle, whereas the active branch of the *Cloaca Maxima* forms an angle of about 60° with the axes of the new tunnels. These run at an average depth of 30.3 m below ground level, midway between the two branches of the *Cloaca Maxima*, with a minimal slope, less than 3.5%, and the North Tunnel 1.25 m deeper than the South Tunnel. Excavation of the North Tunnel started in March 2018 and was followed by the excavation of the South Tunnel, about three months later. The North and South Tunnel passed under the *Cloaca Maxima* in September 2019 and two months later, respectively, at advancement rates as high as 20 m/day.

A number of monitoring sections transverse to the tunnel axes, shown in Figure 2(a), were installed along the stretch before the construction of Line C by the general contractor, Metro C S.C.p.A. These were intended to assess the effects induced by the excavation on the existing archaeological and monumental environment and inform implementation of mitigation and protective measures, if required. Five of these sections (MOR-07 to MOR-11), with an average spacing of 25 m, are located in the area where the new tunnels run under the *Cloaca Maxima*, and are considered in this study. They consist mainly of surface levelling points, with the exception of section MOR-09, located 25 m South-West of the ancient sewers, where a number of Trivec extensimeters-inclinometers were installed down to a maximum depth of 34 m below ground level, to measure all components of displacement below ground surface.

3.2 *Soil profile and groundwater regime*

The ground conditions in the area of the *Cloaca Maxima* can be obtained from the geological longitudinal profile of the South Tunnel, shown in Figure 5, resulting from the repeated and extensive campaigns of ground investigation carried out since 1995, when the preliminary design of the line began.

Starting from the ground surface, the following layers can be identified:

- Anthropogenic, made-ground layer (R) with variable thickness that can be as large as about 14 m in the area under examination. This consists mainly of medium-coarse grained materials, with some silt, often embedding heterogeneous rocky inclusions and occasionally some ancient masonry walls remnants.
- Mid-Pleistocene synvolcanic fluvial-lacustrine deposits (Tb1), with significant vertical and horizontal variability. Two different eteropic facies can be distinguished: the first (Tb1a) composed of lenses and discontinuous levels of fine sands in abundant fine-grained matrix, oxidised yellowish/reddish sandy or clays silts; the second (Tb1b) made of medium-fine, slightly gravelly sand including cobbles with 1 – 2 cm maximum diameter.
- Plio-pleistocene Paleotevere deposits, made of heterometric medium-fine gravel in abundant sandy matrix (SG).
- Pliocene stiff silty clays with occasional thin sandy levels (Apl), known as “Argille Vaticane”, highly overconsolidated, exhibiting very low permeability, high stiffness and shear strength, also due to their significant marl content.

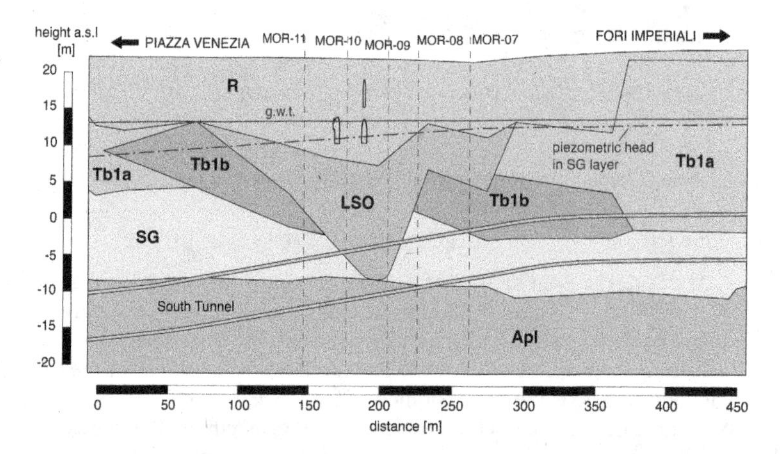

Figure 5. Soil profile and groundwater conditions, along the longitudinal section of South Tunnel.

Occasionally, along the whole of Contract T3, the continuity of the Pleistocene soil layers is broken by channels filled with recent Holocene Tiber alluvial deposits (LSO), consisting of soft heterogeneous silts, either clayey or sandy, with levels of organic matter and isolated tuff inclusions. In particular, in the area of interest, see Figure 5, a thick LSO layer, filling a paleo-valley in the pyroclastic soils, is found below the shallow R layer, from a depth of about 14 m b.g.l.. This extends approximately 100 m ahead and behind the Cloaca Maxima, and has a maximum thickness of about 15 m right under the ancient sewers, where it reaches the top of the Apl base layer.

As shown in Figure 5, in the vicinity of Fori Imperiali station the running tunnels are excavated in mixed face conditions, in the Tb1 and SG layers, and then they plummet into the SG layer, approximately under the monitoring section MOR-07, and finally in the underlying Apl layer towards the end of the stretch, after almost full-face crossing of the LSO layer.

Pore water pressure measurements performed prior to tunnel construction pointed out that the hydraulic head in the Tb1 deposits corresponds to a groundwater table depth of 7.5 – 8.0 m below ground level. An aquifer with a South-East to North-West directed seepage flow, approximately in the same direction as the running tunnels, connected to a hydraulic recharge located far outside the urban centre is confined in the SG layer. In the area under examination, the hydraulic head corresponding to this aquifer is consistently lower than that of the overlying less permeable layers by a few meters. In particular, this difference is larger where there LSO layers are present, showing evidence of a downward groundwater flow in the fine-grained materials, consistently with observations made in other stretches located at the beginning of Contract T3 (Losacco et al. 2020).

Table 1 summarises the indicative geotechnical parameters of all the soil layers described above as obtained from both laboratory and *in-situ* tests carried out for the design of the line.

Table 1. Physical and mechanical parameters of soil layers.

Layer	w (%)	γ (kN/m3)	w_L (%)	I_P (%)	φ' (°)	c' (kPa)	s_u (kPa)	E_{oed} (MPa)	k (m/s)
R	61.5	17.0	–	–	35	–	–	–	5×10^{-5}
LSO	20–32	17.5	35–48	12–29	30	0–16	75	7–9	2.2×10^{-5}
		19.5					200		7.7×10^{-7}
Tb1a	25–27	19.0	25	15	31	7–20	110	17.5	5×10^{-6}
Tb1b		19.4	–	–	34	–	–	–	5×10^{-5}
SG	22–28	20.0	–	–	35	–	–	–	1×10^{-4}
Apl	19–21.5	20–20.4	36–46	18–21.5	26–30	40–60	250–450	21.5–50	5×10^{-7}

4 TUNNELLING-INDUCED SOIL SETTLEMENTS

4.1 *Surface settlements*

Figure 6 shows the transverse surface settlement troughs induced by the construction of each of the running tunnels of Line C. These were obtained from surface precision levelling carried out at monitoring sections MOR-07 to MOR-11 during the advancement of the EPB shields beneath the *Cloaca Maxima*.

In Figure 6, the settlements generated by the second bore – the South Tunnel – are incremental, *i.e.*, they do not include the settlements induced by the excavation of the North Tunnel. The average centreline position of the two tunnels between section MOR-07 (axes spacing $s = 14.7$ m) and MOR-11 ($s = 24.1$ m) is also indicated in the figure. For each monitoring section, the plotted settlement profiles are representative of steady-state conditions, in the sense that they are measured when the excavation face has travelled a sufficient distance beyond the monitoring section (approximately 100 m for the North Tunnel and 120 m for the South Tunnel) and no additional settlements are generated by further excavation.

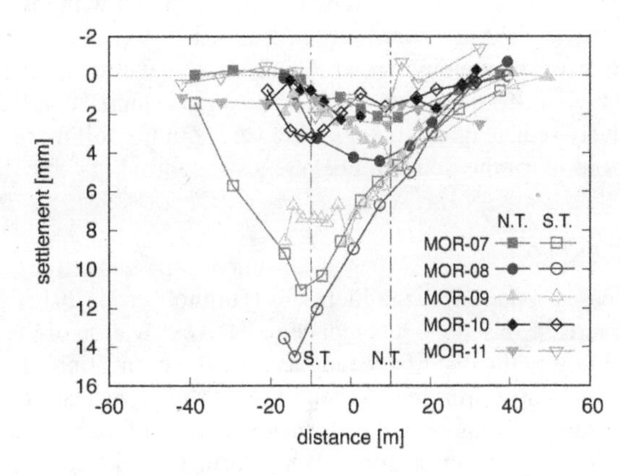

Figure 6. Recorded transverse surface settlement troughs in monitoring sections MOR-07 to MOR-11.

The first notable feature of the measured settlement profiles plotted in Figure 6 is that, consistently with what observed in many case histories (Fargnoli et al. 2015; Wan et al. 2017), the settlement troughs induced by the excavation of twin tunnels are not symmetric, with the second bore inducing significantly larger settlements than the first. Previous research (Losacco & Viggiani 2019) has shown that, when the tunnels are closely spaced, this is related mainly to the disturbance induced in the soil by the first excavation, even when the operational parameters of the TBM are the same for the two bores.

Figure 6 shows also that excavation of the North Tunnel induces small maximum settlement in section MOR-07, of approximately 2.2 mm, and twice this value in sections MOR-08 and MOR-09. Minimal settlements are recorded also for section MOR-10, located between the two branches of the *Cloaca* Maxima, and in section MOR-11. The settlements recorded in the successive monitoring sections were even smaller than those in section MOR-11 and are not reported in the figure. The volume loss values and the trough width parameters reported in Table 2 were obtained interpolating the measured settlements with Gaussian functions (Peck 1969). The values of volume loss follow the same trend as the maximum settlements, with a value of volume loss $V_L = 0.18\%$ in section MOR-07, $V_L = 0.49\%$ in section MOR-08 and progressively reducing in the following monitoring sections down to negligible values beyond the *Cloaca Maxima*.

Table 2. Trough width parameter and volume loss recorded for passage of North and South Tunnel.

	North Tunnel		South Tunnel			
Monitoring section	K	V_L (%)	K_{South}	$V_{L,South}$ (%)	K_{North}	$V_{L,North}$ (%)
MOR-07	0.41	0.18	0.54	1.13	0.63	1.31
MOR-08	0.53	0.49	–	–	0.63	1.81
MOR-09	0.41	0.31	–	–	0.79	1.23
MOR-10	0.46	0.23	0.25	0.17	0.62	0.42

The difference of tunnelling-induced surface settlements observed in the monitoring sections under examination is even more evident during passage of the South Tunnel. In this case, maximum settlements as high as 11.1 mm, *i.e.*, more than five times the settlement induced by the first excavation, were recorded for section MOR-07, increasing to 14.1 mm in section MOR-08, *i.e.*, more than three times the settlement induced by the first excavation in the same section, and then somewhat reducing to 7.4 mm, or almost twice the settlement recorded after the passage of the first tunnel. Consistently with what observed for the excavation of the North Tunnel, relatively small maximum settlements, as low as 3.2 mm for section MOR-10 and 2.1 mm for section MOR-11 were measured with the progress of the South Tunnel beyond the *Cloaca Maxima*. Given the significant asymmetry of the settlement trough generated by the South Tunnel with respect to the centreline, two different interpolations of the measured settlements with Gaussian functions were attempted for the two sides, the North side yielding larger K and V_L, in general. The corresponding volume loss values vary from $V_L = 1.3\%$ in section MOR-07 to a maximum $V_L = 1.8\%$ in section MOR-08, and then progressively reduce down to negligible values in the following sections. Settlement trough parameters V_L and K for the South Tunnel are also reported in Table 2.

4.2 *Effect of face support pressure*

If compared to other cases reported in the literature (Fargnoli et al. 2015; Losacco & Viggiani 2019b), the observed increase in settlements induced by the excavation of the second tunnel is too large to be attributed solely to the disturbance induced by the excavation of the first tunnel and it is likely to be related to variations in the soil conditions or to significant changes in operational parameters of the EPB shield during tunnel advancement. The different response observed in the various monitoring sections during excavation of the North Tunnel also supports this hypothesis. To give an insight of the likely causes of the observed phenomena, Figure 7 shows the average values of face pressure recorded during the construction of each lining ring, at the six face pressure sensors located in the excavation chamber of the EPB shield.

Inspection of the plots in Figure 7 reveals that, during the excavation of the North Tunnel, most of the pressure sensors experience limited oscillations around a mean value of 350 kPa, with some isolated lower recorded values of 250 kPa or less, such as *e.g.*, while crossing monitoring section MOR-08, which might in part explain the slightly larger settlement recorded in this section after the first bore. On the contrary, remarkably low face pressure values were measured during the initial phases of excavation of the South Tunnel, between sections MOR-07 and MOR-09, with the topmost sensors, P1 and P6, measuring pressures consistently below 300 kPa until section MOR-09 and recording minimum values as low as 230 kPa between sections MOR-07 and MOR 08, which is very consistent with the observed large settlements displayed in Figure 6 for the second bore.

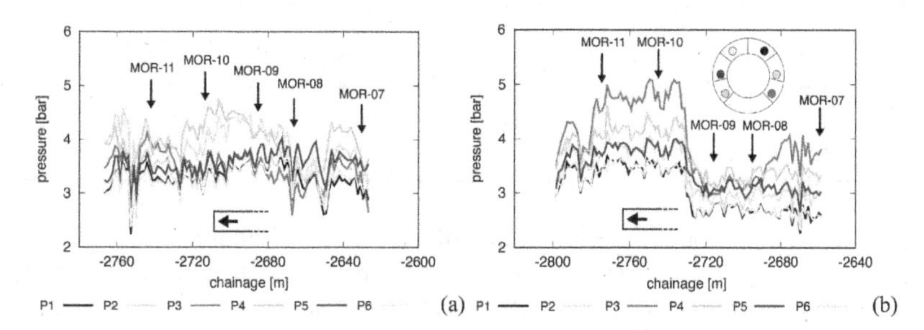

Figure 7. Pressure in the excavation chamber during construction of (a) North and (b) South tunnel.

Between sections MOR-09 and MOR-10, the operational parameters of the EPB shield were changed so that the recorded pressure in the topmost sensors, P1 and P6 was almost constant at approximately 350 kPa, while the recorded pressures at the lowermost sensors were constantly between 400 kPa (P4) and 500 kPa (P3), which justifies the limited settlements measured from section MOR-10 onwards. It is likely that the low face support pressure exerted during the initial phases of excavation are due to the tunnel face approaching the big paleo-valley filled with LSO soil right under the *Cloaca Maxima*. A thorough assessment of the potential role played by the geotechnical conditions and by other operational parameters of the EPB shield (*e.g.*, the tail void grouting pressure) would require further analyses, which are out of the scope of this paper.

4.3 *Subsurface settlements*

Besides surface levelling points, monitoring section MOR-09 was also equipped with five integrated Trivec inclinometers and extensimeters. The measurements of vertical displacements obtained from the Trivec were integrated starting from ground surface, where the values of the settlement obtained from precision levelling could be used as known boundary conditions.

The plots in Figure 8 represent the incremental subsurface settlement troughs obtained at 5 m depth intervals between ground level and a depth of 20 m below, after excavation of each tunnel. The data in Figure 8 show quite clearly that the maximum settlement is almost constant down to a depth of 15 m below ground level, while a sudden increase in maximum settlement, and a simultaneous reduction of the settlement trough width, are recorded at 20 m depth. The relative increase in maximum settlement between ground surface and this depth is +29% for the North Tunnel and +62% for the South Tunnel.

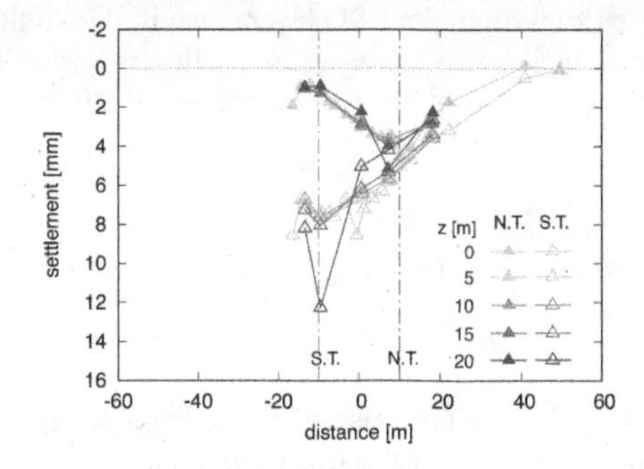

Figure 8. Recorded transverse sub-surface settlement troughs in section MOR-09.

Table 3. Subsurface maximum settlement, trough width parameter and volume loss in section MOR-09.

	North Tunnel			South Tunnel		
Depth (m)	w_{max} (mm)	K	V_L (%)	w_{max} (mm)	K_{North}	$V_{L,North}$ (%)
0	3.4	0.41	0.31	7.4	0.79	1.23
5	3.6	0.53	0.34	7.7	0.93	1.20
10	3.7	0.64	0.34	7.6	1.17	1.18
15	4.0	0.78	0.33	8.0	1.35	1.06
20	5.1	0.71	0.27	12.3	1.09	0.82

Table 3 summarises the values of maximum settlement, trough width parameter K and volume loss obtained at different depths for the two tunnels. Figure 9 shows the variation of the maximum settlement w_{max}, of the relative trough width parameter K, and of the volume loss V_L as a function of the depth normalised by the depth of the tunnel axis, z/z_0, in which $z_0 = 30.1$ m and $z_0 = 28.7$ m for the North and the South Tunnels, respectively. For the South Tunnel, interpolation with a Gaussian function was only attempted for the northern half of the settlement troughs, due to lack of data on the other side.

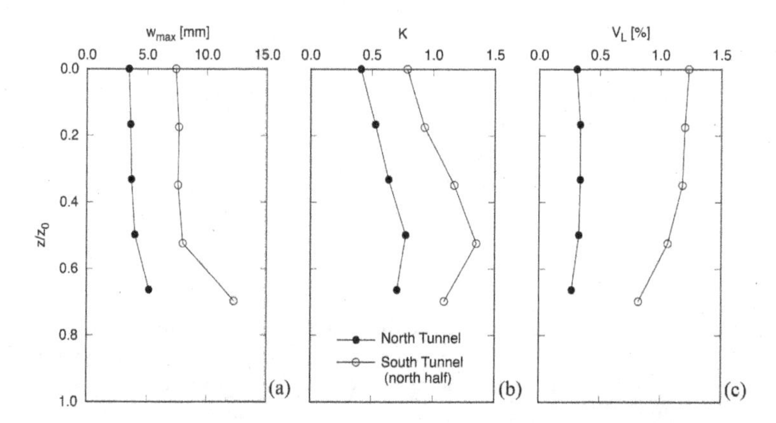

Figure 9. Variation with depth of: (a) maximum settlement, (b) trough width parameter and (c) volume loss.

As mentioned above, the transverse settlement trough induced by the South Tunnel is wider and larger than that induced by the North Tunnel at all depths, and this is reflected in the values of both K and V_L. For both tunnels, the trough width parameter K increases gradually with depth in the made ground layer, to decrease sharply at the transition with the LSO layer. The volume loss does not change significantly over the thickness of the made ground layer, indicating a limited amount of dilation, while it decreases slightly in the LSO layer, even if, due to the significant reduction of K, the settlement above the tunnel crown increases. A qualitatively similar response was observed both in the field, at the beginning of Contract T3 (Losacco et al. 2020), and numerically (Losacco &Viggiani 2019).

5 OBSERVED INTERACTION WITH THE ANCIENT TUNNELS

5.1 *Settlement profiles*

During excavation of the tunnels, the response of the ancient sewers was monitored by means of levelling points installed in the floor of the *Chiavicone* and hydraulic level settlement gauges installed in the floor of the deactivated branch and on both the floor and the vault of the active branch of the *Cloaca Maxima*. Figures 10(a) and (b) illustrate the incremental settlement profiles obtained from the measurements carried out after the excavation of the North and the South Tunnel, respectively. The surface settlement trough obtained for monitoring section MOR-10 is also shown as a reference. The comparison between surface and deep displacements is justified because, as shown by the data in Figure 9, the vertical displacements are practically constant down to a depth of 15 m, and the floor of the *Cloaca Maxima* is located at a depth of approximately 12 m.

Consistently with the limited surface settlements induced by the excavation of the North Tunnel beyond section MOR-09, possibly related to the fairly high face support pressure applied between sections MOR-09 and MOR-11, with mean values in the range 350 – 400 kPa, the settlement profiles observed for all the buried structures and for MOR-10 are practically superimposed, see Figure 10(a). The deformed configuration induced in the ancient tunnels by the excavation of the South Tunnel has the same shape as the surface settlement trough observed in section MOR-10, although some differences exist in the magnitude of measured settlements, see Figure 10(b).

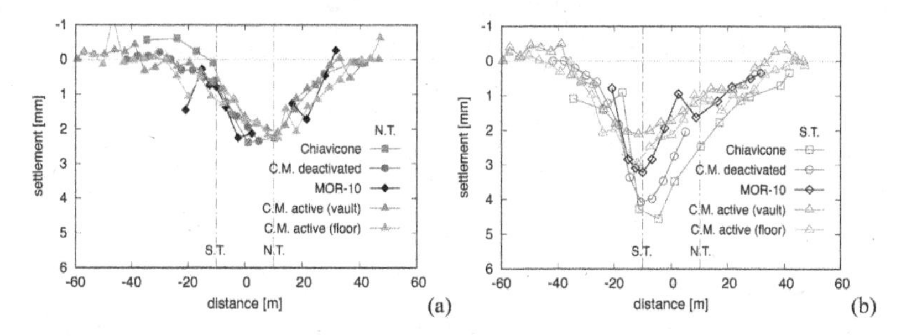

Figure 10. Recorded incremental settlement profiles along the sewers after construction of (a) North Tunnel and (b) South Tunnel.

In particular, the measured maximum settlements progressively reduce moving from the *Chiavicone* ($w_{max} = 4.5$ mm) to the active branch of the *Cloaca Maxima* ($w_{max} = 3.0$ mm at the floor, almost coincident with the maximum surface settlement of MOR-10), which somewhat recalls the reduction of maximum surface settlements observed between monitoring section MOR-09 and monitoring section MOR-11, see Figure 6 and Table 2. A smaller settlement is measured at the crown of the active branch of the Cloaca Maxima ($w_{max} = 2.1$ mm) if compared to the floor, suggesting that the cross-section of the existing tunnel has undergone a slight ovalisation. The observed change in induced settlement is likely to be due to a combination of the changes in applied face pressure – or any other operational conditions of the TBM – and of soil-structure interaction effects due to the finite bending and shear stiffness of the ancient tunnels. The relative importance of these aspects is currently under investigation.

5.2 *Deformed shape of cross-sections of Cloaca Maxima*

As indicated schematically in Figure 2, three cross-sections for each of the two branches of the *Cloaca Maxima* (S1 – S6 in Figure 2(a)) were instrumented with rod extensimeters to measure the convergence induced by excavation of the new tunnels under the ancient sewers. In particular, both vertical and horizontal extensimeters were installed in the deactivated branch (S4 – S6), while horizontal rods only were used in the active one (S1 – S3). Incremental vertical and horizontal strains in the monitoring sections, as induced by the construction of each of the new tunnels, have been calculated by assuming a circular-shaped cross-section, having the same area as the real tunnels of the *Cloaca Maxima*. In the absence of vertical extensimeters, the vertical strains of the cross-section of the active branch of the Cloaca Maxima were inferred from the differential settlements between crown and invert of the tunnel measured by the hydraulic level gauges. The results of these calculations are reported in Figures 11 (a) to (d), in which a qualitative representation of the incremental deformed configuration of the cross-section is also displayed, assuming a circular shape as said above.

For the deactivated branch of the *Cloaca Maxima*, see Figures 11(a) and (b) for the North and South tunnel respectively, the collected data indicate that excavation of the new tunnels induces significant elongation in the vertical direction and some minor elongation in the horizontal direction, in the cross sections located immediately above the tunnel under construction, while it causes squatting of the farthest cross-section, with vertical shortening and horizontal stretching of comparable magnitude. For the active branch, Figures 9(c) and (d), significant ovalisation is induced by excavation, particularly for the South Tunnel, while squatting of the farthest cross-section is not observed. In all cases, the cross-section located between the new tunnels undergoes slight homothetic expansion.

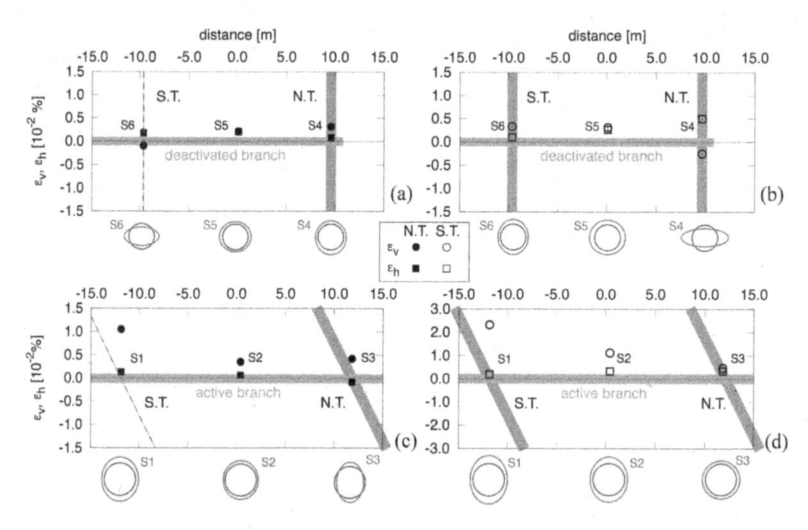

Figure 11. Incremental vertical and horizontal strains in the deactivated, (a) and (b), and active branch, (c) and (d) of the *Cloaca Maxima*, induced by the construction of Line C tunnels.

6 CONCLUDING REMARKS

The *Cloaca Maxima* is one of the earliest sewer systems in the world; its concrete and masonry tunnels channelled Rome refuse beneath the Fora and around the hills, and stood among other extensive drainage networks (Hopkins 2007). The sewer played a key role in the process of urbanization in the area that became the *Forum Romanum,* thus solidifying Rome's economic and military hold on central Italy and was acclaimed in the first century as a work "[...] for which the new magnificence of these days has scarcely been able to produce a match [...]" (Livy, I.56.2). Repaired and restored frequently, has been functioning continuously for more than 2500 years.

As it is often the case with large infrastructural projects, construction of Line C represented a unique opportunity to increase our knowledge of the monument. Aimed at the acquisition of the information required to carry out the interaction analyses between the line and the monument, the extensive investigations carried out by Metro C S.C.p.A. at the design stage, which included coring of the tunnel lining, digital photogrammetry, laser scanning and topographic surveys, made it possible to define precisely the internal geometry of the sewer pipe and to identify the techniques used for its construction and the mechanical properties of the materials. This knowledge was used for a preliminary assessment of the risk of damage and for the design of the monitoring system.

The vertical displacements of the sewers, measured by means of precision levelling on the floor of the *Chiavicone* and hydraulic level settlement gauges installed in the floor and the vault of the

Cloaca Maxima, show that the ancient sewers are relative flexible and substantially conform to the greenfield displacements. The most significant factors affecting the magnitude of the vertical displacements of the sewers are the local variations of the geological profile and the operational parameters of the EPB shield. The measurements of horizontal and vertical convergence provided interesting information on the evolution of the deformed shape of the cross-sections of *Cloaca Maxima* as Line C tunnels were driven in the vicinity of the sewer.

The *Cloaca Maxima* is a very peculiar monument because it is a large active sewer and at the same time a monument of great archaeological value, so that the evaluation of the effects of the construction of Line C had to be carried out under this double profile, even if the lack of decorative elements did not create particular problems of conservation from the point of view of the latter. The constructive simplicity of the structures and the good quality of the materials used have shown high durability and, overall, the condition of burial of the monument has favoured its conservation. As far as the hydraulic and static functionality are concerned, the threshold values on the acceptable deformations for sewers reported in the literature are all much larger than those that were measured and, as a matter of fact, construction of the tunnels was completed without significant impact on the monument.

Recent work has been carried out to address rationally the problem of excavation-induced ground movements affecting existing tunnels and pipelines, and quantify the relative role played by bending and shear deformations (Franza & Viggiani 2021). This depends largely on the ratio of the shear and bending stiffness of the tunnel lining, which in turn affects the soil structure interaction induced by ground settlements. Work is currently under way to obtain characteristic values of this ratio for the various types of lining identified in the *Cloaca Maxima* but also other types of historical linings such as, *e.g*., Victorian brick tunnels or cast iron segmental linings.

REFERENCES

Antognoli L. 2015. Il Chiavicone della Suburra, in Elisabetta Bianchi (ed.) *La* Cloaca Maxima *e i sistemi* fognari *di Roma dall'antichità ad oggi*. Roma: Palombi Editori: 159–166.

Antognoli, L. & Bianchi, E. 2009. La *Cloaca Maxima* dalla suburra al foro romano. *Studi Romani* 12: 89–125.

Corsetti, G. 1925. I collettori bassi delle fogne di Roma, *Annali dei Lavori Pubblici*, Roma: Stabilimento Litografico del Genio Civile.

Di Mucci G. & Miniero N. 2015. Indagini sperimentali su un tratto della Cloaca Maxima nell'ambito degli studi interazione linea-preesistenze della tratta T3 della nuova linea C, in Elisabetta Bianchi (ed.) *La* Cloaca Maxima *e i sistemi fognari di Roma dall'antichità ad oggi*. Roma: Palombi Editori: 207–222.

Fargnoli, V., Boldini, D., & Amorosi, A. 2015. Twin tunnel excavation in coarse grained soils: Observations and numerical back-predictions under free field conditions and in presence of a surface structure. *Tunnelling and Underground Space Technology* 49:454–469.

Franza, A. & Viggiani G.M.B. 2021. Role of shear deformability on the response of tunnels and pipelines to single and twin tunnelling. *J. Geotech. Geoenvironmental Eng. ASCE* 147(12): 04021145.

Hopkins, J.N.N. 2007. The *Cloaca Maxima* and the monumental manipulation of water in archaic Rome. Waters *of Rome Journal* 4:1–15

Losacco, N. & Viggiani, G.M.B. 2019. Class A prediction of mechanised tunnelling in Rome. *Tunnelling and Underground Space Technology* 87:160–173.

Losacco, N. & Viggiani, G.M.B. 2019b. Mechanised tunnel excavation through an instrumented site in Rome, in *National Conference of the Researchers of Geotechnical Engineering*, Springer, Cham: 245–254.

Losacco, N., Romani, E., Viggiani, G.M.B & DiMucci G. 2020. Embedded barriers as a mitigation measure for tunnelling induced settlements: a field trial for the Line C in Rome, in *Tunnels and Underground Cities: Engineering and Innovation Meet Archaeology, Architecture and Art*, Volume 12: Urban Tunnels-Part 2: 5845.

Palombi, D. 2013. Cloaca massima e storia urbana, *Archeologia Classica* 64:133–168

Wan, M.S.P., Standing, J.R., Potts, D.M., & Burland, J.B. 2017. Measured short-term subsurface ground displacements from EPBM tunnelling in London Clay. *Géotechnique*, 67(9):748–779.

Geotechnical Engineering for the Preservation of Monuments and Historic Sites III – Lancellotta, Viggiani, Flora, de Silva & Mele (Eds)

Safeguarding of the *Aurelian Walls* at *Porta Asinaria* from conventional tunnelling

S. Rampello & L. Masini
University of Rome La Sapienza

ABSTRACT: Construction of the line C of Rome underground is being carried out in a complex context due to the presence of archaeological artefacts, historical buildings and monuments of invaluable value. Along the contract T3 of the line, between the shaft 3.3 and *San Giovanni* station, two tunnels have been excavated for a length of about 140 m following a three step procedure: excavation of two small diameter tunnels with a mini TBM; soil improvement via low-pressure cement grouting; and conventional excavation of the two line tunnels in the improved soil. The tunnels, excavated at a depth of about 25 m, reach *San Giovanni* station passing at a short distance from the ancient Aurelian Walls. This paper presents the displacement measured at ground surface during the construction activities, showing the efficiency of a protective barrier made by a line of piles in reducing the movements induced by tunnelling in the Aurelian Walls.

1 INTRODUCTION

The construction of new underground railway lines and the extension of existing ones require deep open excavations and bored tunnels in urban environments, in close vicinity to existing buildings and structures. In these conditions, the main design requirement is to contain ground movements during and after the excavation works to prevent nearby structures from undergoing excessive deformations. This is particularly relevant for existing structures of monumental and historical value for which the limits in allowable movements are very onerous, as it often the case in European cities.

Contract T3 of the new line C of the Rome underground, currently under construction, underpasses the historical centre of the city encountering the historical monuments of Roman age (I to V century): these are the Aurelian Walls at *Porta Asinaria* and *Porta Metronia*, the Church of *Santo Stefano Rotondo*, the *Acquedotto Celimontano*, the *Anfiteatro Flavio* (Coliseum), the *Basilica di Massenzio*, the *Colonnacce* and the *Foro di Cesare*.

Due to the exceptional archaeological and historical value of the structures potentially affected by the construction of the line, the design included a detailed study of the interaction between the construction activities and the monuments. To this aim, the general contractor, Metro C, set-up a multidisciplinary steering technical committee with the assignment of implementing all the necessary procedures to safeguard the monuments and historical buildings.

The main tasks of the steering committee were: (i) to evaluate the influence of the construction of the line C (tunnelling and deep open excavations) on the existing monuments; (ii) to suggest, where necessary, appropriate geotechnical or structural mitigation measures; (iii) to develop a comprehensive and redundant monitoring scheme to follow in real time the response of the monuments to construction; (iv) to assist the general contractor during construction with the evaluation of the monitoring data, to optimise construction sequences and procedures.

The analysis of the interaction between the excavation activities and built environment was carried out following procedures of increasing complexity.

 DOI 10.1201/9781003329756-17

At a first stage, simplified (Level 1) analyses were performed computing surface and near-surface displacements by semi-empirical methods, ignoring the stiffness and weight of the existing monuments (e.g.: Attewell & Woodman 1982; Attewell et al. 1986; Ou & Hsieh 2011); evaluation of the potential damage induced by tunneling and deep excavations was carried out through the interaction diagrams proposed by Burland & Wroth (1974), which relate the deflection ratio and the horizontal tensile strain to given damage categories. Based on the outcome of these evaluations, the study ended if the damage was deemed negligible, or continued to a higher level of complexity (Level 2).

At this second stage, the interaction between the tunnels or the deep excavations and the monuments was studied through 2D or 3D Finite Element (FE) analyses that studied the soil-structure interaction adopting a simplified description of the mechanical behaviour of the monuments. Damage was then re-evaluated using the interaction diagrams. Depending on the computed results, either damage was deemed acceptable, or prospective remedial techniques were suggested (Burghignoli et al. 2013; Rampello et al. 2012).

Structural and geotechnical mitigation interventions have been adopted in the project of the line C, the first being aimed to strengthen the structures, while the second to reduce the ground movements induced by tunnelling or deep excavations.

Definitive structural interventions mainly consisted of reinforcements made by steel wire ropes, or chains made by steel bars (e.g.: Church of *Santo Stefano Rotondo* and *Basilica di Massenzio*), while temporary structural intervention manly consisted of buttresses made of steel tube-joint structures and multiprop towers (e.g.: Aurelian Walls at *Porta Asinaria* and *Porta Metronia*, and *Basilica di Massenzio*).

The geotechnical mitigation interventions were of active or passive type, the first permitting to control the ground settlements during tunnelling, while the second producing a favourable variation of the displacement field induced by tunnelling.

Compensation grouting is an example of active mitigation intervention whose efficiency has been shown by site applications (e.g.: Mair 2008; Mair & Hight 1994) and laboratory tests (e.g.: Masini et al. 2012, 2014). A protective embedded barrier is instead an example of passive mitigation intervention that can be adopted when the structure lies to the side of the tunnel: it is installed before tunnelling, between the tunnel and the structure for which damage must be prevented, providing a restraint to ground movements (e.g.: Bai et al. 2014; Bilotta, 2008; Bilotta & Taylor, 2005; Di Mariano et al. 2007; Fantera et al. 2016; Katzenbach et al. 2013; Masini & Rampello, 2021; Rampello et al. 2019).

During construction of the line C, compensation grouting was adopted to prevent potential damage induced by tunnel excavation under the Aurelian Walls at *Porta Metronia*, while an embedded barrier was preinstalled to protect the Aurelian Walls at *Porta Asinaria*. In both cases, field monitoring was carried out to verify the design assumptions and evaluate the actual performance of the excavation works, having also the opportunity of calibrating the semi-empirical methods.

This paper describes the displacement field monitored during conventional excavation of two tunnels, about 140 m long, connecting the multifunctional shaft 3.3, operating as a launch pit for the TBM/EPB machines which excavated the tunnels in the direction of *Amba Aradam* station, and the existing *San Giovanni* station. Both the shaft 3.3 and the tunnels were excavated close to the Aurelian Walls at *Porta Asinaria* (3rd century A.D.). To prevent any damage eventually induced by tunnelling on the ancient city wall, a protective barrier made by adjacent piles was preinstalled close to the North-bound tunnel, where the walls are closest to the tunnel, at 23 m to 26 m from its axis.

Field monitoring provided the surface ground movements induced by the three-step procedure adopted to construct the tunnels: (i) mechanised excavation of two mini-tunnels; (ii) soil improvement via low-pressure cement grouting; (iii) conventional excavation of the main tunnels in the improved soil.

In this paper, the reduction of the surface settlements obtained behind the barrier, in the portion of soil facing the ancient city wall, is evaluated through the comparison of the monitoring data collected in *green-field* conditions and in the presence of the barrier, demonstrating the efficiency of this kind of mitigation intervention.

2 THE AURELIAN WALLS AT PORTA ASINARIA

The Aurelian Walls are large defensive walls built by Emperor Aurelian between 270 and 275 A.D. with most of their length (12.5 km over 19 km) having survived past centuries in a fair preservation state. Aurelian Walls at *Porta Asinaria* belong to the South-Eastern part of the town wall and are located at a distance of 24 m to 27 m from the diaphragm wall of a 30 m-deep excavation and of 23 m to 34 m from the axis of the North-bound tunnel of the line C of the Rome underground. Figure 1a shows the aerial view of the Aurelian Walls at *Porta Asinaria*, together with the multifunctional pit 3.3. The latter operated as a launch pit for the two TBM/EPB machines that bored the 6.7 m-diameter twin tunnels in the direction of *Amba Aradam* station, while the two conventional tunnels, about 140m long, were excavated towards *San Giovanni* station.

The wall is 4 m thick and about 18 m tall from the foundation plane, with the foundation located at 8–9 m below ground surface. The structure of the Aurelian Walls is made of combined tuff and brick masonry, with an inner core of poorly bonded tuff blocks. Initially, the ground surface was at the same level on both sides of the walls but, over the centuries, material has been accumulated on the side facing the *Basilica di San Giovanni*. Nowadays the city wall retains a backfill of anthropic origin, about 10 m high, cumulated since medieval times without installing any drainage system, which has caused the development of outwards displacements as high as about 0.4 m at the top of the walls and diffused cracks along the masonry surface. The aerial photo of Figure 1b shows the Aurelian Walls and *Porta Asinaria* during the excavation of the multifunctional pit 3.3, together with the temporary safeguarding interventions installed against the wall façade and the back-excavation carried out to reduce the earth trust acting in the wall.

Unlike the wall, *Porta Asinaria* is characterised by the same elevation of the ground surface on both the *extra*- and the *intra-moenia* sides (Figure 1c–d). Its current state largely corresponds to the restorations of the time of the Emperor Honorius. The structure of *Porta Asinaria* is essentially composed of two semi-circular towers connected to each other by two walkways placed on two different levels. The towers today are about 25 m high from the elevation of ground surface on the *extra-moenia* side (+33.9 m asl) with a diameter of about 4.9 m, while the walkway area is about 18 m high.

Thanks to the recent restorations carried out since 1950, *Porta Asinaria* is characterised by a good state of conservation, with extensive portions of the masonry completely rebuilt. However, the restoration mainly concerned the external façade, while the core continues to have rather modest mechanical properties.

According to the FE interaction (Level 2) analyses carried out at the design stage, excavation of the multifunctional pit 3.3, as well as that of the two tunnels, would have induced not negligible effects on the walls, with an increment of wall rotation. Specifically, the numerical simulation of the shaft excavation estimated a wall rotation $\alpha = 0.11°(0.2\%)$ with a maximum settlement of the wall of 27 mm, while the numerical analysis simulating tunnels excavation, with an assumed volume loss $V_L = 2.5\%$, provided a wall rotation $\alpha = 14°$ with a maximum wall settlement of 16 mm.

The above quantities were deemed to be potentially dangerous for an already damaged ancient masonry structure. Therefore, the primary design challenge was that of preventing any damage induced by the excavation activities on the ancient and vulnerable Aurelian Walls, located a short distance away from the pit and the North-bound tunnel.

To this end, in order to attain an estimated reduction of wall rotation of about 36%: (i) about 8 m of the backfill were removed behind the wall prior to the excavation of the pit 3.3, starting from the middle plane of the shaft, located at a distance of 30 m from the Aurelian Walls, thus eliminating the earth thrust acting on it; (ii) a very stiff retaining system was designed for the shaft using 1.2 m-thick diaphragm walls, top-down construction with five levels of props made by cast-in-situ concrete stiff slabs, and a high embedment ratio of the diaphragm walls below the dredge line ($L/H = 1.56$), with the diaphragms extending in the stiff and overconsolidated Pliocene clay, thus preventing any deep-seated movement.

The monitoring data collected during shaft excavation confirmed that the back-excavation carried out to safeguards the ancient structure and the high stiffness of the support system, together with the high embedment ratio of the diaphragm walls below the dredge line, were the key to minimise the effects induced by the excavation of the multifunctional pit on the Aurelian Wall at *Porta Asinaria* (Masini et al. 2021).

Figure 1. Aerial view of Aurelian Walls at *Porta Asinaria* (a-b); front view of *Porta Asinaria* (*extra moenia*) (c); front view of *Porta Asinaria* (*intra moenia*) (d).

To mitigate tunnelling effects, a protective barrier made by a line of piles was instead pre-installed close to the North-bound tunnel, in order to achieve an estimated reduction of about 54% in the wall rotation (see Figure 2). The efficiency of this kind of intervention, preliminary evaluated through a numerical study (Rampello et al. 2019) and a field test carried out in *green-field* conditions (Losacco et al. 2019; Masini & Rampello 2021), is discussed in this paper.

3 PROJECT DESCRIPTION AND CONSTRUCTION SEQUENCE

Figure 2 shows the two tunnels that were excavated using conventional procedures at a depth of about 25 m and for a length of 140 m, between the lunch pit 3.3 and the existing *San Giovanni* station, with the axis of the North-bound tunnel located at a distance of 23 m to 34 m from the ancient city wall.

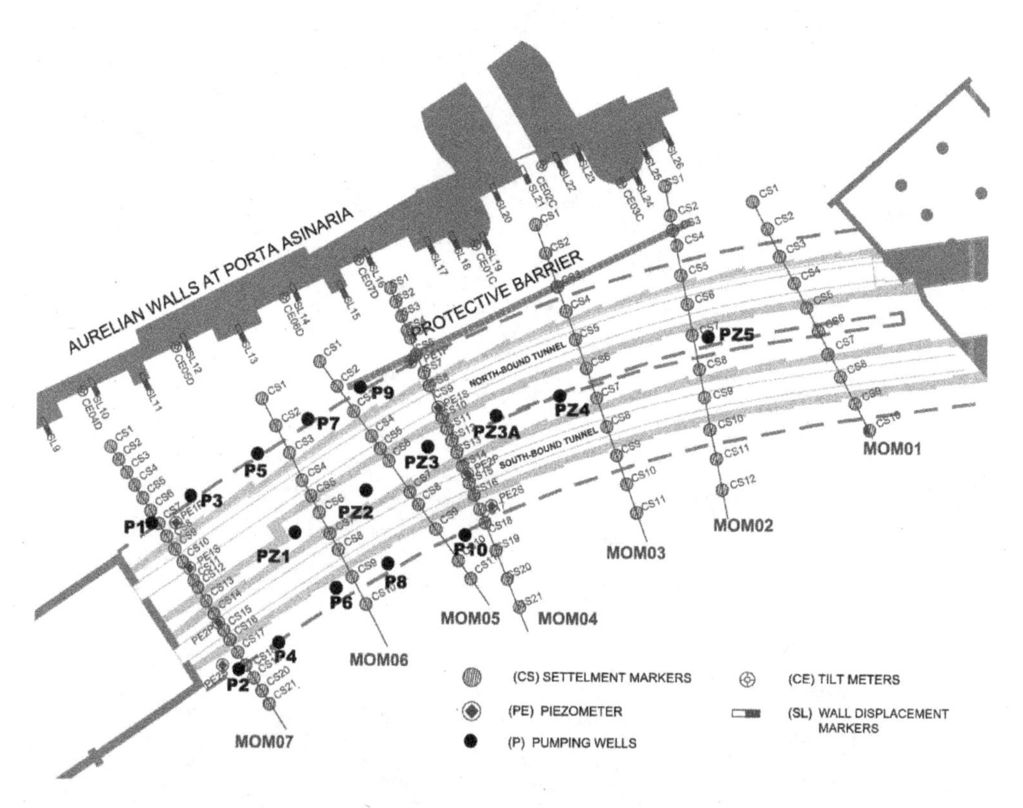

Figure 2. Plan view of the monitoring system.

To mitigate tunnelling-induced effects on the city wall, an embedded barrier made by adjacent piles was installed before tunnelling activities begun. It extends for 63 m in the zone where the North-bound tunnel is closer to the wall: 23 m and 26 m at MOM-02 and MOM-04 alignments, (Figure 2). The piles have a diameter $D = 0.8$ m, a length $L = 28$ m, and are installed at a spacing $s = 1.0$ m. The barrier runs approximately parallel to the wall with a minimum distance of about 8 m from the axis of the North-bound tunnel, about 8 m wide, and 18 m from the ancient city wall. An embedded capping beam (cross-section of $1 \times 1\text{m}^2$) connects the head of the piles. The capping beam and the piles are made of cast-in-place reinforced concrete, with a 28-day compressive strength of 32 MPa and Young's modulus of 31 GPa, while the yield strength of the rebar steel is equal to 235 MPa.

To monitor the ground movements induced by the tunnelling activities, seven arrays of instruments were set up about normal to the tunnels, named sections MOM 01-02-03-04-05-06-07. The instrumentation along the MOM alignments included settlement markers installed at ground surface and vibrating-wire piezometer cells. Arrays MOM-07 and MOM -04 were instrumented with closer displacement markers (spacing 2.5 m) and vibrating-wire piezometer cells, as well as with inclinometer and Trivec casings, the first providing horizontal displacements only, while the second measuring the three orthogonal components Δx, Δy and Δz of the displacement vectors along the vertical measuring line, with a depth spacing of 0.5 m. The displacement markers installed at the ground surface incorporate sockets into which a removable survey plug can be screwed with good positional repeatability for manual surveying.

In this paper, reference is made to surface ground settlements measured by precision levelling only: this was performed using a digital level which can detect the height of the plane of collimation on a suitable bar-coded staff to a resolution of 0.01 mm.

Monitoring of the wall movements during the excavation activities was performed through precision levelling on displacement markers installed along the wall side facing the tunnels, about 0.5 m above the ground level, and by electric tiltmeters installed at wall mid-height to measure the out-of-plane rotation.

Ground conditions at *PortaAsinaria* are described by Fantera et al. (2016), Masini et al. (2019a–b, 2021a–b), Masini and Rampello (2021) and Rampello et al. (2019), to which reference is made for further details. Table 1 reports the strength parameters and the overconsolidation ratio as obtained from laboratory tests.

A section through the instrumented array MOM-04 is plotted in Figure 3, showing soil layering, the two tunnels, and the protective barrier. A 14 m-thick layer of made ground (MG) is first encountered, from ground surface at about +35 m asl, mainly consisting of coarse grained material, sand and gravel; recent alluvial soils of the Tiber river are found underneath, extending down to a depth of 26 m (+8.7 m asl). The alluvia are variable in grading involving slightly overconsolidated clayey silt and sandy silt (CS-SS); they overly a layer of sand and gravel of Pleistocene age (SG), with a thickness of about 14m, followed by a thick layer of stiff and overconsolidated silty clay (OSC), the blue Vatican clay of Pliocene age.

Table 1. Strength parameters and overconsolidation ratio.

Soil	γ (kN/m^3)	c' (kPa)	ϕ' (°)	OCR (–)	S_u (kPa)
Made Ground (MG)	17	5	34	1	–
Clayey silt and Sandy Silt (CS-SS)	19.5	28	27	1.25	120
Sandy Gravel (SG)	20	0.1	40	1	–
Overconsolidated Stiff Clay (OSC)	20.9	41.3	25.7	2.5	400

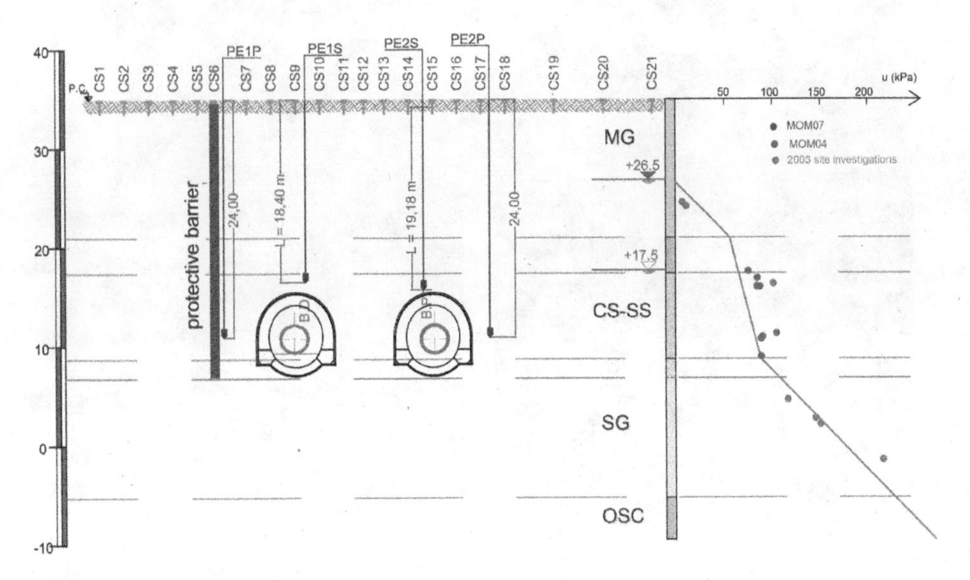

Figure 3. Transversal section through instrumented array MOM-04.

The pore water pressure regime is characterised by downwards seepage in the silty soils from the made ground, where a constant hydraulic head $H = 26.5$ m asl was measured, to the deep layer of sandy gravel, at constant head $H = 17.5$ m asl. This is a typical condition encountered along the line C of Rome underground, induced by pumping from the deep and permeable layer of sandy gravel for anthropic purposes. Tunnels excavation was carried mainly in the alluvia sandy silt whose permeability was preliminary reduced via low-pressure cement grouting.

The two tunnels connecting the multifunctional pit 3.3 to *San Giovanni* station were excavated following a three-step procedure: (i) two small-diameter tunnels ($D = 3$ m) were first excavated using a mini-slurry shield, tunnelling boring machine at a depth of about 25 m (cover to diameter ratio $C/D = 7.8$); (ii) soil improvement via low-pressure cement grouting was then carried out using *tubes à manchettes* installed in boreholes excavated radially to the bored tunnels; (iii) tunnel construction was finally completed through conventional excavation of the main tunnels in the improved soil. Both the tunnels have a curvilinear cross-section with an average diameter of the equivalent circular cross-section $D_{eq} = 8.03$ m. At the alignment MOM-07, the initial portion of the North-bound tunnel has an enlarged cross-section with an equivalent diameter $D_{eq} = 10.3$ m.

Figure 4 shows the mini-TBM at the launching pit (Figure 4a) and at the intermediate jacking station (Figure 4b), a view from the inside of a mini-tunnel after completion of the radial injections (Figure 4c) and an intermediate stage of tunnel excavation in the improved soil (Figure 4d – primary lining installed).

At stage (ii), the *tubes à manchettes* (4 *manchettes*/m) were installed in 20 boreholes per section, drilled at a longitudinal spacing of 0.6 m: the boreholes had a diameter of 80 mm, a length of 5–7 m and a spacing of 18° in the radial direction. Soil improvement was obtained injecting first a Mistrà-type cement grout with 10–15% of bentonite content and a water-cement ratio of 2.5–3.5, and a chemical mixture of silica components in a second stage, to reduce further the permeability of the improved soil that had to be excavated during tunnelling.

Figure 4. Mini-TBM at the launching pit and at the intermediate jacking station (a-b); mini tunnel at the completion of radial injections (c); conventional tunnelling in the improved soil.

In the conventional excavation procedure, fan-like overlapping pipe umbrellas, made by 114.7 mm-diameter steel pipes, were drilled and grouted at the roof of the tunnel, parallel to the direction of advancement of the excavation face. Each roof shield consisted of 41 steel pipes, 12 m long, covering an excavation span of 8m. Full face excavation was carried out along with the installation of the primary support 1m away from the excavation face, which consists of IPN 160 steel ribs and 0.2 m-thick shotcrete. The final concrete lining, 0.8–1.0 m-thick, was installed 35 m away from the excavation face.

At the early stages of excavation of the invert of the South-bound tunnel, the first to be excavated, a local collapse involved a small portion of soil at the tunnel spring line as a result of basal heave of the improved soil. Therefore, before continuing the tunnelling activities, 16 relief wells were activated (see Figure 2), which induced an average drawdown of the hydraulic head of about 9.5 m in the layer of sandy gravel.

4 FIELD MONITORING

In this section the vertical displacements measured at ground surface by the settlement markers installed along the 6 instrumented arrays, MOM-07 to MOM-02, are discussed for each excavation stage, interpreting the transversal displacement profiles attained in plane strain conditions through the empirical relationships currently adopted in applications (e.g.: Moh et al. 1996; O'Reilly & New 1982; Peck 1969).

For evaluating the effects induced by mini-tunnelling, the reference undeformed ground surface was calculated as the average over the time of the displacement readings taken for distances of the excavation face not lower than 30 m from each instrumented section. The effects induced by radial borehole drilling were instead evaluated assuming as base-line the time average of the readings in the time interval between the end of excavation of the North-bound mini-tunnel and the start of drilling (31/03/17–27/06/17), while the reference of each section for the low-pressure grouting injections was taken considering the displacement measured about two months before the start of the injection activities.

Table 2 reports the start and end dates of each construction phase: excavations of both the small-diameter tunnels using the mini TBM was carried out in about two months (70 days), while soil improvements around the mini-tunnels took about 16 months (7 months for boreholes drilling and 9 months for grout injections). Conventional tunnelling required a much longer time, of about one year and a half for the South-bound tunnel due to some problems occurring during the excavation of the tunnel invert, while North-bound tunnel was excavated in about half year.

Typical time histories of the vertical displacements observed at the instrumented arrays are shown in Figure 5 with reference to section MOM-04 (see Figure 2), while Table 3 reports the volume per unit length described by the settlement profiles, the corresponding volume loss (mini-tunnel diameter $D = 3$ m) and the maximum observed settlements (−)/heaves (+) measured during the construction stages preliminary to excavation of the main tunnels.

Table 2. Construction stages.

construction stage		start	end
mini-tunnelling	South-bound tunnel	20/01/17	16/02/17
	North-bound tunnel	13/03/17	31/03/17
soil improvement	borehole drilling	27/06/17	22/01/18
	grouting injections	23/01/18	19/10/18
activation of relief wells		from 20/11/18	
South-bound running tunnel		18/01/19	07/07/19
North-bound running tunnel		26/04/19	21/10/19

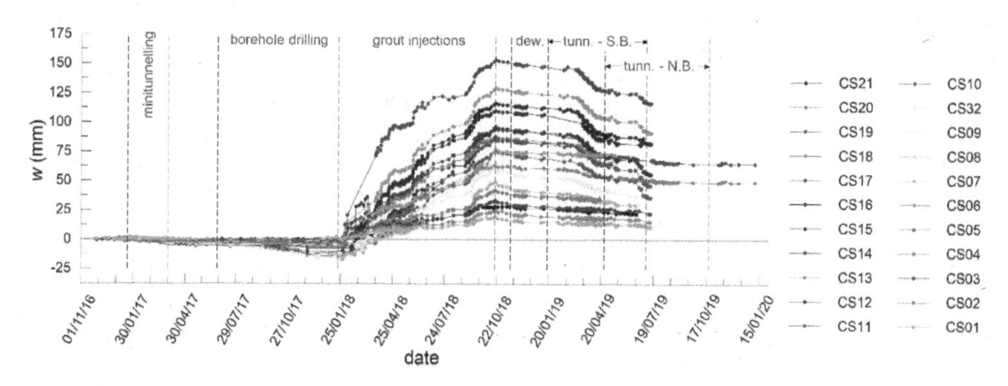

Figure 5. Time histories of the vertical displacements measured at section MOM-04.

Excavation of both the mini-tunnels induced negligible settlements, with average and maximum values $w_{ave} = -3.6$ mm and $w_{max} = -5.6$ mm, respectively, while radial boreholes drilling caused a progressive increase of the settlements as drilling activities approached the instrumented section, with $w_{ave} = -12$ mm and $w_{max} = -20.5$ mm, which are about three times higher than the corresponding values observed at the end of mini-tunnels excavation. The subsequent low-pressure grout injections caused a massive heave, with average and maximum values as high as $w_{ave} = 137.3$ mm and $w_{max} = 165.6$ mm.

Table 3. Effects induced by construction activities preliminary to excavation of main tunnels.

	tunnelling with mini-TBM			drilling of radial boreholes			low-pressure grout injections		
MOM	ΔV (m^3/m)	V_L (%)	w_{max} (mm)	ΔV (m^3/m)	V_L (%)	w_{max} (mm)	ΔV (m^3/m)	V_L (%)	w_{max} (mm)
07	0.110	0.78	–4.01	0.211	1.49	–8.43	–3.035	–21.47	95.75
06	0.202	1.43	–5.61	0.892	6.31	–20.47	–4.032	–28.52	132.7
05	0.095	0.67	–2.98	0.284	2.00	–12.92	–12.29	–86.23	141.5
04	0.099	0.70	–3.26	0.451	3.19	–13.06	–5.65	–39.96	150.0
03	0.084	0.59	–3.41	0.212	1.50	–12.33	–5.89	–41.66	165.6
02	0.062	0.44	–2.61	0.098	0.69	–4.55	–4.59	–32.47	138.4

Part of the measured heave was lost during excavation of the main tunnels, as a result of both the conventional tunnelling activities and the dewatering from the relief wells in the deep layer of sandy gravel.

Measurements provided by precision levelling during mini-tunnelling and borehole drilling are characterised by a significant scatter do not highlighting any clear difference between the volume losses computed in the presence and the absence of the barrier.

Figure 6a–b show the heave profiles induced by the grout injections carried out to improve the soil strength and reduce its permeability. Although the low pressures adopted to inject the grout, the heave measured at ground surface was not negligible in *green-field* conditions (MOM 07-06-05), being equal to about 96 mm to 140 mm, and was even higher in the sections interacting with the barrier (MOM 04-03-02), being in the range of 138 mm to 166 mm. The barrier did not produce any appreciable reduction of the ground uplift behind its location, in the portion of the soil facing the walls.

At the start of construction of the South-bound running tunnel, during the excavation of the tunnel invert, water came into the tunnel due basal heave of the improved soil, so that the excavation was

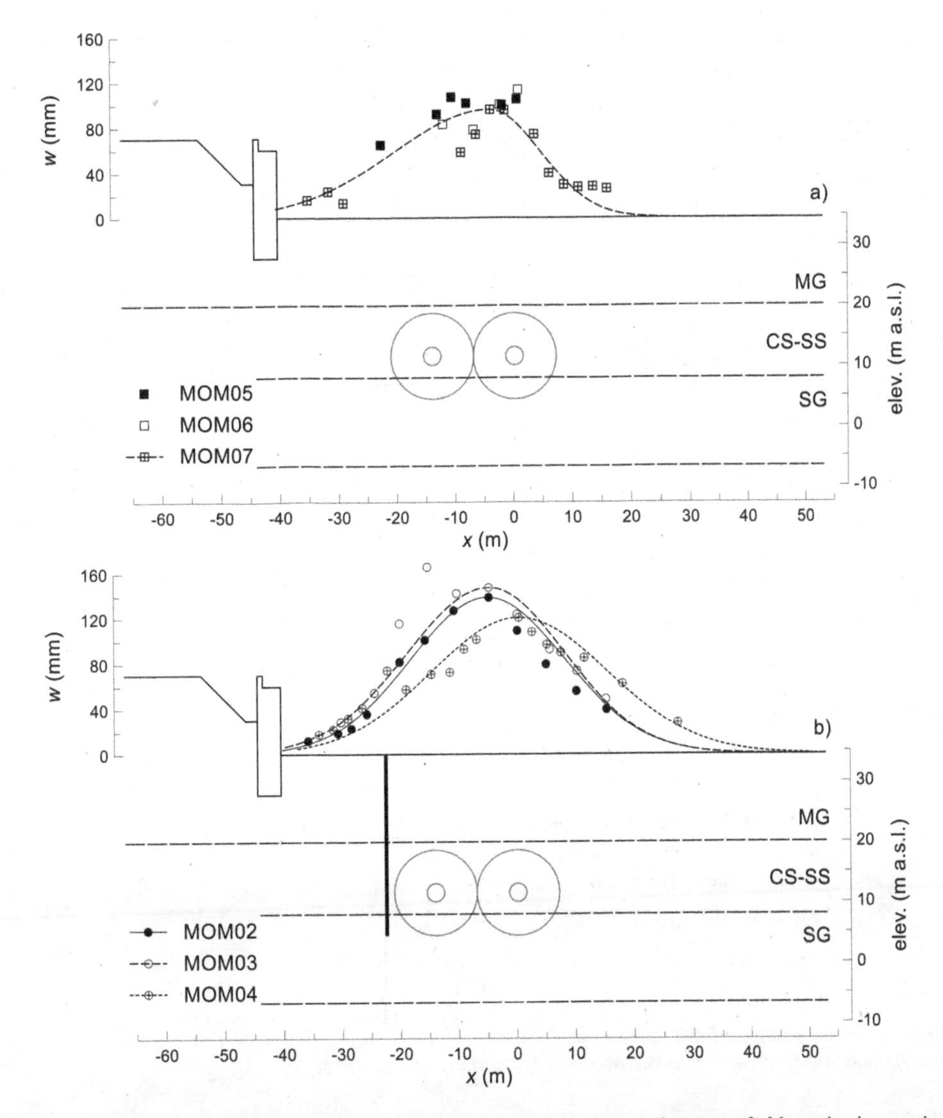

Figure 6. Settlement profiles induced by low-pressure grout injections at the *green-field* monitoring sections MON-05-06-07 (a), and in the presence of the barrier (sections MON-02-03-04) (b).

suspended while activating the pumping wells to reduce the hydraulic head at the base of the improved soil (see Figure 2). The excavation was resumed about two months later.

Pumping from the relief wells lowered the hydraulic head by about 9.5 m in the layer of sandy gravel (SG), with an increase of the effective stresses in the layer of clayey silt and the development of further settlements at ground surface.

To assess the effect of dewatering in the sandy gravel, the time needed to attain the end of consolidation in the layer of clayey silt and sandy silt (CS-SS) was evaluated using Terzaghi's theory of one-dimensional consolidation assuming a consolidation coefficient $c_v = 2 \cdot 10^{-4}$ m^2/s, and a drainage path of 12 m, evaluating an end-of-consolidation time of about 10 days. Therefore, the settlements induced at ground surface by dewatering can be assumed to be nearly fully developed before the start of tunnel excavation.

As an example, Figure 7a shows the time-history of the surface settlement measured by the settlement marker CS14 of the array MOM-07, starting from the beginning of well activation (20/11/2018).

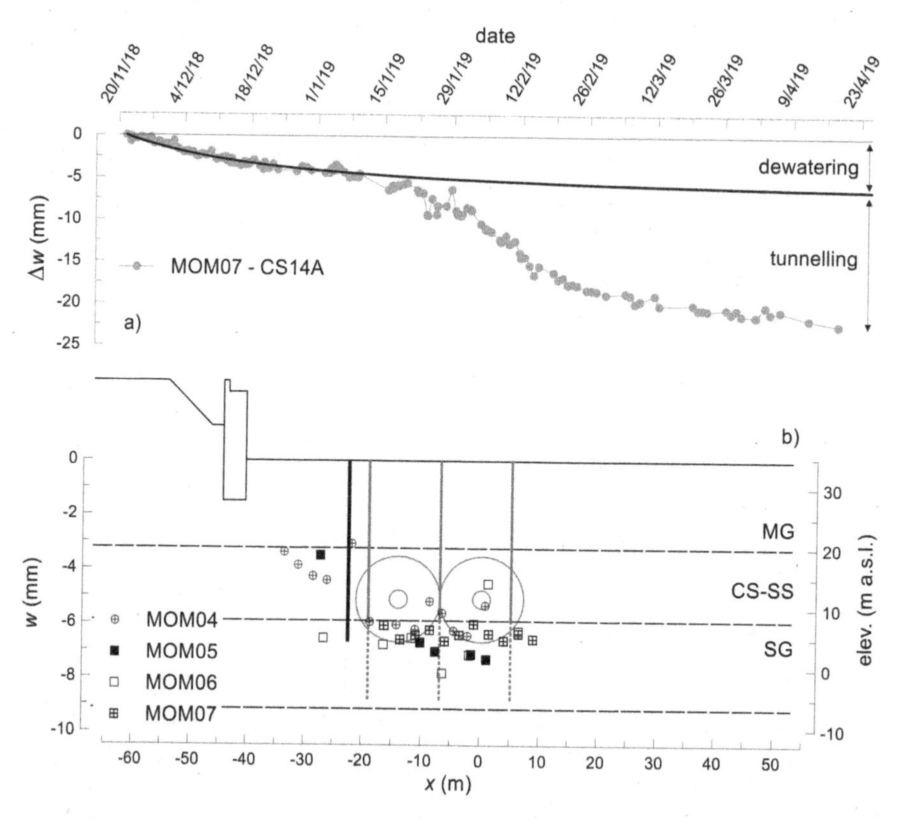

Figure 7. Time-history of the relative settlements observed during dewatering and excavation of the South-bound tunnel (a); settlements induced by the dewatering operations (b).

The first portion of the curve is characterised by a slow increase in the settlements, attributable to the consolidation process induced by dewatering, while the second presents a sharper increase associated to tunnels excavation. The first portion of the curve can be best-fitted using a hyperbole, the origin of which corresponds to the start of the pumping activities, thus estimating the final settlements induced by dewatering.

The ground surface settlements induced by lowering the pore water pressure in the sandy gravel are plotted in Figure 7b in a section transversal to the axes of the mini-tunnels. All the monitoring arrays affected by the dewatering activities exhibit similar behaviour, with a large scatter of the data and maximum settlements of –7 mm to –8 mm between the relief wells, above the two mini-tunnels.

The barrier made of adjacent piles and partially embedded in the layer of sandy gravel had, as expected, no effect in the observed settlement profiles.

Figure 8a–b show the settlement troughs induced by the excavation of both the tunnels in the green-field sections (MOM 07-06-05) and in the sections interacting with the barrier (MOM 04-03-02): the zero abscissa is referred to the axis of the South-bound tunnel.

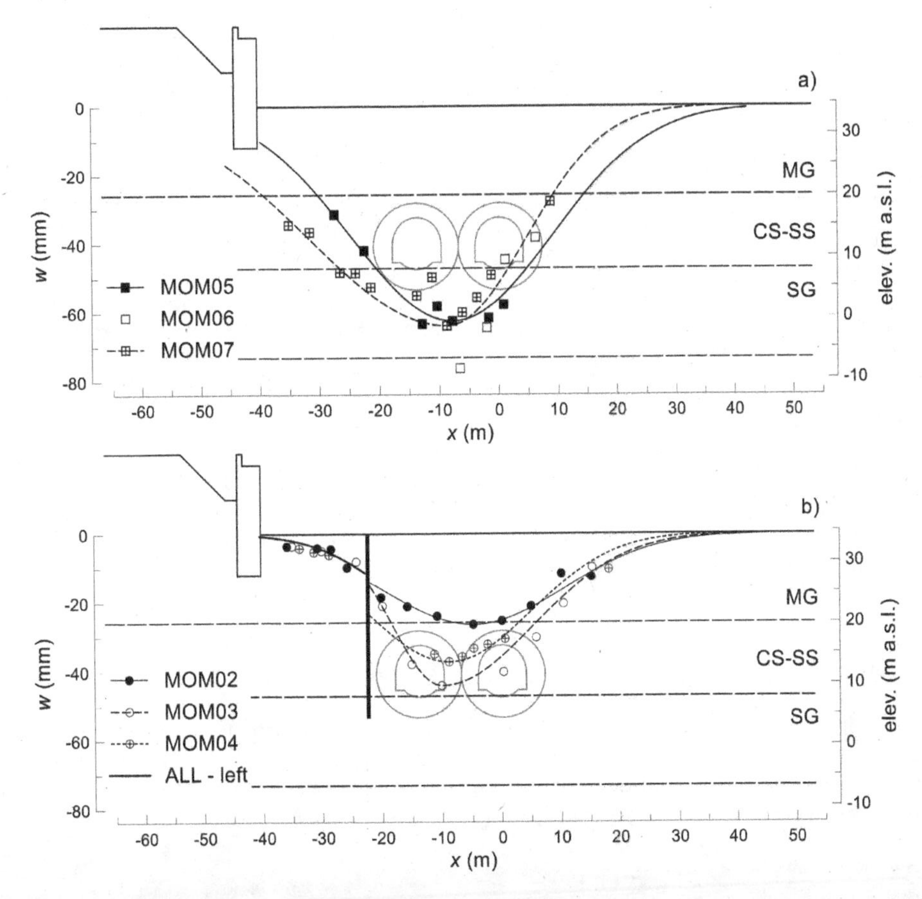

Figure 8. Settlement profiles induced by excavation of South-bound tunnel at the *green-field* monitoring sections MON-06-07 (a), and in the presence of the barrier (sections MON-04-05) (b).

The *green-field* sections show higher settlements, with maximum values of about –65 mm (Table 4), attained at about the mid plane between the tunnels. An asymmetrical settlements trough was also observed at section MOM-07, due to the larger excavated cross-section of the North-bound tunnel, so that separate best-fit Gaussian approximation of the surface settlements was carried out to the wright (+) and the left (−) of the maximum settlement, providing a trough width factor $i^{(-)} = 22.7$ m much higher than $i^{(+)} = 13.4$ m. Surface settlements evaluated at the location of the Aurelian Walls were also higher at section MOM-07, being equal to about −20 mm, if compared with the ones at section MOM-05, that are equal to about −12 mm.

For the sections interacting with the protective barrier (Figure 8b), lower settlements were measured both above the tunnels and close to the Aurelian Walls: maximum settlements above the tunnels were of about –26 mm to –44 mm (Table 4), while at the location of the Aurelian Walls, the surface settlements were not higher than –5 mm. At the wall façade a settlement reduction of 58% was obtained, while at the location of the embedded barrier the measured settlements reduced by about 72% being equal to about –43 mm in *green-field* conditions (sect. MOM-05, Figure 8a) and about −12 mm in the presence of the embedded barrier (MOM 04-03-02, Figure 8b). The efficiency of the adopted mitigation intervention is also appreciable considering the volume loss computed for green-field conditions, $V_L = 2.60\%$ (sect. MOM-05) and in the presence of the protective barrier, $V_L = 0.96\% - 1.31\%$, with an average reduction of about 56%.

It is worth mentioning that the FE interaction analyses predicted slightly lower reductions of the surface settlements at the wall and the barrier locations, equal to 41% and 66%, respectively, that were in a fair agreement with the observed reductions mentioned above, equal to 58% and 72%, respectively.

Table 4. Effects induced by tunnels excavation.

	MOM	$i^{(-)}$	$i^{(+)}$	ΔV (m^3/m)	V_L (%)	w_{max} (mm)
green-field	07	22.7	13.4	3.643	2.71	–64.0
	05	16.8	16.8	2.635	2.60	–62.6
	04	13.9	13.9	1.163	1.15	–37.2
protective barrier	03	8.2	15.5	1.327	1.31	–44.1
	02	15.4	15.4	0.969	0.96	–26.4

It is worth noting that the maximum settlements induced by dewatering plus conventional excavation of the main running tunnels (≈ -75 mm) were sensibly lower than the maximum heave induced by the grout injections carried out at low pressure from by the radial *tubes à manchettes* ($\approx$ 165 mm).

5 MOVEMENTS OF THE AURELIAN WALLS AT PORTA ASINARIA

The Aurelian Walls at *Porta Asinaria* are about parallel to the tunnels in the portion facing sections MOM-06 to MOM-03. Prior to start with the excavation activities, Metro C implemented temporary safeguarding interventions on the city wall, consisting of buttresses made by steel tube-joint structures (Figure 9) to prevent any damage eventually induced by unexpected events.

Figure 9. Temporary safeguarding interventions installed at the Aurelian Walls.

The portion of the wall facing the multifunctional pit 3.3 and the tunnels were also equipped with 10 electric tiltmeters, installed at wall mid-height to measure out-of-plane rotation, and 24 displacement markers installed at about 0.5 m above the ground surface, for monitoring the vertical displacements of the wall (see Figure 2). Precision levelling on the displacement markers and monitoring of tiltmeters started before the tunnelling activities, on November 23rd, 2016.

The effects induced by the TBM excavation of the South-bound and North-bound mini-tunnels, as well as those induced by the radial borehole drilling to install the *tubes à manchettes* were negligible, causing displacements ≤ 2 mm, and are not discussed in the following.

To evaluate the effects of the low-pressure grout injections, the displacement measurements were referred to the start of the injections (23/01/2018), evaluating the base-line displacement profile of each section in the time interval ranging from the end of borehole drilling and the start of injections.

The maximum heave produced by the grout injections, equal to +10.5 mm, occurred at the displacement marker SL13, located close to the *green-field* section MOM-06, while the displacement markers located behind the protective barrier experienced substantially lower heaves, decreasing from about +7.5 mm (SL14–SL19) to +3 mm (SL22-SL26).

Tiltmeters CE 04D-05D-06D, located in front of the *green-field* sections MOM 07-06-05, provided small wall rotation towards the *Basilica di San Giovanni*, equal to about +0.05°, while the tiltmeters installed in the portion of the wall located behind the protective barrier, as well as those located behind the multifunctional shaft 3.3, were not substantially affected by the injection activities.

The settlements induced by conventional excavation of the South-bound and North-bound tunnels were referred to the end of the grout injections, about one month after the start of dewatering.

Figure 10 shows the isochrones of the wall settlements induced by the excavation of both tunnels. Negative abscissas in the figure refer to the displacement markers installed in the portion of the Aurelian Walls located in front of the multifunctional shaft 3.3 (SL3–SL9). The maximum settlement of the wall, equal to −12.3 mm, was measured at the location of the displacement marker SL12, located in the portion of the wall facing the *green-field* section MOM-06.

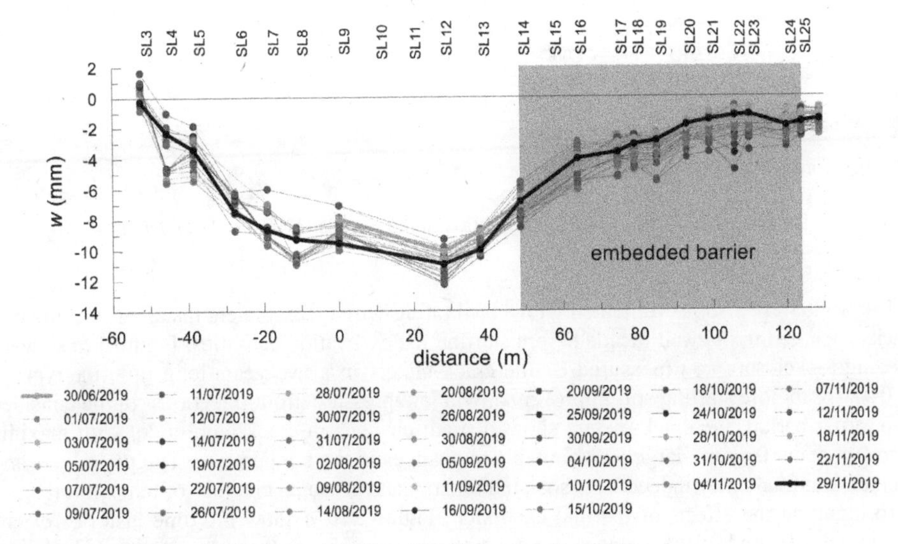

Figure 10. Settlement profiles induced by tunnelling in the Aurelian Walls.

By contrast, the wall settlements are seen to decrease behind the embedded barrier made by adjacent piles (Figure 10): moving from the displacement marker SL14, located behind the left end of the barrier, towards the displacement marker SL25, located behind its right end, the wall settlements reduce from about 7 mm (43%) to about 2 mm (84%), demonstrating the efficiency of this type of mitigation intervention. However, the reduction of wall settlement observed as *Porta Asinaria* is approached is also affected by the slight increasing distance between the North-bound tunnel and the city wall.

The maximum deflection ratios in sagging and hogging were $(\Delta_s/L_s)_{max} = 6.2 \cdot 10^{-5}$ and $(\Delta_h/L_h)_{max} = 2.7 \cdot 10^{-5}$, respectively, both resulting substantially lower than the threshold values proposed by Burland & Wroth (1974): $(\Delta_s/L_s)_{lim} = 8 \cdot 10^{-4}$ and $(\Delta_h/L_h)_{lim} = 4 \cdot 10^{-4}$.

The time histories of the out-of-plane rotations induced in the Aurelian Walls by the tunnels excavation are plotted in Figure 11.

Tiltmeters CE 01D-02D-0D3, installed in the portion of the wall facing the multifunctional shaft 3.3 show nearly constant and negligible rotations towards the excavation, of about −0.03°, with similar values also observed for tiltmeters CE-04D and CED-06D, installed in portion of the wall close to the transversal diaphragm wall of the shaft and the left-end of the barrier, respectively. By contrast, a maximum rotation of about −0.08° was measured on the displacement marker CE-05D, installed close to the displacement marker SL12, in the portion of the wall facing the instruments array MOM-06.

Conversely, the tiltmeters installed in the portion of the wall located behind the protective barrier (CE-07D, CE 01C-02C-03C) were observed to undergo nearly zero rotations, showing once again the efficiency of the embedded barrier in reducing tunnelling effects on the ancient Aurelian Walls at *Porta Asinaria*.

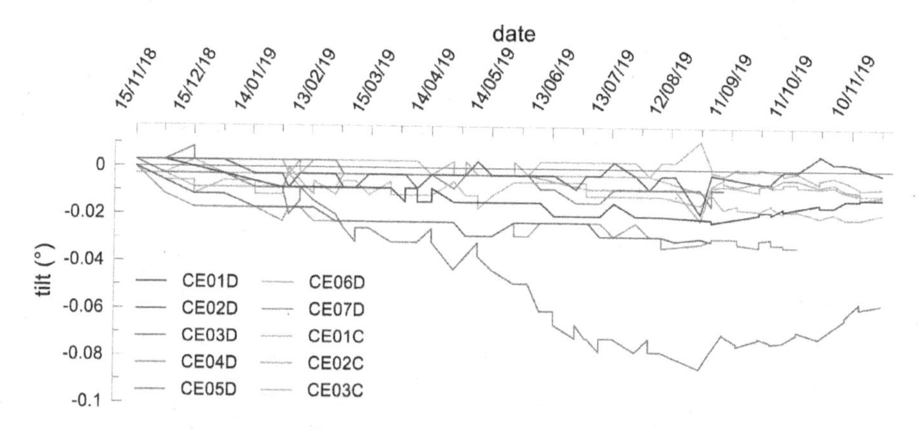

Figure 11. Time histories of the out-of-plane rotation of the Aurelian Walls at *Porta Asinaria*.

The walls were also instrumented with 11 vibrating-wire crack-meters installed over the major cracks observed in the wall façade before starting the excavation activities. Figure 12a shows the opening (−)/closure (+) measured by the crack-meters in a two-years-long time interval from 01/01/2018, before starting soil improvement via low-pressure grout injections. In the considered monitoring period, the crack-meters showed small changes in crack amplitude, with maximum values of about 0.6 mm, largely attributable to changes in the temperature. The periodic changes in crack amplitude may indeed be associated to daily and seasonal changes of temperature.

To highlight the effects of thermal excursion, Figure 12b–c show the time histories of crack opening/closure and of the changes in temperature, respectively: the data refer to a shorter time period (30/05/18–19/07/18), to make clear the strong correlation between the changes in crack-amplitude and temperature in the short (daily) and the long (seasonal) periods. Specifically, the crack-meters show amplitude changes of 0.3–0.6 mm, associated to temperature changes of about 20°C. Correlation between the changes in crack-amplitude and temperature is shown in Figure 12d for the crack-meter MG-06C, in the time period ranging from 1/1/2018 to 27/5/2019: the high computed correlation coefficient $R^2 = 0.989$ demonstrates that the observed variations in crack amplitude are mainly due to changes in the temperature. The slope of the regression line, equal to 0.017 mm/°C, provides an estimate of crack-amplitude variation of about 0.34 mm for a thermal excursion of 20°C, and of about 0.85 mm for the maximum thermal excursion monitored in the time period 20/02/2018–31/07/2018, equal to 50.2°C: both these evaluations are consistent with the data shown in Figure 12.

It can be then concluded that the crack-meters monitored negligible effects on the main cracks present in the façade of the walls, since the changes in their amplitude are essentially attributable to thermal excursions.

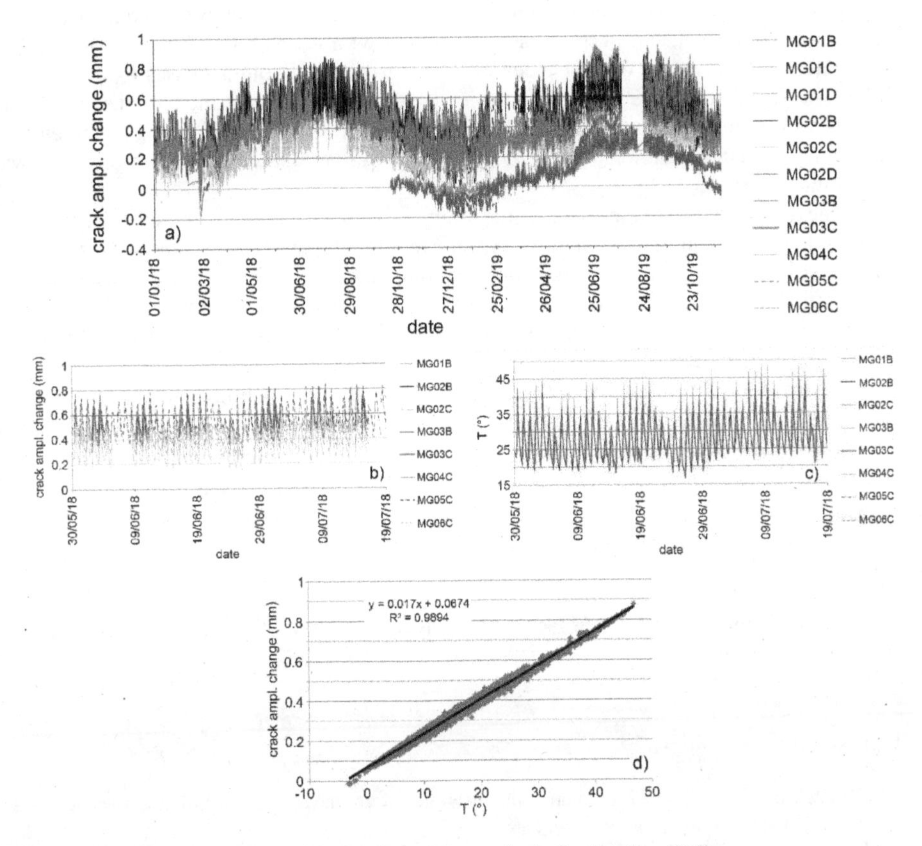

Figure 12. Measured opening (−)/closure (+) of the cracks in the Aurelian Walls.

6 CONCLUSIONS

The analysis of the monitoring data collected in a time period of about 3 years, during the excavation of two tunnels of the Line C of the Rome underground, about 140 m long, permitted to evaluate the effects induced in the Aurelian Walls at *Porta Asinaria* by the construction activities: the low-pressure grout injections preliminary carried out to improve the soil properties of the clayey silt to be excavated during tunnelling and the subsequent excavation of both running tunnels via conventional techniques.

Grout injections, though carried out under low-pressures, induced substantial heave of ground surface above the mini-tunnels, of 80 mm to 150 mm, that were seen to be larger than the maximum settlements induced by subsequent activities: dewatering by the relief wells, which induced maximum settlements of about −10 mm, and tunnels excavation, that produced surface settlement in the range −60 mm to −80 mm.

To prevent any damage in the Aurelian Walls at *Porta Asinaria*, a protective barrier of horizontal extension $L_h = 63$ m, made of adjacent piles 28 m long, partially embedded in the layer of sandy gravel was installed about parallel to the city wall, at a distance of about 8m from the axis of the North-bound tunnel, and of about 18 m from the wall.

The barrier demonstrated to be effective in reducing the settlements induced by tunnels excavation behind its location. Indeed, at the location of the embedded barrier the measured settlements reduced by about 72%, being equal to about −43 mm in *green-field* conditions and to −12 mm in the presence of the embedded barrier. Presence of the barrier was also effective in reducing by about 56% the volume loss evaluated in the arrays interacting with it: values of $V_L = 0.96\% - 1.31\%$ were computed in the presence of the barrier, against a *green-field* volume losses $V_L = 2.60\%$.

The beneficial effects of the protective barrier were also evident for the Aurelian Walls that experiences maximum settlements of about 12 mm in the portion of the wall facing the *green-field* sections, while substantially lower values, of 3 to 7 mm, were measured in the portion of the wall located behind the barrier.

Despite the differential settlements measured along the wall development, the subsidence profile always provided maximum deflection ratios Δ/L lower than the threshold values suggested in the literature, both in sagging and hogging. Moreover, the barrier was effective in reducing the out-of-plane rotation of the wall towards the tunnels: this was equal to a maximum of about −0.08° in the wall portion facing the *green-field* sections, reducing to about −0.03° for the wall portion protected by the embedded barrier.

It may be concluded that the embedded barrier was effective in preventing any damage potentially induced by conventional tunnelling to the Aurelian Walls at *Porta Asinaria* and that the excavation activities were performed within the design prescriptions, without causing any detrimental effect on the ancient city wall.

ACKNOWLEDGEMENTS

The Authors are indebted to Metro C ScPA, particularly to Mr. Eliano Romani, for making available all the monitoring data.

REFERENCES

Attewell, P.B. & Woodman, J.P. 1982. Predicting the dynamics of ground settlement and its derivatives caused by tunnelling in soil. *Ground Engineering*, **15**(8), 13–22, 36.

Attewell P.B., Yeates, J. & Selby, A.R. 1986. Soil movements induced by tunnelling and their effects on pipelines and structures. Glasgow: Blakie.

Bai, Y., Yang, Z. & Jiang, Z. 2014. Key protection techniques adopted and analysis of influence on adjacent buildings due to the Bund Tunnel construction. *Tunnelling and Underground Space Technology*, **41**, 24–34.

Bilotta, E. & Taylor, R.N. 2005. Centrifuge modelling of tunnelling close to diaphragm wall. *Int. J. Phys. Model. Geotech.* **1**, 25–41.

Bilotta, E. 2008. Use of diaphragm walls to mitigate ground movements induced by tunnelling. *Géotechnique* **58** (2), 143–155.

Burghignoli, A., Callisto, L., Rampello, S., Soccodato, F.M. & Viggiani, G.M.B. 2013. The crossing of the historical centre of Rome by the new underground Line C: a study of soil structure-interaction for historical buildings. In *Geotechnics and Heritage: Case Histories*: 97–136. London: CRC Press.

Burland, J.B. & Wroth, C.P. 1974. Settlements of buildings and associated damage. *Proc. Int. Conf. on Settlements of Structures*, Cambridge, 611–654.

Di Mariano, A., Gens, A., Gesto, J.M. Schwartz, H. 2007. Ground deformation and mitigating measures associated with the excavation of the new Metro line. In *Geotechnical Engineering in Urban Environments*, Proc. of the 14th ECSMGE, Millpress Science Publisher, Rotterdam, The Netherlands, vol. **4**: 1901–1906.

Fantera, L., Rampello, S. & Masini, L. 2016. A Mitigation Technique to Reduce Ground Settlements Induced by Tunnelling Using Diaphragm Walls. *Procedia Engineering* **158**, 254–259.

Katzenbach, R., Leppla, S., Vogler, M., Seip, M. & Kurze, S. 2013. Soil-structure interaction of tunnels and superstructures during construction and service time. In: *Proc. of the 11th Int. Conf. on Modern Building Materials, Structures and Techniques*, MBMST 2013. Procedia Engineering, vol. **57**: 35–44.

Losacco N., Romani E., Viggiani G.M.B., Di Mucci G. 2019. Embedded barriers as a mitigation measure for tunnelling induced settlements: A field trial for the line C in Rome. In: *Proc. of the WTC 2019 ITA-AITES World Tunnel Congress*, Naples, Italy

Mair, R.J. & Hight, D. 1994. Compensation grouting. *World Tunnelling Subsurface Excavation* **7** (8).

Mair, R.J. 2008. 46th Rankine Lecture: Tunnelling and geotechnics: new horizons. *Géotechnique* **58** (9), 695–736.

Masini, L., Rampello, S., Viggiani, G.M.B., & Soga, K. 2012. Experimental and numeri-cal study of grout injections in silty soils. Proc. *7th International Symposium on Geotechnical Aspects of Underground Construction in Soft Ground*, Rome, 495–503.

Masini, L., S. Rampello, & K. Soga. 2014. An approach to evaluate the efficiency of compensation grouting. *J. Geotechnical and Geoenvironmental Engineering* **140** (12): 04014073.

Masini, L., Rampello, S. & Romani, E. 2019a. Performance of a deep excavation for the new Line C of Rome underground. In *Geotechnical Research for Land Protection and Development* – Proc. of CNRIG 2019 Springer Nature Switzerland AG 2020 F. Calvetti et al. (Eds.): CNRIG 2019, Lecture Notes in Civil Engineering (LNCE) **40**: 575–582.

Masini, L., Rampello, S., Carloni, S. & Romani, E. 2019b. Ground response to mini-tunnelling plus ground improvement in the historical city centre of Rome. *In Tunnels and Underground Cities: Engineering and Innovation meet Archaeology, Architecture and Art,* WTC2019, Naples 3–9 May: 5876–5885.

Masini, L., Gaudio, D., Rampello, S. & Romani, E. 2021a. Observed Performance of a Deep Excavation in the Historical Center of Rome. *Journal of Geotech. Geoenviron. Eng.*, ASCE, **147**(2): 05020015.

Masini, L., Rampello, S., Fantera, L. & Romani, E. 2021b. Mitigation of tunnelling effects via pre-installed barriers: the case of Line C of Rome underground. In *Challenges and Innovations in Geomechanics, Proc. 16th Int. Conf. of IACMAG*, Springer Nature Switzerland AG 2021, M. Barla et al. (Eds.): IACMAG 2021, Lecture Notes in Civil Engineering (LNCE), Torino 2021, **126** (2): 197–205.

Masini, L. & Rampello, S. 2021. Predicted and observed behaviour of pre-installed barriers for the mitigation of tunnelling effects. *Tunnelling and Underground Space Technology*, 118: 104200.

Moh, Z.C., Huang, R.N. & Ju, D.H. 1996. Ground movements around tunnels in soft ground. *Proc. Int. Symp. on Geotechnical Aspects of Underground Construction in Soft Ground*, London, 725–730.

O'Reilly, M.P. & New, B.M. 1982. Settlements above tunnels in the United Kingdom – Their magnitudes and prediction. *Proc. Tunnelling '82 Symposium*, London: 173–181.

Ou C.Y. & Hsieh P.G. 2011. A simplified method for predicting ground settlement profiles induced by excavation in soft clay. *Comput. Geotech.* **38** (8): 987–997.

Peck, R. B. 1969. Deep excavation and tunnelling in soft ground. State-of-the-art-report, Mex-ico City, State of the Art Volume, *Proc. 7th Int. Conf. on Soil Mech and Found. Engng* (ICSMFE): 225–290.

Rampello, S., Callisto, L., Viggiani, G.M.B. & Soccodato, F.M. 2012. Evaluating the effects of tunnelling on historical buildings: the example of a new subway in Rome. *Geomechanics and tunnelling* **5** (3): 275–299.

Rampello, S., Fantera, L. & Masini, L. 2019. Efficiency of embedded barriers to mitigate tunnelling effects. *Tunnelling and Underground Space Technology* **89**: 109–124.

Geotechnical Engineering for the Preservation of Monuments and Historic Sites III – Lancellotta, Viggiani, Flora, de Silva & Mele (Eds)
© 2022 Copyright the Editor(s), ISBN 978-1-032-35998-4

Author index